MG MGA
WORKSHOP MANUAL
1500, 1600 and 1600 Mk 2

Issued by

THE BRITISH MOTOR CORPORATION LIMITED
MG CAR DIVISION

Exported by

THE BRITISH MOTOR CORPORATION LIMITED
EXPORT SALES DIVISION

M.G. 'MGA'. Issue 5. 40954

INTRODUCTION

This Manual has been prepared to provide the service operator with the necessary information for the maintenance and repair of the M.G. Series MGA 1500, MGA 1600, and MGA 1600 Mk. II.

The Manual also serves as a ready-reference book for service supervision and covers items of procedure for the guidance of both the fully qualified and the less-experienced mechanic.

UNIT ARRANGEMENT

In the Manual the complete vehicle is divided into Sections each of which deals with an assembly or major component and carries a reference letter. Where necessary, a Section is divided into two or three parts, having a single, a double, or a treble reference letter. In such cases the double-letter Section refers specifically to Series MGA 1600 and MGA 1600 Mk. II cars and the treble-letter Section refers to the Series MGA 1600 and MGA 1600 Mk. II fitted with centre-lock wheels and Dunlop disc brakes. These Sections should always be used in conjunction with the corresponding single-letter Section. Where there is no double-letter or treble-letter Section the information contained in the single-letter Section refers to all models.

NUMBERING OF PAGES AND ILLUSTRATIONS

The pages and illustrations are numbered consecutively within each Section, and the Section title and letter(s) are shown at the top of each page.

SERVICE TOOLS

Use of the correct tools contributes to an efficient, economic, and profitable repair. References have therefore been made to such tools throughout the Manual.

CONTENTS

THE M.G. (Series MGA) TWO-SEATER

THE M.G. (Series MGA 1600) COUPÉ

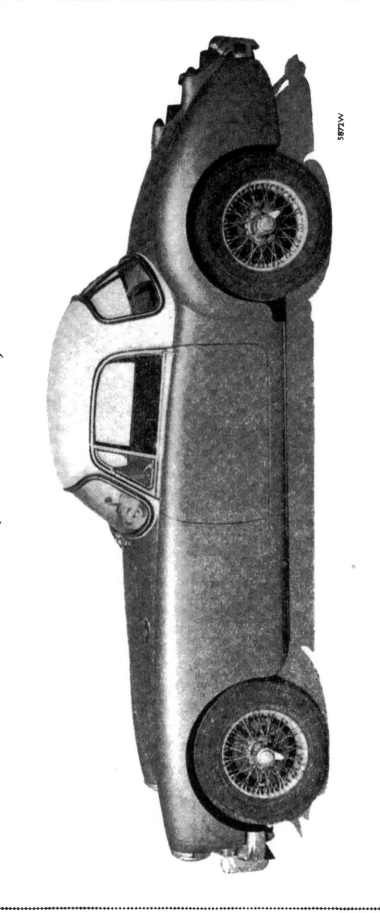

5872W

THE M.G. (Series MGA 1600—Mk. II) TOURER

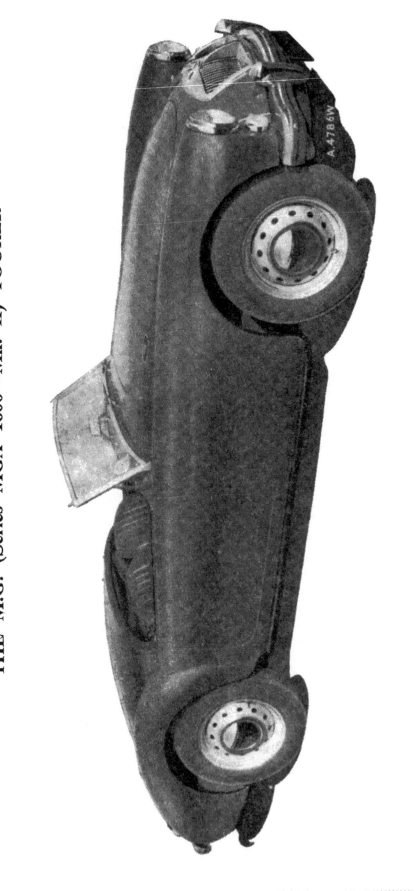

GENERAL DATA
(Series MGA)

ENGINE

Type	15GB.
(From Car No. 61504)	15GD.
Number of cylinders	4.
Bore	2·875 in. (73·025 mm.).
Stroke	3·5 in. (89 mm.).
Capacity	90·88 cu. in. (1489 c.c.).
Firing order	1, 3, 4, 2.
Compression ratio	8·3 : 1.
Capacity of combustion chamber (valves fitted) ..	2·3 to 2·4 cu. in. (38·2 to 39·2 c.c.).
Valve operation	Overhead by push-rod.
B.M.E.P.	130 lb./sq. in. at 3,500 r.p.m.
Torque	77·4 lb. ft. at 3,500 r.p.m.
Cooling system	Thermo-siphon, pump- and fan-assisted.
Oversize bore : 1st	·010 in. (·254 mm.).
Max.	·040 in. (1·016 mm.).

CRANKSHAFT

Main journal diameter	2 in. (50·8 mm.).
Minimum regrind diameter	1·96 in. (49·78 mm.).
Crankpin journal diameter	1·8759 to 1·8764 in. (47·65 to 47·66 mm.)
Crankpin minimum regrind diameter	1·8359 in. (46·64 mm.).
Main bearings	
Number and type	3. Shell-type.
Material: Top and bottom halves	Steel-backed white metal.
Length	1·375 in. (34·925 mm.).
End-clearance	·002 to ·003 in. (·051 to ·076 mm.).
End-thrust	Taken by thrust washers at centre main bearing.
Running clearance	·0005 to ·002 in. (·0127 to ·0508 mm.).
Undersizes	−·010 in., −·020 in., −·030 in., −·040 in. (−·254 mm., −·508 mm., −·762 mm., −1·016 mm.).

CONNECTING RODS

Length between centres	6·5 in. (165·1 mm.).
Big-end bearings	
Material: Top and bottom halves	Steel-backed lead-indium or lead-tin.
Bearing side-clearance	·008 to ·012 in. (·203 to ·305 mm.).
Bearing diametrical clearance	·0001 to ·0016 in. (·002 to ·04 mm.).
Undersizes	−·010 in., −·020 in., −·030 in., −·040 in. (−·254 mm., −·508 mm., −·762 mm., −1·016 mm.).

PISTONS

Type	Aluminium alloy.
Clearances: Bottom of skirt	·0017 to ·0023 in. (·043 to ·051 mm.).
Top of skirt	·0035 to ·0042 in. (·090 to ·106 mm.).
Oversizes	+·010 in., +·020 in., +·030 in., +·040 in. (+·254 mm., +·508 mm., +·762 mm., +1·016 mm.).

PISTON RINGS

Compression: Plain	Top ring.
Tapered	2nd and 3rd rings.
Width	·0615 to ·0625 in. (1·56 to 1·58 mm.).
Thickness	·111 to ·118 in. (2·81 to 3·0 mm.) to Engine No. 40824. ·119 to ·126 in. (3·02 to 3·2 mm.) from Engine No. 40825.

Fitted gap	·008 to ·013 in. (·20 to ·33 mm.).
Clearance in groove	·0015 to ·0035 in. (·038 to ·089 mm.).
Oil control type	Slotted scraper.
Width	·1552 to ·1562 in. (3·94 to 3·99 mm.).
Thickness	·111 to ·118 in. (2·81 to 3·0 mm.) to Engine No. 40824.
	·119 to ·126 in. (3·02 to 3·2 mm.) from Engine No. 40825.
Fitted gap	·008 to ·013 in. (·20 to ·33 mm.).
Clearance in groove	·0016 to ·0036 in. (·040 to ·091 mm.).

GUDGEON PIN

Type	Clamped.
Fit	·0001 to ·00035 in. (·0025 to ·009 mm.). Hand push fit at 68° F. (20° C.).
Diameter	·6869 to ·6871 in. (17·447 to 17·4523 mm.).

VALVES AND VALVE GEAR
Valves

Seat angle: Inlet and exhaust	45°.
Head diameter: Inlet	1½ in. (38·1 mm.).
Exhaust	1 9/32 in. (32·54 mm.).
Stem diameter: Inlet and exhaust	·342 in. (8·68 mm.).
Valve lift	·357 in. (9·06 mm.).
Valve stem to guide clearance: Inlet	·00155 to ·00255 in. (·0394 to ·0635 mm.).
Exhaust	·00105 to ·00205 in. (·027 to ·052 mm.) to Engine No. 4044.
	·002 to ·003 in. (·051 to ·076 mm.) from Engine No. 4045.
Valve rocker clearance: Running	·017 in. (·432 mm.) (hot).
Timing	·060 in. (1·52 mm.).
Timing markings	Dimples on timing wheels.
Chain pitch and number of pitches	⅜ in. (9·52 mm.), 52 pitches.
Inlet valve: Opens	16° B.T.D.C.
Closes	56° A.B.D.C.
Exhaust valve: Opens	51° B.B.D.C.
Closes	21° A.T.D.C.

VALVE GUIDES

Length: Inlet	1⅞ in. (47·63 mm.).
Exhaust	2 9/32 in. (57·94 mm.).
Diameter: Inlet: Outside	·5635 in. (14·31 mm.).
Inside	·3438 in. (8·73 mm.).
Exhaust: Outside	·5635 in. (14·31 mm.).
Inside	·3438 in. (8·73 mm.).
Fitted height above head	·625 in. (15·87 mm.).

VALVE SPRINGS

Free length: Inner	1 31/32 in. (50 mm.).
Outer	2 3/64 in. (51·99 mm.).
Fitted length: Inner	1 7/16 in. (36·51 mm.).
Outer	1 9/16 in. (39·69 mm.).
Number of working coils: Inner	6½.
Outer	4½.
Pressure: Valve open	Inner 50 lb. (22·7 kg.). Outer 105 lb. (47·6 kg.).
Valve closed	Inner 30 lb. (13·6 kg.). Outer 60½ lb. (27 kg.).

TAPPETS

Type	Barrel with flat base.
Diameter: Body	$\frac{13}{16}$ in. (20·64 mm.).
Length	2·293 to 2·303 in. (58·25 to 58·5 mm.).

ROCKERS

Outside diameter before fitting	·751 in. (19·07 mm.).
Inside diameter (reamed in position)	·616 to ·620 in. (15·65 to 15·74 mm.).
Bore of rocker arms	·7485 to ·7489 in. (19·01 to 19·02 mm.).
Rocker ratio	1·426 : 1.

CAMSHAFT

Journal diameters: Front	1·78875 to 1·78925 in. (45·43 to 45·44 mm.).
Centre	1·72875 to 1·72925 in. (43·91 to 43·92 mm.).
Rear	1·62275 to 1·62325 in. (41·22 to 41·23 mm.).
End-float	·003 to ·007 in. (·076 to ·178 mm.).
Bearing: number and type	3. Thinwall steel-backed white metal.
Outside diameter (before fitting)	Front 1·920 in. (48·76 mm.), centre 1·860 in. (47·24 mm.), rear 1·754 in. (44·55 mm.).
Inside diameter (reamed in position)	Front 1·790 in. (45·47 mm.), centre 1·730 in. (43·94 mm.), rear 1·624 in. (41·25 mm.).
Clearance	·001 to ·002 in. (·0254 to ·0508 mm.).

ENGINE LUBRICATION SYSTEM

Oil pump

Type	Eccentric rotor.
Relief pressure valve operates	75 to 80 lb./sq. in. (5·3 to 5·6 kg./cm.²).
Relief valve spring: Free length	3 in. (76·2 mm.).
Fitted length	2$\frac{5}{32}$ in. (54·77 mm.) at 16 lb. (7·26 kg.) load.
Identification colour	Red spot.

Oil filter

Type	Tecalemit (element Part No. 1H779) or Purolator (element Part No. 1H1054) up to Engine No. 26932. Tecalemit or Purolator (element Part No. 8G683) from Engine No. 26933.
Capacity	$\frac{1}{2}$ pint (·28 litre).

Oil pressure

Normal running: Minimum	10 to 25 lb./sq. in. (·7 to 1·7 kg./cm.²).
Maximum	50 to 75 lb./sq. in. (3·5 to 5·2 kg./cm.²).

TORQUE WRENCH SETTINGS

Cylinder head nuts	50 lb. ft. (6·91 kg. m.).
Main bearing nuts	70 lb. ft. (9·7 kg. m.).
Connecting rod set screws	35 lb. ft. (4·83 kg. m.).
Clutch assembly to flywheel	25 lb. ft. (3·46 kg. m.).
Road wheel nuts	60 to 62·5 lb. ft. (8·3 to 8·65 kg. m.).
Gudgeon pin clamp	25 lb. ft. (3·45 kg. m.).
Manifold stud nuts	25 lb. ft. (3·45 kg. m.).
Water pump securing bolts	25 lb. ft. (3·45 kg. m.).
Clutch to flywheel bolts	35 to 40 lb. ft. (4·8 to 5·5 kg. m.).
Oil filter centre-bolt	15 lb. ft. (2·07 kg. m.).
Brake calliper securing bolts	45 to 50 lb. ft. (6·22 to 6·91 kg. m.).

FUEL SYSTEM

Carburetter

Make and type	S.U. twin H4 semi-downdraught.
Diameter	1½ in. (38·1 mm.).
Needle	GS.
Jet	·090 in. (2·29 mm.).
Piston spring	Red.

AIR CLEANER

Make and type	Vokes—oil-wetted.

FUEL PUMP

Make and type	S.U. electric—high pressure.
Delivery test	10 gal. per hr. (45·4 litres per hr.).
Suction lift	33 in. (83·8 cm.).
Output lift	48 in. (121·9 cm.).

COOLING SYSTEM

Type	Pressurized radiator. Thermo-siphon, pump- and fan-assisted.
Thermostat setting	70 to 75° C. (158 to 167° F.).
Quantity of anti-freeze: 15° frost	1 pint (57 litre).
25° frost	1½ pints (·85 litre).
35° frost	2 pints (1·1 litres).

IGNITION SYSTEM

Sparking plugs	Champion N5, was NA8.
Size	14 mm.
Plug gap	·024 to ·026 in. (·625 to ·660 mm.).
Coil	Lucas HA12.
Distributor	Lucas. Type DM2. Later models DM2.P4.
Distributor contact points gap	·014 to ·016 in. (·35 to ·40 mm.).
Suppressors	Lucas No. 78106A fitted on each H.T. cable.
Timing	7° B.T.D.C.

CLUTCH

Make and type	Borg & Beck A6–G. Single dry plate.
Diameter	8 in. (20·3 cm.).
Facing material	Wound yarn—Borglite.
Pressure springs	6.
Colour	Black and yellow.
	Cream and light green: commencing Engine No. 16225.
Damper springs	6.
Colour	White with light-green stripes.
Release lever ratio	9 : 1.

GEARBOX

Number of forward speeds		4.
Synchromesh		Second, third, and fourth gears.
Ratios:	Top	1·0 : 1.
	Third	1·374 : 1.
	Second	2·214 : 1.
	First	3·64 : 1.
	Reverse	4·76 : 1.

Overall ratios: Top	4·3 : 1.
Third	5·908 : 1.
Second	9·520 : 1.
First	15·652 : 1.
Reverse	20·468 : 1.
Speedometer gears ratio	5 : 12.

STEERING

Type	Rack and pinion.
Steering-wheel turns—lock to lock	$2\frac{2}{3}$.
Steering-wheel diameter	$16\frac{1}{2}$ in. (419·10 mm.).
Camber angle	1° positive to $\frac{1}{2}$° negative on full bump.
Castor angle	4°.
King pin inclination	9° to $10\frac{1}{2}$° on full bump.
Toe-in	Wheels parallel.
Track: Front	Disc wheels $47\frac{1}{2}$ in. (1·203 m.). Wire wheels $47\frac{7}{8}$ in. (1·216 m.).
Rear	Disc wheels $48\frac{3}{4}$ in. (1·238 m.). Wire wheels $48\frac{3}{4}$ in. (1·238 m.).

FRONT SUSPENSION

Type	Independent coil.	
Spring detail:	*To Car No.* 15151	*From Car No.* 15152
Coil diameter (mean)	3·238 in. (82·24 mm.).	3·28 in. (82·25 mm.).
Diameter of wire	·498 in. (12·66 mm.).	·54 in. (13·72 mm.).
Free height	$9{\cdot}28\pm\frac{1}{16}$ in. (23·49 cm. ±1·6 mm.).	$8{\cdot}88\pm\frac{1}{16}$ in. (22·55 cm. ±1·6 mm.).
Number of free coils	7·5.	7·2.
Static laden length	$6{\cdot}60\pm\frac{1}{32}$ in. (16·76 cm. ±·8 mm.).	
Static laden length at load of	1,095±20 lb. (497±9·1 kg.).	
Maximum deflection	4 in. (10·16 cm.).	
Dampers (front)	Piston type.	

REAR SUSPENSION

Type	Semi-elliptic.
Spring detail:	
Number of leaves	6.
Width of leaves	$1\frac{3}{4}$ in. (44·45 mm.).
Gauge	$\frac{7}{32}$ in. (5·56 mm.).
Working load	450 lb. (203·7 kg.).
Free camber	3·60 in. (91·44 mm.).
Dampers (rear)	Piston type.

PROPELLER SHAFT

Type	Tubular. Reverse spline.
Make and type of joints	Hardy Spicer. Needle roller.
Propeller shaft length (between centres of joints)	$31\frac{3}{8}$ in. (79·69 cm.).
Overall length	$38\frac{13}{32}$ in. (97·44 cm.).
Diameter	2 in. (50·8 mm.).
Type (with 15GD series power unit)	Tubular, incorporating sliding spline joint.
Overall length (fully extended)	$32\frac{11}{16}$ in. (83·03 cm.).
Overall length (fully compressed)	$31\frac{3}{4}$ in. (80·65 cm.).
Length between centres of joints (fully extended)	$30\frac{5}{16}$ in. (77 cm.).
Length between centres of joints (fully compressed)	$29\frac{3}{8}$ in. (74·65 cm.).
Diameter (main tube)	2 in. (50·8 mm.).

GENERAL DATA—*continued*

REAR AXLE

Make and type	B.M.C. 'B' type, three-quarter-floating.
Ratio: Standard	10/43.
Optional	9/41.
Adjustment	Shims.

ELECTRICAL EQUIPMENT

System	12-volt. Positive earth.
Charging system	Compensated voltage control.
Battery: Type	Lucas SG9E.
Type (Export only)	Lucas STGZ9E (dry-charged).
Voltage	6-volt (2 off).
Capacity (20-hr. rate)	58-amp.-hr.
Starter motor	Lucas 4-brush M35G.
Dynamo	Lucas C39PV2.
	Lucas C40/1 ('MGA 1600' after Engine No. 16GA6272).

BRAKES

Type	Lockheed hydraulic (front and rear).
Size	10 in. × 1¾ in. (25·4 cm. × 44·45 mm.).
Front	2 leading shoes.
Rear	Single leading shoe.
Drum size	10 in. (254 mm.) (front and rear).
Lining dimensions	9·6 in. × 1¾ in. (24·38 cm. × 44·45 mm.).
Lining area: Front	67·2 sq. in. (433·55 cm.²).
Rear	67·2 sq. in. (433·55 cm.²).
Material	Ferodo DM12.

WHEELS

Type: Ventilated disc	4J × 15.
Wire (optional)	4J × 15, 48-spoke.

TYRES

Size	5·60—15.
Tyre pressures: Normal: Front	17 lb./sq. in. (1·2 kg./cm.²).
Rear	20 lb./sq. in. (1·4 kg./cm.²).
Fast motoring: Front	21 lb./sq. in. (1·48 kg./cm.³).
Rear	24 lb./sq. in. (1·69 kg./cm.²).
Competition work and sustained high-speed motoring: Front	23 lb./sq. in. (1·62 kg./cm.²).
Rear	26 lb./sq. in. (1·83 kg./cm.²).

CAPACITIES

	Imp.	U.S.	Litres
Engine sump (including filter)	7½ pts.	9 pts.	4·25
Gearbox	4½ pts.	5·4 pts.	2·56
Rear axle	2¼ pts.	2·7 pts.	1·28
Cooling system	10 pts.	12 pts.	5·67
Steering rack	½ pt.	·6 pt.	·28
Fuel tank	10 gal.	12 gal.	45·4
Brake system	1 pt.	1·2 pts.	·568

GENERAL DIMENSIONS

Wheelbase	94 in. (238·8 cm.).
Overall length	156 in. (396·2 cm.).
Overall width	58 in. (147·3 cm.).
Overall height	50 in. (127·0 cm.).
Ground clearance..	6 in. (15·24 cm.).
Weight: fully equipped with tools, spare wheel, oil, water, and	
2 gallons of fuel (2·5 U.S. gal., 9·1 litres)	1,988 lb. (901·81 kg.).
Turning circles	28 ft. (8·534 m.).

GENERAL DATA
(MGA 1600)

ENGINE

Type	16GA.
Number of cylinders	4.
Bore	2·968 in. (75·39 mm.).
Stroke	3·5 in. (89 mm.).
Capacity	96·9 cu. in. (1588 c.c.).
Firing order	1, 3, 4, 2.
Compression ratio	8·3 : 1.
Capacity of combustion chamber (valves fitted)	2·36 cu. in. (38·7 c.c.).
Valve operation	Overhead by push-rod.
B.M.E.P.	135 lb./sq. in. (9·5 kg./cm.^{2}) at 4,000 r.p.m.
Torque	87 lb. ft. (12·03 kg. m.) at 3,800 r.p.m.
Cooling system	Thermo-siphon, pump- and fan-assisted.
Oversize bore: 1st	·010 in. (·254 mm.).
Max.	·040 in. (1·016 mm.).
Maximum b.h.p. (standard)	79·5 at 5,600 r.p.m.

CRANKSHAFT
Main bearings } Refer to Series MGA data on preceding pages.

CONNECTING RODS

Length between centres	6·5 in. (165·1 mm.).

Big-end bearings

Material: Top and bottom halves	Steel and lead-indium.
Bearing side-clearance	·008 to ·012 in. (·203 to ·305 mm.).
Bearing diametrical clearance	·0010 to ·0025 in. (·025 to ·063 mm.).
Undersizes	−·010 in., −·020 in., −·030 in., −·040 in. (−·254 mm., −·508 mm., −·762 mm., −1·016 mm.).

PISTONS

Refer to Series MGA data on preceding pages.

PISTON RINGS

Compression: Plain	Top ring.
Tapered	2nd and 3rd rings.
Width	·0615 to ·0625 in. (1·56 to 1·58 mm.).
Thickness	·141 to ·148 in. (3·57 to 3·76 mm.).
Fitted gap	·009 to ·014 in. (·229 to ·356 mm.).
Clearance in groove	·0015 to ·0035 in. (·038 to ·089 mm.).
Oil control type	Slotted scraper.
Width	·1552 to ·1562 in. (3·94 to 3·99 mm.).
Thickness	·135 to ·142 in. (3·43 to 3·61 mm.).
Fitted gap	·009 to ·014 in. (·23 to ·36 mm.).
Clearance in groove	·0016 to ·0036 in. (·040 to ·091 mm.).

GUDGEON PIN

Refer to Series MGA data on preceding pages.

VALVES AND VALVE GEAR

Valves

Seat angle: Inlet and exhaust	45°.
Head diameter: Inlet	1½ in. (38·1 mm.).
Exhaust	1$\frac{9}{32}$ in. (32·54 mm.).

Stem diameter: Inlet .. .. ·342 in. (8·68 mm.).

Exhaust .. .. ·342 in. (8·68 mm.).

Valve lift .. .. ·350 in. (8·89 mm.).

Valve stem to guide clearance: Inlet .. .. .. ·00155 to ·00255 in. (·0394 to ·0635 mm.).

Exhaust .. .. .. ·002 to ·003 in. (·051 to ·076 mm.).

Valve rocker clearance: Running .. .. ·015 in. (·38 mm.) (cold).

Timing .. .. ·021 in. (·53 mm.).

Timing markings .. .. Dimples on timing wheels.

Chain pitch and number of pitches .. .. $\frac{3}{8}$ in. (9·52 mm.), 52 pitches.

Inlet valve: Opens .. .. 16° B.T.D.C.

Closes .. .. 56° A.B.D.C.

Exhaust valve: Opens .. .. 51° B.B.D.C.

Closes .. .. 21° A.T.D.C.

VALVE GUIDES

Length: Inlet .. .. 1⅞ in. (47·63 mm.).

Exhaust .. .. 2¹¹⁄₆₄ in. (55·95 mm.).

Diameter: Inlet and exhaust: Outside .. .. .. ·5635 to ·5640 in. (14·31 to 14·32 mm.).

Inside .. .. ·34425 to ·34475 in. (8·744 to 8·757 mm.).

Fitted height above head .. .. ·625 in. (15·87 mm.).

VALVE SPRINGS ⎫
TAPPETS   ⎬ Refer to Series MGA data on preceding pages.
ROCKERS   ⎪
CAMSHAFT ⎭

ENGINE LUBRICATION SYSTEM

Oil pump

Type .. .. Eccentric rotor.

Relief pressure valve operates .. .. 50 lb./sq. in. (3·5 kg./cm.²).

Relief valve spring: Free length .. .. .. 3 in. (76·2 mm.).

Fitted length .. .. .. 2⁸⁄₃₂ in. (54·77 mm.) at 16 lb. (7·26 kg.) load.

Identification colour .. .. .. Red spot.

Oil filter

Type .. .. Tecalemit or Purolator.

Capacity .. .. 1 pint (1·2 U.S. pints, ·57 litre).

Oil pressure

Normal running: Minimum .. .. 15 lb./sq. in. (1·05 kg./cm.²).

Maximum .. .. 50 lb./sq. in. (3·5 kg./cm.²).

TORQUE WRENCH SETTINGS. Refer to Series MGA data on preceding pages.

FUEL SYSTEM

Carburetter

Make and type .. .. S.U. twin H4 semi-downdraught.

Diameter .. .. 1½ in. (38·1 mm.).

Needle .. .. No. 6.

Jet .. .. ·090 in. (2·29 mm.).

Piston spring .. .. Red.

AIR CLEANER AND FUEL PUMP. Refer to Series MGA data on preceding pages.

COOLING SYSTEM. Data as for Series MGA on preceding pages except:

Thermostat opening temperature: Crack open .. .. 68° C. (154° F.) ⎫ from Engine No. 16GA4788.

Fully open .. .. 83° C. (181° F.) ⎭

Filler cap spring pressure .. .. 7 lb. (3·18 kg.) from Car No. 71832.

GENERAL DATA—*continued*

IGNITION SYSTEM

Sparking plugs	Champion N5.
Size	14 mm.
Plug gap	·024 to ·026 in. (·625 to ·660 mm.).
Coil	Lucas HA12.
Distributor	Lucas Type DM2. Later models DM2.P4.
Distributor contact points gap	·014 to ·016 in. (·35 to ·40 mm.).
Suppressors	Lucas No. 78106A fitted on each H.T. cable.
Static timing	7° B.T.D.C.

CLUTCH

Make and type	Borg & Beck A6–G. Single dry plate.
Diameter	8 in. (20·3 cm.).
Facing material	Wound yarn—Borglite.
Pressure springs	6.
Colour	Black and yellow.
	Cream and light green: from Engine No. 16225.
Damper springs	6.
Colour	White with light-green stripes.
Release lever ratio	9 : 1.

GEARBOX

Refer to Series MGA data on preceding pages.

STEERING

Type	Rack and pinion.
Steering-wheel turns—lock to lock	2⅜.
Steering-wheel diameter	16½ in. (419·10 mm.).
Camber angle	1° positive to ½° negative on full bump.
Castor angle	4°.
King pin inclination	9° to 10½° on full bump.
Toe-in	Wheels parallel.
Track (MGA 1600):	
Front	Disc wheels 47½ in. (1·203 m.).
	Wire wheels 47⅞ in. (1·216 m.).
Rear	Disc wheels 48¾ in. (1·238 m.).
	Wire wheels 48¾ in. (1·238 m.).
Track (MGA 1600 with Dunlop disc brakes):	
Front	47 $\frac{49}{53}$ in. (1·217 m.).
Rear	48⅞ in. (1·242 m.).

FRONT SUSPENSION

Type	Independent coil.
Spring detail:	
Coil diameter (mean)	3·28 in. (82·25 mm.).
Diameter of wire	·54 in. (13·72 mm.).
Free height	8·88 ± $\frac{1}{16}$ in. (22·55 cm. ± 1·6 mm.).
Number of free coils	7·2.
Static laden length	6·60 ± $\frac{1}{32}$ in. (16·76 cm. ± ·8 mm.).
Static laden length at load of	1,095 ± 20 lb. (497 ± 9·1 kg.).
Maximum deflection	4 in. (10·16 cm.).
Dampers (front)	Piston type.

REAR SUSPENSION

Type .. .. Semi-elliptic.
Spring detail:
  Number of leaves .. .. 6.
  Width of leaves .. .. 1¾ in. (44·45 mm.).
  Gauge .. .. 7/32 in. (5·56 mm.).
  Working load .. .. 450 lb. (203·7 kg.).
  Free camber .. .. 3·60 in. (91·44 mm.).
Dampers (rear) .. .. Piston type.

PROPELLER SHAFT

Type .. .. Tubular, flanged type.
Propeller shaft length .. .. 30½ in. (77·47 cm.).
Overall length .. .. 32 11/16 in. (82·98 cm.).
Diameter .. .. 2 in. (50·8 cm.).
Make and type of joints .. .. Hardy Spicer needle roller.

REAR AXLE
ELECTRICAL EQUIPMENT ⎤ Refer to Series MGA data on preceding pages.

BRAKES

Type .. .. Lockheed hydraulic; disc front, drum rear.
Lining material .. .. DON24.
Disc material .. .. DON55.
Lining dimensions .. .. 9·63 in. × 1·7 in. (244·6 mm. × 43·2 mm.).
Total lining area (rear) .. .. 65·48 sq. in. (422·36 cm.²).
Number of rivets (per shoe) .. .. 12.
Disc diameter .. .. 11 in. (27·9 cm.).

BRAKES (MGA 1600 with Dunlop disc brakes)

Type .. .. Dunlop disc (front and rear).
Disc diameter .. .. 11 in. (27·9 cm.).
Fluid .. .. Wakefield Crimson (S.A.E. 70.R3).

WHEELS ⎤
TYRES ⎦ Refer to Series MGA data on preceding pages for Series MGA 1600.

WHEELS (MGA 1600 with Dunlop disc brakes)

Type .. .. Ventilated disc, 4J × 15. Centre lock.

TYRES (MGA 1600 with Dunlop disc brakes)

Size .. .. 5·90—15. Road Speed.
Tyre pressures:
  Normal: Front .. .. 18 lb./sq. in. (1·27 kg./cm.²).
      Rear .. .. 20 lb./sq. in. (1·4 kg./cm.²).
  Fast motoring: Front .. .. 22 lb./sq. in. (1·55 kg./cm.²).
      Rear .. .. 24 lb./sq. in. (1·69 kg./cm.²).
  Competition work and sustained high-speed motoring ⎤ Front .. .. 24 lb./sq. in. (1·69 kg./cm.²).
    ⎦ Rear .. .. 26 lb./sq. in. (1·83 kg./cm.²).

CAPACITIES

	Imp.	U.S.	Litres
Engine sump (including filter)	7½ pts.	9 pts.	4·25
Gearbox	4½ pts.	5·6 pts.	2·56
Rear axle	2¼ pts.	2·7 pts.	1·28
Cooling system	10 pts.	12 pts.	5·67
Steering rack	½ pt.	·6 pt.	·28
Fuel tank	10 gal.	12 gal.	45·4
Brake system	1 pt.	1·2 pts.	·568
Oil cooler	¾ pt.	·9 pt.	·426

GENERAL DIMENSIONS

Wheelbase	94 in. (238·8 cm.).
Overall length	156 in. (396·2 cm.).
Overall width	58 in. (147·3 cm.).
Overall height	50 in. (127·0 cm.).
Ground clearance..	6 in. (15·24 cm.).
Turning circles (MGA 1600)	30 ft. 6 in. (9·296 m.).
Turning circles (MGA 1600 with Dunlop disc brakes) ..	Right-hand 32 ft. 1 in. (9·78 m.).
	Left-hand 32 ft. 6 in. (9·91 m.).

WEIGHTS

Fully equipped with tools, spare wheel, oil, water, and 2 gallons (2·5 U.S. gal., 9·1 litres) of fuel	2,016 lb. (914 kg.).
Engine (dry)	359 lb. (162·84 kg.).
Gearbox (dry)	67¼ lb. (30·50 kg.).
Rear axle (dry)	117½ lb. (53·32 kg.).

GENERAL DATA
(Series MGA 1600—Mk. II)

ENGINE

Type	16GC.
Number of cylinders	4.
Bore	3·0 in. (76·2 mm.).
Stroke	3·5 in. (89 mm.).
Capacity	99·5 cu. in. (1622 c.c.).
Firing order	1, 3, 4, 2.
Compression ratio: High	8·9 : 1.
Low	8·3 : 1.
Capacity of combustion chamber (valves fitted) ..	2·624 cu. in. (43·0 c.c.).
Valve operation	Overhead by push-rod.
Maximum horse-power (standard): High compression ..	90 at 5,500 r.p.m.
Low compression ..	85 at 5,500 r.p.m.
B.M.E.P.: High compression	148 lb./sq. in. (10·4 kg./cm.²) at 4,000 r.p.m.
Low compression	140 lb./sq. in. (9·84 kg./cm.²) at 3,000 r.p.m.
Torque: High compression	97 lb. ft. (13·1 kg. m.) at 4,000 r.p.m.
Low compression	92 lb. ft. (12·72 kg. m.) at 3,000 r.p.m.
Cooling system	Thermo-siphon, pump- and fan-assisted.
Oversize bore: 1st	·010 in. (·254 mm.).
Max.	·040 in. (1·016 mm.).

CRANKSHAFT

Refer also to Series MGA data on preceding pages.

Main bearings

Journal length: Front	1·528 to 1·544 in. (38·817 to 39·224 mm.).
Intermediate	1·471 to 1·473 in. (37·363 to 37·414 mm.).
Rear	1·494 to 1·498 in. (37·940 to 38·049 mm.).
Bearing length	1·25 in. (31·75 mm.).
Diametrical clearance	·001 to ·0027 in. (·0254 to ·0685 mm.).

CONNECTING RODS

Refer also to Series MGA data on preceding pages.

Small-end bore	·750 to ·7512 in. (19·05 to 19·08 mm.).

Big-end bearings

Diametrical clearance	·001 to ·0025 in. (·0254 to ·063 mm.).

PISTON RINGS

Compression: Top ring	Plain.
Second and third rings	Tapered.
Width	·0615 to ·0625 in. (1·56 to 1·58 mm.).
Thickness	·125 to ·132 in. (3·175 to 3·35 mm.).
Fitted gap	·009 to ·014 in. (·229 to ·356 mm.).
Clearance in groove	·0015 to ·0035 in. (·038 to ·089 mm.).
Oil control type	Slotted scraper.
Width	·1552 to ·1562 in. (3·94 to 3·99 mm.).
Thickness	·125 to ·132 in. (3·175 to 3·35 mm.).
Fitted gap	·009 to ·014 in. (·23 to ·36 mm.).
Clearance in groove	·0016 to ·0036 in. (·040 to ·091 mm.).

GENERAL DATA—*continued*

GUDGEON PIN

Type	Clamped.
Fit	·0001 to ·0006 in. (·0025 to ·0152 mm.). Hand push fit at 68° F. (20° C.).
Diameter: Outer	·7499 to ·7501 in. (19·047 to 19·050 mm.).
Inner	·3215 in. (7·94 mm.).
Length	2·693 to 2·703 in. (68·402 to 68·656 mm.).

VALVES AND VALVE GEAR

Valves

Seat angle: Inlet	45°.
Exhaust	45°.
Head diameter: Inlet	1·562 to 1·567 in. (39·6 to 39·8 mm.).
Exhaust	1·343 to 1·348 in. (34·11 to 34·23 mm.).
Stem diameter: Inlet	·342 in. (8·68 mm.).
Exhaust	·342 in. (8·68 mm.).
Valve lift	·350 in. (8·89 mm.).
Valve stem to guide clearance: Inlet	·00155 to ·00255 in. (·0394 to ·0635 mm.).
Exhaust	·002 to ·003 in. (·051 to ·076 mm.).
Valve rocker clearance: Running	·015 in. (·38 mm.) (cold)
Timing	·021 in. (·53 mm.).
Timing markings	Dimples on timing wheels.
Chain pitch and number of pitches	⅜ in. (9·52 mm.), 52 pitches.
Inlet valve: Opens	16° B.T.D.C.
Closes	56° A.B.D.C.
Exhaust valve: Opens	51° B.B.D.C.
Closes	21° A.T.D.C.

VALVE GUIDES

Length: Inlet	1⅝ in. (41·275 mm.).
Exhaust	2¾ in. (55·95 mm.).
Diameter: Inlet and exhaust: Outside	·5635 to ·5640 in. (14·31 to 14·32 mm.).
Inside	·34425 to ·34475 in. (8·744 to 8·757 mm.).
Fitted height above head	·625 in. (15·87 mm.).

VALVE SPRINGS

Free length: Inner	1³¹⁄₃₂ in. (50 mm.).
Outer	1²³⁄₂₄ in. (48·8 mm.).
Fitted length: Inner	1·449 in. (36·8 mm.).
Outer	1·575 in. (40 mm.).
Number of working coils: Inner	6½.
Outer	4½.
Load: Full lift: Inlet and exhaust	Inner 50 lb. (22·7 kg.). Outer 113 lb. (51·2 kg.).
No lift: Inlet	Inner 28 to 32 lb. (12·7 to 14·51 kg.). Outer 53 to 57 lb. (24 to 25·8 kg.).
Exhaust	Inner and outer 53 to 57 lb. (24 to 25·8 kg.).

TAPPETS
ROCKERS   Refer to Series MGA data on preceding pages.
CAMSHAFT

General Data 14

M.G. 'MGA'. Issue 2. 48539

LUBRICATION

Refer also to Series MGA data on preceding pages.

Normal pressure: Running	70 lb./sq. in. (4·9 kg./cm.²) at 30 m.p.h.
Idling	15 lb./sq. in. (1·05 kg./cm.²) at 500 r.p.m.

TORQUE WRENCH SETTINGS
FUEL SYSTEM
AIR CLEANER AND FUEL PUMP

Refer to Series MGA data on preceding pages.

COOLING SYSTEM

Thermostat opening temperature	150·8° F. (66° C.).

IGNITION SYSTEM

Static ignition timing: High compression	10° B.T.D.C. (up to Engine No. 4003).
	5° B.T.D.C. (from Engine No. 4004).
Low compression	10° B.T.D.C.

CLUTCH

Make and type	Borg & Beck 8A6–G single dry plate.
Facing material	Wound yarn.
Friction plate damper springs	6. Maroon and light green.
Pressure springs	6. Light grey.
Minimum free length	2·27 in. (57·658 mm.).
Rate	282 lb. in. (3·24 kg. mm.).
Total spring load (mean)	1,200 lb. (544·3 kg.).
Test length	1·56 in. (39·624 mm.).
Load	195 to 205 lb. (88·45 to 92·98 kg.).

GEARBOX

Number of forward speeds	4.
Synchromesh	Second, third, and fourth gears.
Ratios: Top	1·0 : 1.
Third	1·374 : 1.
Second	2·214 : 1.
First	3·64 : 1.
Reverse	4·76 : 1.
Overall ratios: Top	4·1 : 1.
Third	5·633 : 1.
Second	9·077 : 1.
First	14·924 : 1.
Reverse	19·516 : 1.
Speedometer gears ratio	5: 12.

STEERING
FRONT SUSPENSION
REAR SUSPENSION
PROPELLER SHAFT

Refer to Series MGA data on preceding pages.

REAR AXLE

Make and type	B.M.C. 'B' type, three-quarter-floating.
Ratio	10/41.

ELECTRICAL EQUIPMENT
BRAKES } Refer to Series MGA data on preceding pages.
WHEELS

TYRES

Size .. .. 5·60—15 Gold Seal nylon (tubed).

Pressures:

Normal use including motorways up to 100 m.p.h.: Front   21 lb./sq. in. (1·47 kg./cm.²).

Rear   24 lb./sq. in. (1·68 kg./cm.²).

Maximum or near-maximum performance: Front .. .. 24 lb./sq. in. (1·68 kg./cm.²).

Rear .. .. 27 lb./sq. in. (1·89 kg./cm.²).

Size .. .. 5·90—15 Road Speed RS5 (tubed).

Pressures:

Normal use: Front .. .. 17 lb./sq. in. (1·19 kg./cm.²).

Rear .. .. 20 lb./sq. in. (1·40 kg./cm.²).

Maximum or near-maximum speeds sustained for lengthy

periods or for competition use: Front .. .. .. 24 lb./sq. in. (1·68 kg./cm.²).

Rear .. .. .. 27 lb./sq. in. (1·89 kg./cm.²).

CAPACITIES
GENERAL DIMENSIONS } Refer to Series MGA data on preceding pages.

WEIGHTS

Kerbside weight	2,016 lb. (914·4 kg.).
Shipping weight	1,987 lb. (901·2 kg.).
Engine and clutch (dry)	355 lb. (136·0 kg.).

GENERAL INFORMATION

CONTROLS

Hand brake

The hand brake lever is located alongside the driver's seat and operates the rear wheel brakes only.

To operate, pull up the lever and press the knob in the end with the thumb to lock the lever in position. To release the brakes, pull upwards on the lever to automatically release the lock and then push downwards.

Always apply the hand brake when parking.

Brake pedal

The pedal operates the hydraulic brakes on all four wheels and will also operate the twin stop warning lamps when the ignition is switched on.

Gear lever

The four forward gears and the reverse gear are engaged by moving the lever to the positions indicated in the illustration.

To engage the reverse gear move the lever to the left of the neutral position until resistance is felt, apply side pressure to the lever to overcome the resistance and then pull it backwards to engage the gear.

Synchromesh engagement is provided on second, third and fourth gears.

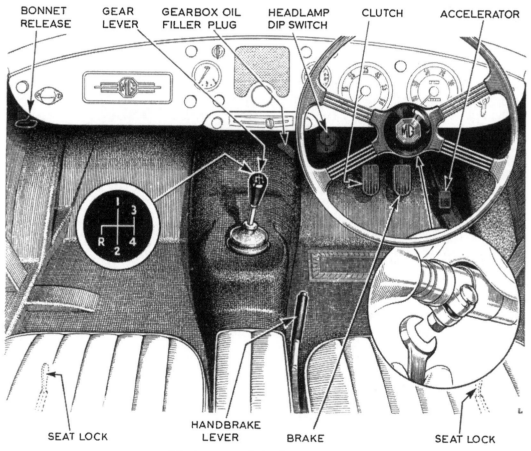

The location of the driving controls

Seat adjustment

A lever is provided at the front of each seat and this must be pressed outwards to release the catches and allow the seat to slide.

Steering column adjustment

This enables the steering wheel to be placed in the most comfortable driving position after slackening a clamp bolt below the wheel hub.

Headlamp beam dip switch

This is situated on the left of the clutch pedal and is foot operated. The switch will dip the headlamp beams on one depression and raise them on the next.

Bonnet lock release

The bonnet is hinged at the rear and the lock is released by pulling on the ring below the instrument panel on the extreme left-hand side of the car.

The bonnet is still held by the safety catch, which must be released before the bonnet can be raised.

To re-lock the bonnet in the fully closed position after opening, press downwards on the front of the bonnet until the lock is heard to engage.

INSTRUMENT PANEL

Speedometer

The speedometer also records the trip and total distances. The trip recorder is reset to zero by pushing upwards the knob below the instrument and turning it anti-clockwise.

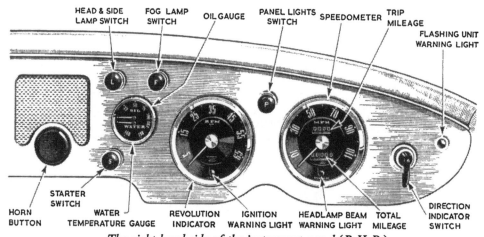

The right-hand side of the instrument panel (R.H.D.)

Main beam warning light

The warning light at the bottom of the speedometer dial glows red when the headlamp main beams are in use, as a reminder to dip the beams when approaching other traffic.

Engine revolution indicator

This dial is calibrated in hundreds of revolutions per minute. Normal use of the engine will not require speeds over 5,000 r.p.m. and great care must be taken if the needle does approach the amber sector of the dial, which commences at 5,500 r.p.m. Under favourable conditions the needle may be allowed to enter the amber sector but under no circumstances must it enter the red sector.

Ignition warning light

The warning light at the bottom of the revolution indicator dial glows red when the ignition is switched on and will go out again when the engine is started and its speed is increased sufficiently for the dynamo to charge the battery. Should the light glow at all engine speeds, the dynamo is not charging the battery.

Oil pressure gauge

The pressure of the oil should be between 30 lb./sq. in. and 80 lb./sq. in. (2·1 kg./cm.^{2} and 5·6 kg./cm.^{2}) under normal running conditions. Approximately 10 lb./sq. in. (·7 kg./cm.^{2}) should be shown when the engine is idling.

Water temperature gauge

The temperature of the cooling water leaving the cylinder head is indicated by this gauge and should be approximately 160° F. when the engine is running normally.

Starter switch

Pull out the knob marked 'S' to operate the starter motor. The switch must be pushed in immediately the engine starts.

Lamp switch

To switch on the sidelamps, tail-lamps, and number-plate illumination lamp pull out the knob marked 'L'.
Turn the knob clockwise and pull out again to switch on the headlamps.
See **'Headlamp beam dip switch'** and **'Main beam warning light'.**

Fog lamp switch

A fog lamp is not fitted as standard equipment, but the switch marked 'F' on the instrument panel is connected to the battery and is ready for use when a fog lamp is connected to it.
Pull out the knob to switch on the fog lamp.

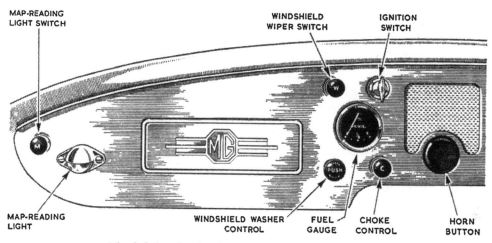

The left-hand side of the instrument panel (R.H.D.)

Panel lamp switch

To illuminate the instruments turn the control knob 'P' clockwise. The first movement of the knob will switch on the lamps and further turning to the right will dim the lamps.
The panel lamps will only operate when the sidelamps are also switched on.

Direction indicator switch

The lever-type switch on the outer edge of the panel controls the flashing indicator unit. The unit will operate only while the ignition is switched on and flashes the sidelamp and tail lamp, on the side of the car to which the switch lever is moved, until it is automatically switched off.
While the flashing unit is switched on, the warning light next to the switch will show green.

Fuel gauge

This operates only when the ignition is switched on.

Choke or mixture control

To enrich the mixture and assist starting when the engine is cold, pull out the knob marked 'C' and lock it in position by turning it anti-clockwise. Turn the knob clockwise and push it inwards to the normal running position as soon as the engine is warm enough to run without the rich mixture.
Never allow the engine to run for any length of time with the knob pulled out.

Ignition switch

The fuel pump and gauge are brought into action by this switch, which is also the master switch for the windshield wipers and direction indicators.

Windshield wiper switch

The windshield wipers are self-parking and operate only when the ignition is switched on.

Pull out the control 'W' to set the wiper blades in motion. Push in the knob to switch off the motor and park the blades.

Map-reading lamp

The map-reading lamp is controlled by the adjacent knob, which must be pulled out to switch on the light. The lamp will only operate while the sidelamps are switched on.

Windshield washer

When windshield-washing equipment is fitted it is operated by the knob marked 'Push' below the fuel gauge.

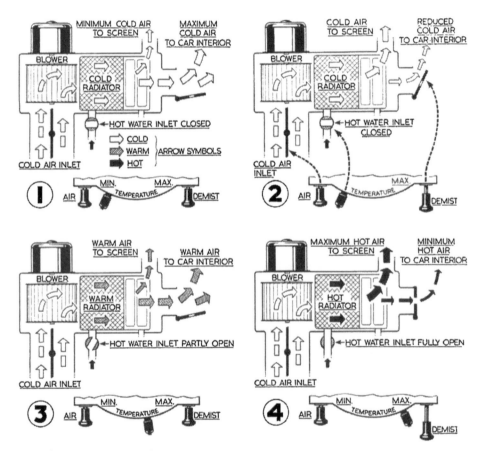

The circulation of the air through the heater unit with the controls positioned as recommended on page General Information 5

HEATING AND DEMISTING EQUIPMENT

The 2·75-kw. heating and demisting unit is fitted as an extra to standard equipment (see Section S.9).

Fresh air is ducted from the radiator grille to the heating element and blower motor mounted below the bonnet. Water from the engine cooling system is used to heat the element.

Warmed air issues from the toeboard or the windshield demisting vents according to the position of the controls mounted below the instrument panel

Air

The left-hand knob controls the air supply. When the knob is pushed in the air duct is open and air at atmospheric temperature will enter the car when it is in motion and will issue from the toeboard or demisting vents.

While the control is pushed in it may be turned clockwise to switch on the blower motor, if the ignition is switched on also, and this will increase the flow of air into the car unit and may be used to give a supply of air when the car is stationary.

If the blower motor is switched off by the air control, the knob can be pulled outwards to close the air duct and prevent fresh air entering the car from the toeboard or windshield vents. The blower cannot be switched on while the knob is pulled out.

NOTE.—The heating and demisting equipment control panel fitted to some cars has the blower motor operating switch incorporated in the temperature control lever. These control panels may be identified by the temperature lever knob, which is round and marked 'B'. Pull out the knob to switch on the blower motor.

The left-hand control on these panels will regulate the air supply only.

Demist

The right-hand knob on the heater unit control panel operates a shutter in the panel above the gearbox cover When the control is pushed into the normal position the shutter is open and most of the air from the unit will enter the car at the toeboard while some will issue from the vents below the windshield. As the knob is pulled out the shutter closes and more air is delivered to the car from the demisting vents, giving the maximum supply of air to the windshield. This is the demist position of the control if the blower is switched on and also the defrost position if the heater is operating.

Temperature

The temperature lever operates the water valve on the engine. When the lever is in the left-hand position the hot water supply is cut off and air entering the car through the unit will not be heated. As the lever is moved to the right the water supply is increased and the maximum temperature is obtained.

As a general guide, here are some of the more frequently required positions:

(1) *No additional ventilation or heating.* Pull out the air control, push the temperature control to the left.

(2) *Hot weather.* Push in the air and demist controls. Move the temperature control to the left. To increase the supply of air switch on the blower motor.

(3) *Warm weather.* Set the controls as for hot weather. To increase the supply of air switch on the blower motor. To prevent mist forming on the windshield pull out the demist control partially.

(4) *Cold weather.* Place the air control in its normal position. Place the temperature lever according to the degree of heating required. Switch on the blower to increase the air supply. (If demisting is required pull out the demist control).

(5) *Severe cold.* Move the temperature control to the right for maximum heating and pull out the demist control fully to give a maximum supply of hot air to the screen. Switch on the blower motor to increase the air flow.

WINDSHIELD WASHER

The washing equipment supplied as an optional fitting is operated by pumping the knob on the instrument panel. As the knob moves towards the panel a jet of cleaning fluid is ejected onto the windshield from nozzles on the scuttle.

Set the windshield wipers in motion before operating the cleaning jets.

Fluid for the windshield is stored in an unbreakable bottle clipped to the engine bulkhead. When refilling with fluid, lift the bottle from its clip and unscrew the cap.

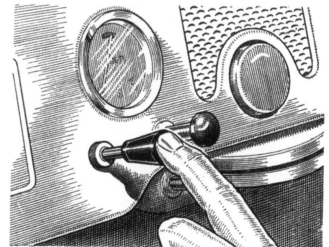

Operating the windshield washer

M.G. 'MGA'. Issue 1. 22142

General Information 5

FOLDING THE HOOD

Never fold the hood if it is wet or damp; wait until it is dry.

(1) Release the hood from the pillars at the top of the windshield by unscrewing the wing bolts.

(2) Release the rear bottom edge of the hood from the three buttons and the turnbuckle at each side. Pull on the centre knob of each button to release them from their attachment pins.

(3) Raise the front of the hood slightly to release the tension in the canvas and pull to the rear the bottom of the hood where it is attached to the tonneau panel to release it from the two anchor brackets on the panel.

(4) Tip the seats forward, unfasten the sidescreen container, and turn it over onto the tonneau panel to expose the hood stowage compartment.

(5) Leave the rear window panel suspended over the tonneau panel and collapse the hood into the stowage compartment, pulling the canvas clear of the hood irons and folding it forward over the front hood rail.

(6) Fold the rear window forward over the hood, pulling out the spare canvas at each side and folding it neatly over the front of the window.

(7) Push the hood into the stowage compartment and turn the sidescreen container forward to cover the hood

(8) Remove the sidescreens and stow them in the container pockets with the cranked bracket of each screen at opposite ends and facing towards the rear.

(9) Secure the sidescreen container over the folded hood with the six buttons (three on each side).

(10) Tighten the sidescreen clamping nut on each door to prevent its possible loss.

SERIAL NUMBERS

The major components of the vehicle have serial numbers and these will be found in the positions illustrated on pages General Information 6 and 7. When in communication with the Company or your Dealer always quote the engine number and car number complete with prefixes. The registration number is of no assistance and is not required. Write your name and address clearly.

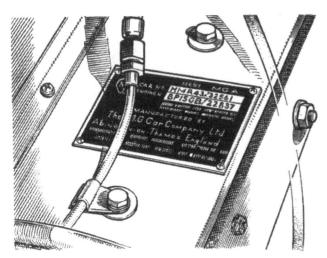

Chassis Number. This is stamped on the identification plate and should be quoted with its prefix. The plate is secured to the top left-hand side of the dash panel

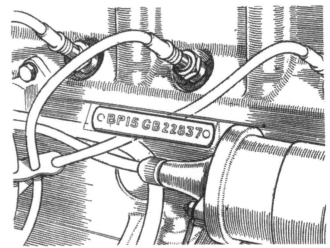

Engine Number (earlier models). This is stamped on the identification plate and also on a plate on the right-hand side of the cylinder block

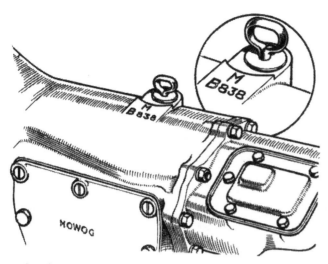

Gearbox Number. Stamped on top of the gearbox casing adjacent to the dipstick

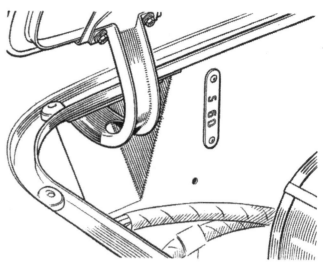

Body Number. Stamped on a plate secured to the right-hand side of the dash panel

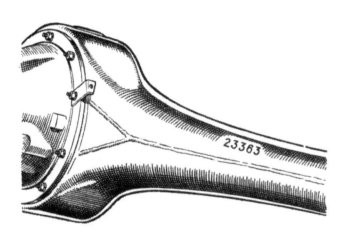

Rear Axle Number. Stamped on the front of the left-hand rear axle tube

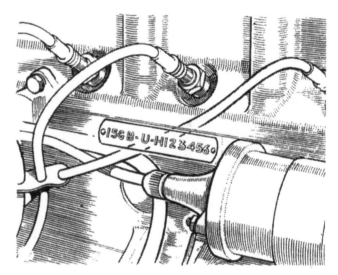

Engine Number (later models). This is stamped on a plate on the right-hand side of the cylinder block

POWER UNIT SERIAL NUMBER CODING

The engine number on later engines comprises a series of letters and numbers, presenting in code the capacity, make, and type of unit, gearbox and ancillaries fitted, and the type of compression together with the serial number of the unit.

1st PREFIX GROUP—Cubic capacity, make, and type

1st Prefix number 8—803 c.c.
9—950 c.c.
12—1200 c.c.
15—1500 c.c.
16—1600 c.c.
22—2200 c.c.
25—2500 c.c.
26—2600 c.c.

1st Prefix letter A—Austin
B—B.M.C. Industrials
G—M.G.
H—Miscellaneous special
J—Commercial
M—Morris
R—Riley
W—Wolseley

2nd Prefix letter A–Z used for the variations of engine type

2nd PREFIX GROUP—Gearbox and ancillaries

A—Automatic gearbox
M—Manumatic clutch
N—Steering-column gear change gearbox
O—Overdrive (Borg-Warner)
P—Police specification
U—Centre gear change gearbox

3rd GROUP—Compression and serial number

H—High compression ⎱ and serial number of unit
L—Low compression ⎰

CODE EXAMPLE

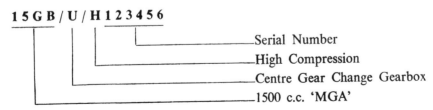

15GB/U/H123456
Serial Number
High Compression
Centre Gear Change Gearbox
1500 c.c. 'MGA'

CAR NUMBER IDENTIFICATION CODE

In order to provide comprehensive information concerning the vehicle the identification plate is stamped with symbols.

The symbols consist of three letters and two figures followed by the usual serial number of the vehicle for cars of one colour, and four letters and two figures followed by the vehicle serial number for vehicles with a duotone finish.

The first letter when related to the code provides indication of the make and model of the vehicle—Morris Minor, M.G. 'MGA', etc.

The second letter provides an indication of the type of vehicle—Saloon, Tourer, Van, etc.

The third letter indicates the colour in which the vehicle is finished or the top colour on duotone vehicles.

The fourth letter indicates the bottom colour in which the duotone vehicle is finished. For duotone vehicles the third and fourth letters are read in the same (colour) column.

The first figure indicates the class to which the vehicle belongs—R.H.D. Home, L.H.D., etc.

The second figure indicates the type of paint used to finish the car—Cellulose, Synthetic, etc.

From this it will be clear that when an owner quotes the code number of his vehicle it is a relatively simple matter to obtain a comprehensive picture of the vehicle concerned by reference to the following tabulated code of symbols.

Model	Code	Type	Code	Colour	Code	Class	Code	Paint	Code
Wolseley 6/80	A	Saloon 4-door	A	Black	A	R.H.D. Home	1	Synthetic	1
Wolseley 4/50	B	Saloon 2-door	B	Light Grey	B	R.H.D. Export	2	Synobel	2
Morris Six	C	Tourer	C	Dark Red	C	L.H.D.	3	Cellulose	3
Morris Oxford	D	2-Seater	D	Dark Blue	D	North America	4	Metallic	4
Morris Cowley	E	Van	E	Mid Green	E	C.K.D.—R.H.D.	5	Primed	5
Morris Minor	F	Truck	F	Beige	F	C.K.D.—L.H.D.	6	Cellulosed body and synthetic wings	6
Morris 5-cwt.	G	Cab	G	Brown	G				
M.G. 'MGA'	H	Mail	H	C.K.D. Finish	H				
M.G. 1½-litre	J	Engineers	J	Dark Grey	J				
M.G. Magnette	K	Chassis	K	Light Red	K				
Riley 1½-litre	L	Traveller	L	Light Blue	L				
Riley 2½-litre	M								
Wolseley 4/44	N								
Quarter-ton	O								
Half-ton	P			Ivory	P				
Wolseley 6/90	R			White	R				
Isis	S			Mid Grey	S				
Wolseley 15/50	T			Light Green	T				
				Dark Green	U				

As an example:

The symbols HDA 43/10101 when decoded give—M.G. 'MGA', 2-seater, Black, North America, Cellulose, Car No. 10101.

Owing to the fact that the technique required to effect repairs to the different paint finishes varies considerably and that the correct paint **must** be used for such purpose, it is to be noted that the last number(s) of the symbols is of particular importance as it defines the nature of the paint used in the Factory to finish the car.

GENERAL INFORMATION—*continued*

IDENTIFICATION OF UNIFIED SCREW THREADS

The general standardization of Unified screw threads makes it necessary to identify all nuts, bolts, and set screws with these threads in order to ensure their being matched with correspondingly threaded components and the fitting of correct replacements.

Identification has been standardized and is effected in the following manner:

Nuts. By a circular groove turned on the end face of the nut or by connected circles stamped on one flat of the hexagon.

Bolts and set screws. By a circular depression turned on the head or by connected circles stamped on one flat of the hexagon.

Wheel stud nuts. By a notch cut in all the corners of the hexagon.

It is of the utmost importance that any nuts, bolts, or set screws marked with the above identifications are used only in conjunction with associated components having Unified threads and that only replacement parts with Unified threads are used, as these are *not* interchangeable with Whitworth, B.S.F., or Metric threads.

The Unified thread is, however, interchangeable with the American National Fine (A.N.F.) thread for all practical purposes.

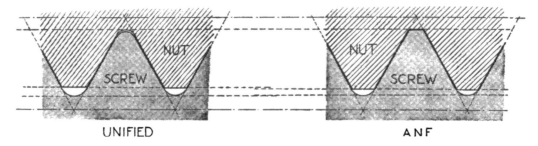

This illustration of the Unified thread and the A.N.F. thread to the same scale indicates their close relationship

Spanners. It is to be noted that all A.N.F.- and Unified-threaded nuts and hexagon-headed bolts are made to the standard American hexagon sizes and that spanners of the appropriate size must be used when tightening or loosening them.

KEY TO SPANNER SIZES (Nominal widths between jaws)

Diameter of screw thread (inches)	$\frac{1}{4}''$	$\frac{5}{16}''$	$\frac{3}{8}''$	$\frac{7}{16}''$	$\frac{1}{2}''$	$\frac{9}{16}''$	$\frac{5}{8}''$	$\frac{3}{4}''$	$\frac{7}{8}''$	$1''$
For B.S.F. screws and nuts	·448	·529	·604	·705	·825	·925	1·016	1·207	1·309	1·489
For A.N.F. screws and nuts	·440	·504	·566	·629	·755	·880	·944	1·132	1·320	1·508
For Unified screws	·440	·504	·566	**·630**	·755	**·817**	·943	1·132	1·321	1·509
For Unified nuts (normal)	·440	·504	·566	**·692**	·755	**·880**	·943	1·132	1·321	1·509
For Unified nuts (heavy)	—	—	—	—	—	—	**1·069**	**1·258**	**1·446**	—

NOTE.—In the case of some Unified-threaded components the size of the hexagon for the nut is different from that of the bolt. Where this occurs the spanner size is shown in heavy type in the above table

General Information 10

M.G. 'MGA'. Issue 2. 40954

PART NAME ALTERNATIVES

	M.G. part name	*Alternatives*
ENGINE	Gudgeon pin	Piston pin. Small-end pin. Wrist pin.
	Scraper ring	Oil control ring.
	Core plug	Expansion plug. Welch plug. Sealing disc.
	Oil sump	Oil pan. Oil reservoir.
CONTROLS	Mixture control	Choke. Strangler.
GEARBOX	Gear lever	Shift lever.
	Change speed fork	Shift fork. Selector fork.
	First motion shaft	Clutch shaft. First reduction pinion. Main drive pinion. Drive gear.
	Layshaft	Countershaft.
AXLE	Crown wheel	Ring gear. Spiral drive gear.
	Bevel pinion	Small pinion. Spiral drive pinion.
	'U' bolts	Spring clips.
	Axle shaft	Half-shaft. Hub driving shaft. Jack driving shaft.
	Differential gear	Sun wheel.
	Differential pinion	Planet wheel.
STEERING	Swivel pin	Pivot pin. Steering pin. King pin.
	Stub axle	Swivel axle.
	Track-rod	Cross-tube.
	Draglink	Side-tube. Steering connecting rod.
ELECTRICAL	Dynamo	Generator.
	Control box	Voltage regulator. Cut-out. Voltage control.
EXHAUST	Silencer	Muffler.
BODY	Bonnet	Hood.
	Wing	Mudguard. Fender.

CLAIMS UNDER WARRANTY

Claims for the replacement of material or parts under Warranty must always be submitted to the supplying Distributor or Dealer, or, when this is not possible, to the nearest Distributor or Dealer, informing them of the Vendor's name and address.

FROST PRECAUTIONS

If the car is not stored in a warmed building, steps must be taken to prevent the cooling water from freezing during frosty weather. As a precautionary measure when frost is anticipated an anti-freezing solution must be used in the cooling system. The heater unit fitted to the M.G. (Series MGA) cannot be drained completely by the cooling system drain taps and the use of anti-freeze is essential on this model in freezing weather.

The cooling system is of the sealed type and relatively high temperatures are developed in the radiator upper tank. For this reason anti-freeze solutions having an alcohol base are unsuitable owing to their high evaporation rate producing a rapid loss of coolant and consequent interruption of circulation.

Only anti-freeze of the ethylene glycol type incorporating the correct type of corrosion inhibitor is suitable and owners are recommended to use Bluecol, Shell, Esso Anti-freeze, or any other anti-freeze conforming to Specification B.S.3151 or B.S.3152.

The recommended quantities of anti-freeze for different degrees of frost resistance are:

15° frost (17° F. or −8° C.)	1 pint (·57 litre, 1·2 U.S. pints)
25° frost (7° F. or −14° C.)	1½ pints (·85 litre, 1·8 U.S. pints)
35° frost (−3° F. or −19° C.)	2½ pints (1·42 litres, 3 U.S. pints)

Where temperatures below 0° F. or −18° C. are likely to be encountered a solution of at least 25 per cent. of anti-freeze must be used to ensure immunity from trouble. Consult your local Dealer on this matter.

First decide what degree of frost protection is required before adding anti-freeze to the radiator.

Make sure that the cooling system is watertight and examine all joints, replacing any defective rubber hose with a new one.

Before introducing anti-freeze mixture to the radiator it is advisable to clean out the cooling system thoroughly by draining out the water and swilling out the water passages with a hose inserted in the radiator filler, keeping the drain taps open.

Avoid excessive topping up, otherwise there is a risk of losing valuable anti-freeze due to expansion of the solution. Only top up when the cooling system is at its normal running temperature.

Generally speaking, anti-freeze is not injurious to cellulose paint, provided it is wiped off in reasonable time. It must not, however, be allowed to remain on the paintwork.

Radiator anti-freeze should not be used in windshield-washing equipment.

RUNNING-IN SPEEDS

The treatment given to a new car will have an important bearing on its subsequent life, and engine speeds during this early period must be limited. The following instructions should be strictly adhered to.

During the first 500 miles (800 km.)

DO NOT exceed 45 m.p.h. (72 km.p.h.).

DO NOT operate at full throttle in any gear.

DO NOT allow the engine to labour in any gear.

FILLING UP WITH FUEL

Considerable loss of fuel can occur as a result of filling the fuel tank until the fuel is visible in the filler tube. If this is done and the vehicle is left in the sun, expansion due to heat will cause leakage, with consequent loss of and danger from exposed fuel.

(1) Avoid overfilling the tank until the fuel is visible in the filler tube.

(2) If the tank is inadvertently overfilled, take care to park the vehicle in the shade with the filler as high as possible.

CAR NUMBER IDENTIFICATION CODE (MGA 1600 MODELS)

The car number symbol consists of three letters and one figure followed by a fifth prefix letter (L) if the vehicle is left-hand drive, and then by the serial number of the vehicle.

The first letter when related to the code provides an indication of the make of the vehicle—M.G., etc.

The second letter provides an indication of the model's cubic capacity.

The third letter indicates the type of body—2-seat Tourer, etc.

The first figure indicates the series of model—1, 2, etc.

1st Prefix Letter—Name		2nd Prefix Letter—Model (cubic capacity)	
A—Austin	M—Morris	A—800–999 c.c.	G—1000–1399 c.c.
C—Austin Cooper	R—Riley	B—2000–2999 c.c.	H—1400–1999 c.c.
G—M.G.	V—Vanden Plas	D—3000 c.c. upwards	L—Up to 799 c.c.
H—Healey	W—Wolseley		
K—Morris Cooper			

3rd Prefix Letter—Body type

A—Ambulance
B—Buckboard
C—Chassis
D—Coupé
E—G.P.O. Engineers
G—G.P.O. Mail

H—Hearse
J—Convertible
K—Truck
L—Hire-car
M—Limousine
N—2-seat Tourer

P—Hard Top
O—Chassis and Cab
R—Chassis and Scuttle
S—4-door Saloon
2S—2-door Saloon

T—4-seat Tourer
U—Pick-up
V—Van
W—Dual-purpose
X—Taxi

4th Prefix—Series of model (2, 3, etc., used to record a major change).

5th Prefix (used when vehicles differ from standard R.H.D.) D—De-luxe.
L—Left-hand drive.
S—Super De-luxe.

Code Example
GHN 68851

B.M.C. SEAT BELTS

The body of the car incorporates anchorage points to facilitate the fitting of B.M.C. seat belts.

To use the seat belt, position the buckle tongue on the long belt approximately in the centre of the belt and ensure that the upper part of the belt passes over the shoulder; pass the tongue across the body. Adjust the short belt until the buckle is located just in front of the hip and push the tongue into the buckle until it clicks in the locked position. Finally, adjust the long belt until the user is held firmly but comfortably in the seat.

To release the seat belt lift up the buckle lever. After releasing the seat belt the long belt must be stowed in such a way as to give clear access to the doors.

Notes

MAINTENANCE ATTENTION

PERIODICAL

Daily

Check oil level in crankcase, and top up if necessary.
Check coolant level in radiator, and top up if necessary.

Weekly

Check tyre pressures, and regulate if necessary.
Check batteries and top up to correct level.

3,000 miles (5000 km.) service

1. *Engine*
   Top up carburetter piston dampers.
   Lubricate carburetter controls.
   Top up coolant in radiator.
   Check fan belt tension.
   Clean and re-oil air cleaner elements.

2. *Clutch*
   Check level of fluid in hydraulic clutch supply tank, and top up if necessary.
   Check free pedal movement (where applicable).

3. *Brakes*
   Check brakes, and adjust if necessary.
   Make a visual inspection of brake lines and pipes.
   Check fluid level in hydraulic brake supply tank, and top up if necessary.

4. *Body*
   Lubricate door locks and hinges, safety catches, bonnet lock operating mechanism, and safety catch.
   Inject oil through key slot.
   Lightly smear dovetails and striking plates with a suitable grease.

5. *Electrical*
   Check batteries and top up to correct level.

6. *Lubrication*
   Change engine oil.
   Top up gearbox and rear axle.
   Lubricate all nipples except steering rack and pinion.

7. *Wheels and tyres*
   Change road wheels round diagonally, including spare, to regularize tyre wear.
   Check tyre pressures.

6,000 miles (10000 km.) service

1. *Engine*
   Top up carburetter piston dampers.
   Lubricate carburetter controls.
   Top up coolant in radiator.
   Check fan belt tension.
   Check valve rocker clearances, and adjust if necessary.
   Clean and re-oil air cleaners.

2. *Ignition*
   Check functioning of automatic advance and retard mechanism.
   Lubricate all distributor parts as necessary.
   Check and adjust distributor contact points.
   Clean and adjust sparking plugs.

3. *Clutch*
   Check level of fluid in hydraulic clutch supply tank, and top up if necessary.
   Check free pedal movement (where applicable).

4. *Brakes*
   Check brakes, and adjust if necessary.

Make a visual inspection of brake lines and pipes.
Check fluid level in hydraulic brake supply tank, and top up if necessary.
Inspect disc brake friction pads and report if attention is required.

5. *Steering*
   Check wheel alignment, and adjust if necessary.

6. *General*
   Tighten rear road spring seat bolts.

7. *Body*
   Lubricate door locks and hinges, safety catches, bonnet lock operating mechanism, and safety catch.
   Inject oil through key slot.
   Lightly smear dovetails and striking plates with a suitable grease.

8. *Electrical*
   Check specific gravity of battery cells and top up to correct level.
   Lubricate dynamo bearing.
   Check all lamps for correct functioning.

9. *Lubrication*
   Change oils in engine, gearbox, and rear axle.
   Fit new oil filter element.
   Lubricate all nipples except steering rack and pinion.

10. *Wheels and tyres*
    Change road wheels round diagonally, including spare, to regularize tyre wear.
    Check tyre pressures.

9,000 miles (15000 km.) service

Carry out the **3,000 miles (5000 km.) service**.

12,000 miles (20000 km.) service

1. *Engine*
   Remove carburetter suction chambers and pistons, clean, reassemble, and top up the damper pistons.
   Lubricate the carburetter controls.
   Check valve rocker clearances, and adjust as necessary.
   Check fan belt tension.
   Clean and re-oil air cleaners.

2. *Ignition*
   Check functioning of automatic advance and retard mechanism.
   Lubricate all distributor parts as necessary.
   Clean and adjust distributor contact points.
   Fit new sparking plugs.

3. *Clutch*
   Check level of fluid in hydraulic clutch supply tank, and top up if necessary.
   Check free pedal movement (where applicable).

4. *Steering*
   Check steering and front suspension moving parts for wear.
   Check wheel alignment, and adjust as necessary.

5. *Brakes*
   Check brakes, and adjust as necessary.
   Make a visual inspection of brake lines and pipes.
   Check fluid level in hydraulic brake supply tank, and top up if necessary.
   Inspect disc brake friction pads and report if attention is required.

6. *Radiator*
   Drain, flush out, and refill the radiator.

7. *General*
   Tighten rear road spring seat bolts.

M.G. 'MGA'.   Issue 5.   48539

MAINTENANCE ATTENTION—*continued*

12,000 miles (20000 km.) service—*continued*

8. *Body*
Lubricate door locks and hinges, safety catches, bonnet lock operating mechanism, and safety catch.
Inject oil through key slot.
Lightly smear dovetails and striking plates with a suitable grease.

9. *Electrical*
Check specific gravity of battery cells and top up to correct level.
Lubricate dynamo bearing.
Check all lamps for correct functioning.
Check headlamp beam setting, and adjust if necessary.

10. *Lubrication*
Drain off old engine oil, flush out engine, and refill with fresh oil.

Change oil in gearbox and rear axle.
Fit new oil filter element.
Lubricate all grease nipples.
Lubricate steering rack and pinion.

11. *Wheels and tyres*
Change road wheels round diagonally, including spare, to regularize tyre wear.
Check tyre pressures.

24,000 miles (40000 km.) service
Carry out the **12,000 miles (20000 km.) service**, with the following amendment:

1. *Lubrication*
Remove the engine sump and pick-up strainer, clean the sump, strainer, and crankcase, reassemble, and refill with fresh oil.

Maintenance Attention 2

M.G. 'MGA'. Issue 5. 48539

38

SECTION A

THE ENGINE

THE ENGINE COMPONENTS

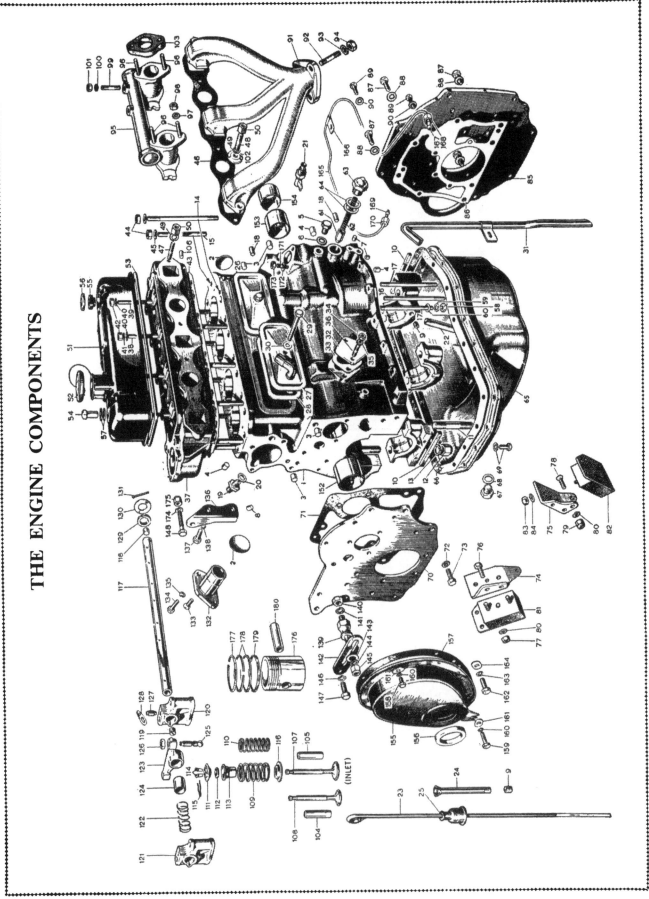

KEY TO THE ENGINE COMPONENTS

No.	Description	No.	Description	No.	Description	No.	Description
1.	Block assembly—cylinder.	45.	Washer—stud.	89.	Set screw—plate to block.	133.	Screw—to block.
2.	Plug—core hole.	46.	Joint—manifold to head.	90.	Lock washer—set screw.	134.	Screw—distributor to housing.
3.	Plug—oil gallery.	47.	Stud—exhaust manifold to head.	91.	Manifold—exhaust.	135.	Washer—spring—screw.
4.	Plug—taper—crankcase oil hole.	48.	Stud—manifolds to head.	92.	Stud—exhaust pipe flange.	136.	Bracket—dynamo—rear.
5.	Plug—screwed—transverse oil hole.	49.	Washer—spring.	93.	Washer.	137.	Screw—bracket to crankcase.
6.	Washer—plug.	50.	Nut—stainless steel.	94.	Nut.	138.	Washer—spring.
7.	Plug—oil relief valve hole.	51.	Cover assembly—rocker gear.	95.	Manifold—induction.	139.	Adjusting link pillar.
8.	Plug—oil filter boss.	52.	Cap—oil filler and cable.	96.	Stud—carburetter.	140.	Nut—pillar to front plate.
9.	Plug—redundant dipper boss (and rear main bearing cap).	53.	Joint—cover to head.	97.	Washer—spring.	141.	Washer.
10.	Joint—front/rear main bearing cap.	54.	Cap nut—cover.	98.	Nut—carburetter stud.	142.	Adjusting link.
11.	Stud—main bearing cap.	55.	Bush—rubber—cap nut.	99.	Stud—inlet manifold (accelerator abutment bracket).	143.	Washer—plain.
12.	Nut—main bearing cap stud.	56.	Cup washer—bush.	100.	Washer—spring.	144.	Washer—spring.
13.	Washer—spring.	57.	Washer—packing—cover stud.	101.	Nut—inlet manifold stud.	145.	Nut—link to pillar.
14.	Stud (long)—cylinder head.	58.	Nut—oil pump stud.	102.	Yoke—manifold.	146.	Washer—spring—link to dynamo.
15.	Stud (short)—cylinder head.	59.	Washer—spring.	103.	Washer—carburetter insulating.	147.	Screw—link to dynamo.
16.	Stud (long)—oil pump.	60.	Washer—plain.	104.	Guide—valve—exhaust.	148.	Bolt—dynamo to bracket.
17.	Stud (short)—oil pump.	61.	Valve—oil relief.	105.	Guide—valve—inlet.	152.	Liner—front camshaft bearing.
18.	Dowel—gearbox mounting plate.	62.	Spring—relief valve.	106.	Plug—oil hole.	153.	Liner—centre camshaft bearing.
19.	Union—oil gauge pipe.	63.	Cap nut—relief valve.	107.	Valve—inlet.	154.	Liner—rear camshaft bearing.
20.	Washer—union.	64.	Washer—cap nut.	108.	Valve—exhaust.	155.	Cover complete—crankcase (front).
21.	Tap—water drain.	65.	Reservoir—oil.	109.	Spring—valve (outer).	156.	Felt ring.
22.	Pipe—drain—rear bearing cap.	66.	Joint—oil reservoir.	110.	Spring—valve (inner).	157.	Joint—crankcase front cover.
23.	Dipper rod.	67.	Plug—drain.	111.	Cup—valve spring.	158.	Set screw—cover to engine plate.
24.	Tube—dipper rod.	68.	Washer—drain plug.	112.	Packing ring—valve.	159.	Set screw—cover and plate to bearing cap.
25.	Dust cap.	69.	Set screw—reservoir (with captive washer).	113.	Shroud—valve guide.	160.	Washer—spring.
26.	Cover—side—rear.	70.	Plate—front mounting.	114.	Cotters—cotter.	161.	Washer—plain.
27.	Cover—side—front—with elbow.	71.	Joint—plate to block.	115.	Circlip—cotter.	162.	Set screw—cover and plate to crankcase.
28.	Joint—side covers.	72.	Washer—spring.	116.	Collars—valve spring (bottom).	163.	Washer—spring.
29.	Set screw—covers.	73.	Set screw—plate to block.	117.	Shaft.	164.	Washer—set screw.
30.	Washer—set screw.	74.	Bracket—front R/H.	118.	Plug—plain.	165.	Pipe—ignition control.
31.	Pipe—vent with clip—crankcase.	75.	Bracket—front L/H.	119.	Plug—screwed.	166.	Clip—pipe.
32.	Plate—cylinder block blanking—N/S.	76.	Screw—R/H bracket to mounting plate.	120.	Bracket—tapped hole.	167.	Nipple.
33.	Joint—blanking plate.	77.	Nut—R/H bracket to plate screw.	121.	Bracket—plain.	168.	Nut—carburetter end.
34.	Nut—blanking plate stud.	78.	Screw—countersunk—L/H bracket to plate.	122.	Spring—rocker shaft spacing.	169.	Olive—distributor end.
35.	Washer—blanking plate stud.	79.	Nut—countersunk screw.	123.	Rocker.	170.	Nut—distributor end.
36.	Stud—L/H side crankcase blanking plate.	80.	Washer—spring.	124.	Bush.	171.	Stud—tachometer pinion housing.
37.	Cylinder head.	81.	Mounting—R/H—front.	125.	Screw—tappet adjusting.	172.	Washer—housing stud.
38.	Stud—rocker bracket—short.	82.	Mounting—L/H—front.	126.	Locknut—screw.	173.	Nut—housing stud.
39.	Stud—rocker bracket—long.	83.	Nut—engine mounting to bracket.	127.	Screw—shaft—locating.	174.	Washer—dynamo bolt.
40.	Washer—stud.	84.	Washer—spring.	128.	Plate—locking—locating screw.	175.	Nut—dynamo bolt.
41.	Washer—spring—stud.	85.	Plate—gearbox mounting.	129.	Washer—double coil.	176.	Piston assembly.
42.	Nuts—rocker bracket studs.	86.	Joint—plate to block.	130.	Washer—plain.	177.	Ring—compression—first.
43.	Joint—head to block.	87.	Set screw—plate to block.	131.	Split pin.	178.	Ring—compression—second and third.
44.	Nut—cylinder head stud.	88.	Lock washer—set screw.	132.	Housing—distributor.	179.	Ring—scraper.
						180.	Gudgeon pin.

GENERAL DESCRIPTION

The M.G. (Series MGA) overhead-valve engine is built in unit construction, with an 8 in. (20·3 cm.) single-plate dry clutch.

The valves are set in line in the detachable cylinder head and are operated by rockers and push-rods from the camshaft in the left-hand side of the engine. Oil seals are fitted to the valves and there is the normal provision for clearance adjustment. The camshaft, running in three steel-backed white-metal bearings, is chain-driven and has a synthetic rubber chain silencer. The oil pump, distributor and engine revolution indicator are driven from the camshaft; each component has its own drive shaft.

The solid-skirt pistons are of aluminium alloy with anodized finish, and carry three compression rings and a slotted oil control ring. The gudgeon pins are clamped in connecting rods, which have steel-backed indium-lead renewable big-end bearings. Three steel-backed white-metal renewable bearings support the forged-steel counterbalanced crankshaft. The thrust is taken by special washers at the centre main bearing. The renewable element full-flow oil filter is secured by its centre bolt to the right-hand side of the engine.

A centrifugal water pump and fan are driven from the crankshaft by the dynamo belt.

Two semi-downdraught S.U. carburetters are supplied with fuel by an electric high-pressure S.U. pump. Air enters the carburetters through twin filters.

LUBRICATION SYSTEM

An eccentric-type oil pump inside the crankcase is driven from the camshaft by a short vertical shaft. Oil is drawn into the pump through a gauze strainer and is delivered through crankcase drillings to a non-adjustable plunger-type relief valve located at the rear of the engine on the left-hand side. From the relief valve the oil passes through an internal drilling across the rear of the block and through an external oil pipe to the main oil filter. From the filter the oil passes to the main oil gallery and drillings supply oil to the main, big-end and camshaft bearings. The connecting rod ends are drilled and supply oil to the cylinder walls.

From the rear camshaft bearing oil passes through the block and a drilling in the rear rocker shaft bracket to lubricate the rockers, returning to the sump via the push-rod holes.

Two grooves in the front camshaft journal register with small holes in the camshaft thrust plate as the camshaft turns, allowing a small amount of oil for chain and wheel lubrication to pass into the timing case twice in each revolution of the camshaft. A drain is provided to return oil from the timing case to the sump.

The filter bowl is filled with oil at full pressure which passes through the element into the annular space around the centre bolt and from there into the main oil gallery.

Section A.1

DRAINING THE SUMP

The sump on new and reconditioned engines must be drained and then filled with new oil after the first 500 miles (800 km.) and at intervals of every 3,000 miles (4800 km.). The hexagon-headed drain plug is at the rear of the sump on the right-hand side. The sump should be drained when the engine is hot as the oil will flow more readily; allow to drain for at least 10 minutes before the drain plug is replaced.

Every 6,000 miles (9600 km.) or every alternate oil change remove and wash out the filter bowl before filling the sump with new oil. Fit a new filter element and replace the bowl.

When the sump has been drained, approximately 7½ pints (9 U.S. pints, 4·26 litres) of oil are required to fill it.

The capacity of the filter is approximately ½ pint (·6 U.S. pint, ·28 litre), giving a total of 8 pints (9·6 U.S. pints, 4·54 litres).

Section A.2

OIL PRESSURE

Under normal running conditions the oil pressure should not drop below 30 lb./sq. in. (2·1 kg./cm.²) on the gauge at normal road speeds, whilst approximately 10 lb./sq. in. (·7 kg./cm.²) should be shown when the engine is idling. New engines with new oil will give considerably higher readings at low speeds.

Should there be a noticeable drop in pressure, the following points should be checked:

(1) That there is a good supply of the correct grade of oil in the engine sump.

(2) That the strainer in the sump is clean and not choked with sludge.

(3) That the bearings, to which oil is fed under pressure, have the correct working clearances. Should the bearings be worn and the clearances excessive, the oil will escape more readily from the sides of the bearings, particularly when the oil is warm and becomes more fluid. This will cause a drop in pressure on the gauge as compared with that shown when the bearings are in good order.

The automatic relief valve in the lubrication system deals with any excessive oil pressure when starting from cold. When hot the pressure drops as the oil becomes more fluid.

THE M.G. (Series MGA) ENGINE

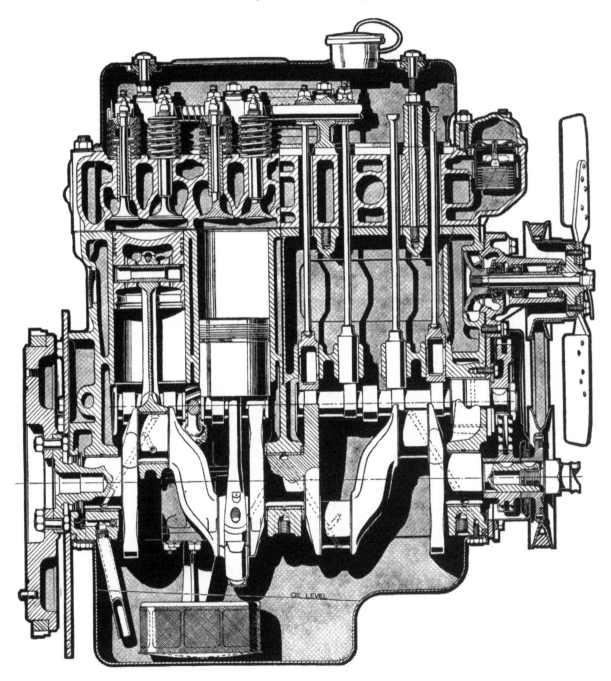

OIL LEVEL

Should the oil filter become blocked, two relief valves in the filter blow off to enable the oil to by-pass the filter and pass direct into the main gallery.

Continuous cold-running and unnecessary use of the mixture control are often the cause of serious oil dilution by fuel with a consequent drop in pressure.

Particular attention is called to the recommended change of oil every 3,000 miles (4800 km.).

Section A.3

OIL PRESSURE RELIEF VALVE

The non-adjustable oil pressure relief valve is situated at the rear of the left-hand side of the cylinder block and is held in position by a domed hexagon nut sealed by two fibre washers. The relief valve spring maintains a valve cup against a seating machined in the block.

The valve should be examined to ensure that the cup is seating correctly and that the relief spring has not lost its tension. The latter can be checked by measuring the length of the spring. To give the correct relief pressure of 75 to 80 lb. per sq. in. (5·3 to 5·6 kg./cm.²) this should not be less than 3 in. (7·6 cm.). Fit a new cup and spring if necessary.

Section A.4

REMOVING AND REPLACING THE WATER PUMP

The water pump is of the centrifugal impeller type and is mounted on a common spindle with the fan in a casting bolted to the front of the cylinder block.

The water pump and fan assembly is attached to the front of the cylinder block by four bolts and may be withdrawn and serviced as detailed in Sections C.7, C.8 and C.9.

If the gasket is damaged as the pump body is withdrawn from the cylinder block, ensure that all traces of it are removed before a new gasket is fitted and the pump replaced.

Section A.5

REMOVING AND REPLACING THE CARBURETTERS

Release the spring clips and detach the breather hose from the air cleaner and rocker cover.

Disconnect the fuel supply pipe at the rear carburetter banjo union.

Remove the split pin and flat washer and release the mixture cable and clevis pin from the mixture control linkage and release the mixture outer cable abutment from its bracket.

Detach the throttle return spring and release the throttle cable.

Unscrew the union nut and disconnect the ignition vacuum control pipe from the rear carburetter.

Remove the nut and flat washer to release the vent pipe from the top of each float-chamber.

Remove the four nuts securing the carburetter flanges and withdraw the carburetters and air cleaners as an assembly.

Replacement is a reversal of the above instructions.

Section A.6

REMOVING AND REPLACING THE MANIFOLDS

Remove the air cleaners and carburetters as detailed in Section A.5.

Remove the nut, bolt and washers securing the exhaust pipe to the steady bracket on the rear engine plate.

Release the distributor suction advance pipe and, if fitted, the heater pipe, from the manifold.

Remove the three exhaust pipe to manifold clamp bolts and spring washers and release the pipe.

Six studs and nuts secure the manifolds to the cylinder head.

The four centre nuts have large washers enabling them to secure both the inlet and exhaust manifolds. The two remaining nuts, one at each end of the manifolds, have small washers and secure the exhaust manifold only.

Replacement of the manifolds is a reversal of these instructions.

Use a new gasket.

Section A.7

REMOVING AND REPLACING THE ROCKER ASSEMBLY

Drain the cooling system, using a clean container for the coolant if it contains anti-freeze which is to be used again.

Release the breather pipe from the front of the rocker cover.

Slacken the clamping nut and withdraw the throttle cable from the lever and outer cable abutment. Unscrew the two nuts and lift off the rocker cover, taking care not to damage the cork gasket or lose the washers and rubber seals. Notice that under the right-hand rear rocker stud nut is a special locking plate. Unscrew the eight rocker-shaft bracket fixing nuts gradually, a turn at a time, until all load has been released.

It is necessary to drain the radiator and slacken the seven external cylinder head securing nuts because four of the rocker-shaft bracket fixing nuts also secure the

cylinder head, and if the seven external cylinder head fixing nuts are not slackened distortion may result and water find its way from the cooling system into the cylinders and sump.

Completely unscrew the eight rocker shaft bracket nuts and remove the rocker assembly, complete with brackets and rockers. Withdraw the eight push-rods, storing them carefully so that they may be replaced in the same positions. To dismantle the rocker shaft assembly, first remove the grub screw which locates the

Fig. A.1

The rocker shaft locating screw is locked by this plate on the rear rocker shaft bracket

rocker shaft in the rear rocker mounting bracket and remove the split pins, flat washers and spring washers from each end of the shaft. Slide the rockers, brackets and springs from the shaft.

Unscrew the plug from the front end of the shaft and clean out the oilways.

Reassembly and replacement is a reversal of the above procedure, replacing the rockers and springs in their original positions on the shaft. Remember to replace the rocker shaft locating screw lock plate. Replace the rocker cover with the vent pipe to the front. Check that the two cap nut rubber bushes and the rocker cover cork gasket are undamaged; if they are found to be faulty, fit new ones or oil leaks may result.

Section A.8

REMOVING AND REPLACING THE CYLINDER HEAD ASSEMBLY

Drain the water from the cooling system. One drain tap is at the base of the radiator on the right-hand side and the other is at the rear of the engine on the right-hand side. If anti-freeze mixture is being used it should be drained into a suitable clean container and carefully preserved for future use.

Remove the top water hose. Remove the three thermostat housing securing nuts and washers and remove the housing and thermostat.

Remove the air cleaners and carburetters as detailed in Section A.5.

Remove the inlet and exhaust manifolds as detailed in Section A.6.

Remove the rocker assembly as detailed in Section A.7 and remove the seven external cylinder head nuts at the same time. Withdraw the push-rods, keeping them in the order of their removal.

Detach the high-tension cables and remove the sparking plugs, taking care not to damage the porcelain insulators. If fitted, remove the heater hose from the water valve on the right-hand side of the cylinder head by slackening the retaining clip.

Unscrew the thermal transmitter from the front of the cylinder head and release the conductor from its supporting clip.

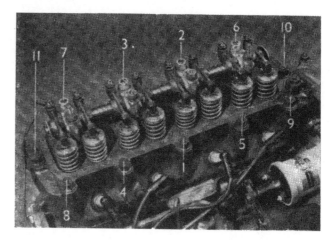

Fig. A.2

The correct order of tightening and slackening the cylinder head retaining nuts

Slacken the clips and disconnect the hoses from the water pipe on the left-hand side of the cylinder head and remove the pipe. Release the ignition vacuum control pipe from the rear cylinder head stud and remove the cylinder head.

To facilitate breaking the cylinder head joint, tap each side of the head with a hammer, using a piece of wood interposed to check the blow. When lifting the head a direct pull should be given to withdraw it evenly up the studs.

Refitting the cylinder head

Make sure that the surfaces of both the cylinder head and the cylinder block are clean. It will be noticed that the cylinder head gasket is marked 'FRONT' and 'TOP' to assist in replacing it correctly with the copper side uppermost. Having slipped the gasket over the studs, next lower the cylinder head into position. Replace the vacuum control pipe clip and fit the seven cylinder head external nuts finger-tight.

Replace the push-rods in the positions from which they were taken. Replace the rocker assembly and securing nuts, finger-tight. Tighten the 11 cylinder head nuts, a turn at a time, in the order given in Fig. A.2. Finally tighten the four rocker assembly nuts.

Reassembly continues in the reverse order to the dismantling procedure.

Fig. A.3

The use of the spring compressor (special tool 18G45) while replacing the cotters is shown in this illustration. The arrow indicates the position of the valve packing ring in the cotter groove below the valve cotters

Switch on the ignition and check the fuel system for leaks. Start the engine and run it until the normal working temperature is reached. Remove the rocker cover and check the valve clearances (see Section A.14). Replace the rocker cover and connect the breather hose.

Section A.9

REMOVING AND REPLACING VALVES

Remove the cylinder head as in Section A.8.

Remove the valve circlip. Compress the double valve springs, using special tool 18G45, and remove the two valve cotters. Release the valve springs and remove the compressor, valve spring cap, shroud, inner and outer springs, and bottom collar.

Remove the valve packing ring from the cotter groove and withdraw the valve from the guide.

Keep the valves in their relative positions when

removed from the head to ensure replacement in their original valve guides. The exhaust valve heads are concave and are smaller in diameter than the inlet valves.

To replace the valves place each valve into its guide and fit the bottom collars, inner and outer valve springs, shrouds, and caps. Compress the valve springs and push a new synthetic rubber packing ring over the tip of the valve stem down to the bottom of the cotter groove (see Fig. A.3). Refit the two valve cotters and remove the compressor. Replace the valve circlip.

NOTE.—Do not fit old valve packing rings, or oil sealing may suffer. The rings are fitted more easily if they have been soaked in clean engine oil for a short period before use.

Section A.10

DECARBONIZING

Remove the cylinder head as described in Section A.8. Withdraw the valves as described in Section A.9.

Remove the cylinder head gasket and plug the waterways with a clean rag.

If special equipment is not available for decarbonizing it will be necessary to scrape the carbon deposit from the piston crowns, cylinder block and cylinder head, using a blunt scraper.

A ring of carbon should be left round the periphery of the piston crown and the rim of carbon round the top of the cylinder bore should not be touched. To facilitate this, an old piston ring can be sprung into the bore so that it rests on top of the piston.

The cylinder head is next given attention. The sparking plugs must be cleaned and adjusted. Clean off the carbon deposit from the valve stems, valve ports and combustion spaces of the cylinder head. Remove all traces of carbon dust with compressed air or by the vigorous use of a tyre pump and then thoroughly clean with paraffin (kerosene) and dry off.

Fit a new cylinder head gasket when replacing the head if the old has been damaged, noting that the gasket is marked to indicate the top face and the front end.

Section A.11

GRINDING AND TESTING VALVES AND SEATINGS

Remove the valves as in Section A.9.

Each valve must be cleaned thoroughly and carefully examined for pitting. Valves in a pitted condition should be refaced with a suitable grinder or new valves should be fitted.

If valve seats show signs of pitting or unevenness they should be trued by the use of a suitable grinder or

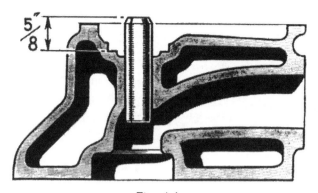

Fig. A.4

Showing the correct fitting of a valve guide

special cutter. When using a cutter, care must be exercised to remove only as little metal as is necessary to ensure a true surface.

When grinding a valve onto its seating the valve face should be smeared lightly with fine or medium grade carborundum paste and then lapped in with a suction grinder (special tool, Part No. 18G328). Avoid the use of excessive quantities of grinding paste and see that it remains in the region of the valve seating only.

A light coil spring placed under the valve head will assist considerably in the process of grinding. The valve should be ground to its seat with a semi-rotary motion and occasionally allowed to rise by the pressure of the light coil spring. This assists in spreading the paste evenly over the valve face and seat. It is necessary to carry out the grinding operation until a dull, even, matt surface, free from blemish, is produced on the valve seat and valve face.

On completion, the valve seat and ports should be cleaned thoroughly with a rag soaked in paraffin (kerosene), dried, and then thoroughly cleaned by compressed air. The valves should be washed in paraffin (kerosene) and all traces of grinding paste removed.

Fit a new oil seal when refitting the valves (see Section A.9) and ensure that the chamfered side of the seal is downwards.

Section A.12

REMOVING AND REPLACING VALVE GUIDES

Remove the cylinder head as shown in Section A.8.

Remove the appropriate valve and springs as in Section A.9. Rest the cylinder head with its machined face downwards on a clean surface and drive the valve guide downwards into the combustion space with a suitable-sized drift. This should take the form of a hardened steel punch $\frac{9}{16}$ in. (14 mm.) in diameter and not less than 4 in. (10 cm.) in length, with a locating

spigot $\frac{5}{16}$ in. (7·9 mm.) diameter machined on one end for a length of 1 in. (2·5 cm.) to engage the bore of the guide.

When fitting new valve guides, these should be pressed in from the top of the cylinder head. The inlet valve guides must be inserted with the end having the largest chamfer at the top, and the exhaust valve guides should have their counterbored ends at the bottom. The valve guides should be driven into the combustion spaces until they are $\frac{5}{8}$ in. (15·8 mm.) above the machined surface of the valve spring seating (see Fig. A.4).

Section A.13

REMOVING AND REPLACING TAPPETS

Remove the carburetters (see Section A.5) and the rocker cover.

Remove the manifolds (see Section A.6).

Disconnect the high-tension leads from the sparking plugs.

Remove the rocker assembly as in Section A.7, and withdraw the push-rods, keeping them in their relative positions to ensure their replacement onto the same tappets. Release the breather pipe and remove the tappet covers and lift out the tappets, also keeping them in their relative positions.

New tappets should be fitted by selective assembly so that they just fall into their guides under their own weight when lubricated.

Assembly is a reversal of the above procedure, but care should be taken to see that the tappet cover joints are oiltight and that the rockers are adjusted to give the correct valve clearance.

Section A.14

ADJUSTING VALVE ROCKER CLEARANCES

If the engine is to give its best performance and the valves are to retain their maximum useful life, it is essential to maintain the correct valve clearance. Accordingly it is recommended that the clearance be checked at regular intervals and any necessary adjustments made.

For the correct valve rocker clearance refer to 'GENERAL DATA'. The engine has been designed to operate with this clearance and no departure from it is permissible. An additional ·001 in. (·025 mm.) must be allowed when the engine is cold.

Provision for adjusting the valve clearance is made in the rocker arm by an adjustable screw and locknut.

The rocker adjusting screw is released by slackening the hexagon locknut with a spanner, while holding the screw against rotation with a screwdriver. The valve clearance can then be set by carefully rotating the rocker screw while checking the clearance with a feeler gauge. This screw is then re-locked by tightening the hexagon locknut while again holding the screw against rotation.

It is important to note that while the clearance is being set the tappet of the valve being operated upon is on the back of its cam, i.e. opposite to the peak.

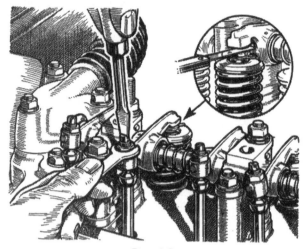

Fig. A.5

The method of adjusting the valve rocker clearance and the correct position for measuring it

As this cannot be observed accurately, the rocker adjustment is more easily carried out in the following order, and this also avoids turning the engine over more than is necessary:

Adjust No. 1 rocker with No. 8 valve fully open.
"	No. 3	"	" No. 6	"	"	"
"	No. 5	"	" No. 4	"	"	"
"	No. 2	"	" No. 7	"	"	"
"	No. 8	"	" No. 1	"	"	"
"	No. 6	"	" No. 3	"	"	"
"	No. 4	"	" No. 5	"	"	"
"	No. 7	"	" No. 2	"	"	"

Section A.15

CHECKING VALVE TIMING

Set No. 1 cylinder inlet valve to ·060 in. (1·52 mm.) clearance with the engine cold, and then turn the engine until the valve is about to open.

The indicating notch in the flange of the crankshaft pulley should then be opposite the longest of the three pointers on the timing cover, i.e. the No. 1 valve should be about to open at T.D.C. and No. 4 piston will be at T.D.C. on its compression stroke.

Do not omit to reset the inlet valve clearance to the recommended clearance (see 'GENERAL DATA') when the timing check has been completed. The clearance of ·60 in. is necessary to bring the opening position of the

valve to T.D.C. **It is not possible to check the valve timing accurately with the normal running valve clearance.**

Section A.16

REMOVING AND REPLACING THE TIMING CHAIN COVER

Drain the cooling system and remove the radiator (see Section C.4).

Slacken the dynamo attachment bolts and remove the belt.

Bend the tab on the starting dog nut locking washer. Unscrew the starting dog nut and remove the locking washer.

Pull off the crankshaft pulley.

The timing cover is secured by nine bolts. Each bolt has a shakeproof washer and a plain washer. Remove all nine bolts with their washers and remove the timing cover.

Care should be taken not to damage the timing cover gasket. If it is damaged, clean the face of the cover flange and the front engine mounting plate and fit a new gasket when reassembling.

The felt washer situated in the timing cover should also be renewed if necessary.

It should be noted that the oil thrower, which is located behind the crankshaft pulley, is fitted with its concave side facing away from the engine.

Replacement of the timing cover is a reversal of the above procedure.

Section A.17

REMOVING AND REPLACING THE TIMING CHAIN

Remove the timing cover and oil thrower as in Section A.16.

Unlock and remove the camshaft chain wheel nut and remove the nut and lock washer. Note that the locating tag on the lock washer fits into the keyway of the camshaft chain wheel.

The camshaft and crankshaft chain wheels may now be removed, together with the timing chain, by easing each wheel forward, a fraction at a time, with suitable smaller levers.

As the crankshaft gear wheel is withdrawn care must be taken not to lose the gear packing washers immediately behind it. When reassembling replace the same number of washers as was found when dismantling, unless new camshaft or crankshaft components have been fitted which will disturb the alignments of the two gear wheels. To determine the thickness of washers required, place a straight-edge across the sides of the camshaft wheel teeth and measure with a feeler gauge the gap between the straight-edge and the crankshaft gear. Subtract ·005 in. (·13 mm.) from the feeler gauge reading and add the resultant thickness of crankshaft gear packing washers.

When replacing the timing chain and gears, set the crankshaft with its keyway at T.D.C., and the camshaft with its keyway approximately at the one o'clock position when seen from the front. Assemble the gears into the timing chain with the two marks on the gear wheels opposite to each other, as in Fig. A.6. Keeping the gears in this position, engage the crankshaft gear keyway with the key on the crankshaft and rotate the camshaft until

Fig. A.6
When replacing the chain wheels the timing marks indicated by the arrows must be in line to give correct valve timing

the camshaft gear keyway and key are aligned. Push the gears onto the shafts as far as they will go and secure the camshaft gear with the lock washer and nut.

Replace the oil thrower, concave side forward, and the remaining components as detailed in Section A.16.

Section A.18

REMOVING AND REPLACING THE POWER UNIT

Drain the oil from the engine and gearbox. Mark the propeller shaft and rear axle drive flange and disconnect the propeller shaft from the axle.

Mark the hand brake cable operating lever and splined shaft to assist replacing the lever in its original position.

Remove the clamping nut and bolt from the lever and prise the lever from the splined shaft.

Remove the nuts, bolts and spring washers and with-

draw the reinforcement bracket from inside the propeller shaft tunnel.

Remove both seats and frames. Remove all the floor covering from the toeboards, floorboards and gearbox cover. Remove the floorboards and toeboards. Remove the gear lever knob and draught excluder cover and the propeller shaft and gearbox covers.

Disconnect the speedometer drive cable.

Remove the banjo bolt to release the flexible supply pipe from the clutch slave cylinder.

Detach the bonnet from the bonnet hinges.

Drain the water from the radiator and disconnect and remove the top and bottom water hoses. Remove the three bolts with spring and flat washers, securing each side of the radiator, and withdraw the radiator.

Remove the carburetters as detailed in Section A.5.

Disconnect the engine revolution indicator drive from the left-hand side of the engine. Release the exhaust pipe from the exhaust manifold and from the steady bracket on the engine rear mounting plate.

Unscrew the thermal transmitter from the cylinder head and release the conductor from its support clip.

Disconnect the flexible oil gauge pipe from the union at the rear of the cylinder block on the right-hand side.

Disconnect the cables from the dynamo, ignition coil, distributor, and starter motor. Remove the gearbox remote control assembly from the gearbox extension.

Place a rope sling around the power unit and attach the lifting tackle. Arrange the sling so that the unit may be lifted slightly and moved forward, and finally lifted from the frame at a sharp angle with the front considerably higher than the rear.

Take the weight of the unit, release the engine from the two front mounting rubbers and remove the rubbers. Remove the nut, bolt and spring washer to release the gearbox from the mounting bracket on the frame cross-member.

Replacement is a reversal of the above instructions, not forgetting to refill the engine and gearbox with oil to Ref. A (page P.2).

Section A.19

REMOVING AND REPLACING THE SUMP AND OIL PUMP STRAINER

Drain the oil from the engine sump.

Remove the bolts and withdraw the sump from the crankcase.

To remove the oil strainer, remove the two bolts securing it to the pump cover.

The strainer may be dismantled for cleaning by removing the centre nut and bolt and the two delivery

A

THE CAMSHAFT, CRANKSHAFT, AND OIL PUMP COMPONENTS

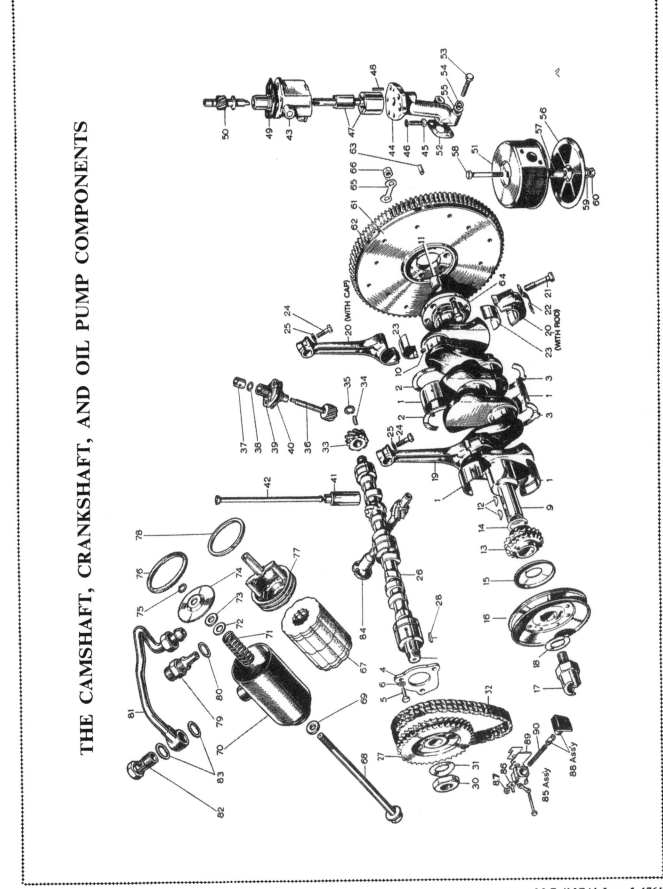

KEY TO THE CAMSHAFT, CRANKSHAFT, AND OIL PUMP COMPONENTS

No.	Description	No.	Description	No.	Description
1.	Bearing—main—standard.	33.	Gear—tachometer drive.	63.	Dowel—clutch.
2.	Thrust washer—upper.	34.	Key—gear.	64.	Bolt—flywheel to crankshaft.
3.	Thrust washer—lower.	35.	Spring ring.	65.	Lock washer—bolt.
4.	Plate—camshaft locating.	36.	Pinion—tachometer drive.	66.	Nut—bolt.
5.	Screw—plate to engine.	37.	Oil seal—pinion.	67.	Element—oil filter.
6.	Washer—spring—plate screw.	38.	Retaining ring.	68.	Bolt—centre.
9.	Crankshaft.	39.	Housing pinion.	69.	Washer—sealing—small.
10.	Restrictor—oil.	40.	Joint washer—housing.	70.	Container.
11.	Bush—first motion shaft.	41.	Tappet—valve.	71.	Spring.
12.	Key—gear/fan pulley.	42.	Push-rod.	72.	Washer.
13.	Gear.	43.	Body and plug.	73.	Washer—felt.
14.	Washer—gear packing.	44.	Cover.	74.	Pressure plate.
15.	Oil thrower—front.	45.	Set screw.	75.	Circlip.
16.	Pulley—crankshaft fan.	46.	Spring washer.	76.	'O' ring.
17.	Nut—starting dog.	47.	Shaft—driving—with rotors.	77.	Valve assembly.
18.	Lock washer—nut.	48.	Dowels—cover.	78.	Washer—sealing—large.
19.	Rod and cap—Nos. 1 and 3.	49.	Joint—pump to block.	79.	Adaptor—oil filter connection.
20.	Rod and cap—Nos. 2 and 4.	50.	Spindle—oil pump driving.	80.	Joint washer—pipe to crankcase.
21.	Set screw—cap.	51.	Body complete—oil strainer.	81.	Pipe assembly—filter to crankcase.
22.	Lock washer.	52.	Joint—strainer to pump.	82.	Screw—banjo union.
23.	Bearings—standard.	53.	Set screw—strainer to pump.	83.	Washer—banjo union screw.
24.	Screw—gudgeon pin clamp.	54.	Spring washer.	84.	Spindle—distributor drive.
25.	Spring washer—screw.	55.	Plain washer.	85.	Tensioner—timing chain.
26.	Camshaft.	56.	Cover—oil strainer.	86.	Lock washer—plug.
27.	Gear—camshaft.	57.	Distance piece—cover.	87.	Plug—body.
28.	Key—gear.	58.	Bolt—cover.	88.	Slipper head and cylinder.
29.	Tensioner ring—gear.	59.	Washer—shakeproof.	89.	Back plate—body.
30.	Nut—gear.	60.	Nut—cover bolt.	90.	Spring.
31.	Lock washer—nut.	61.	Flywheel.		
32.	Chain—camshaft timing.	62.	Ring—starter.		

pipe flange bolts. Note that there is a locating tongue on the side of the cover which must be positioned correctly when replacing. Remember also to replace the distance tube.

Clean out the sump and strainer with paraffin (kerosene) and a stiff brush; never use rag.

When refitting the sump to the engine give particular attention to the sealing gaskets for the crankcase face and the two oil seal packings for the crankcase which fit into recesses in the crankcase.

If the gaskets are in good condition and have not been damaged during removal of the sump they may be used again, but it is always advisable to fit new ones. Before fitting new gaskets, remove all traces of the old ones from the sump and crankcase faces. Smear the faces of the crankcase joint with grease and fit the two halves of the large gasket. Lift the sump into position on the crankcase, insert the 19 bolts and tighten them evenly.

Section A.20

REMOVING AND REPLACING MAIN AND BIG-END BEARINGS

Unless the bearing journals are badly worn the big-end bearings may be renewed without removing the crankshaft. To renew the main bearings it is necessary to withdraw the crankshaft. Liners are used both for the main and the big-end bearings, which are of the shimless type and therefore non-adjustable.

Big-end bearings

Drain the engine oil and remove the sump as in Section A.19.

As the bearings are of the shimless type it is essential that no attempt should be made to adjust bearings which are worn. Always fit new bearings in place of worn parts. If the crankshaft journals are found to be in a worn condition it is advisable to fit a service reground crankshaft, complete with main and big-end bearings, as supplied by the Factory.

Both the big-end and main bearings liners are located in the bearing housings by a small tag on one side of each half-bearing; it should be noted that the bearings are fitted so that the tags come on the same joint edge of the bearing housing although on opposite corners.

To detach the big-end bearings, bend down the locking strips so that the bolts may be removed. Remove the connecting rod caps and extract the bearings. Care should be taken to see that the bearing journals are thoroughly cleaned before installing new bearings. No scraping is required, as the bearings are machined to give the correct diametrical clearance of ·0016 in. (·04 mm.).

Main bearings

Remove the engine from the car and remove the flywheel and clutch, the timing chain (Section A.17), the sump and strainer (Section A.19), and the rear engine mounting plate (Section A.31).

Remove the self-locking nuts securing the main bearing caps to the cylinder block and the two bolts securing the front cap to the front engine bearer plate.

Note that a thrust washer is fitted on each side of the centre main bearing to take the crankshaft end-thrust. These thrust washers each consist of two semicircular halves, one having a lug which is located in a recess in the detachable half of the bearing and the other being plain.

When fitting new bearings no scraping is required as the bearings are machined to give the correct diametrical clearance of ·0005 in. (·0127 mm.).

In the case of a 'run' bearing it is always essential to clean out thoroughly all the oilways in the crankshaft and block, wash out the engine sump with paraffin (kerosene) and clean the oil pump and sump strainer to ensure that no particles of metal are left anywhere in the lubricating system. The rear main bearing cap horizontal joint surfaces should be thoroughly cleaned and lightly covered with Wel-Seal (manufactured by Messrs. Wellworthy Ltd.) sealing compound before the cap is fitted to the cylinder block. This will ensure a perfect oil seal when the cap is bolted down to the block. Replace each main bearing and cap, replacing the thrust washers in their correct positions at the centre main bearing with the oil grooves towards the bearing. Refit the locking strip or locking plates to each bearing cap and bend them to lock the bolts after tightening. Note that the two bolts securing the front main bearing cap to the front bearer plate are locked by a common plate.

Section A.21

REMOVING AND REPLACING PISTONS AND CONNECTING RODS

Remove the cylinder head as in Section A.8. Drain and remove the sump and oil strainer as in Section A.19.

The pistons and connecting rods must be withdrawn from the top of the cylinder block.

Unlock and remove the big-end bolts and remove the bearing caps. Release the connecting rod from the crankshaft.

Withdraw the piston and connecting rod from the top of the cylinder block and refit the bearing cap. The big-end bearing caps are offset, and the caps on the big-ends in Nos. 1 and 3 cylinders are interchangeable when new, as are those for Nos. 2 and 4 cylinders. When used parts are replaced after dismantling it is essential that they should be fitted in their original positions. In order to ensure this, mark the caps and connecting rods on their sides which are fitted together with the number of the cylinder from which each was taken.

Replacement of the piston and connecting rod is a direct reversal of the above, but the piston ring gaps should be set at 180° to each other.

If the piston rings have been removed from the piston they must be replaced as detailed in Section A.24.

It is essential that the connecting rod and piston assemblies should be replaced in their own bores and fitted the same way round, i.e. with the gudgeon pin clamp screw on the camshaft side of the engine. The piston crowns are marked 'FRONT' to assist correct assembly to the connecting rods.

Refit the big-end bearings in their original positions.

Section A.22

DISMANTLING AND REASSEMBLING PISTON AND CONNECTING ROD ASSEMBLIES

The gudgeon pin is rigidly held in the split little-end of the connecting rod by a clamp bolt engaging the central groove of the gudgeon pin.

Before the piston and gudgeon pin can be dismantled from the connecting rod it is necessary to remove the clamp screw. To enable the assembly to be held in a vice for this operation without damage, holding plugs should be inserted in each end of the gudgeon pin.

Unscrew the gudgeon pin clamp and remove it completely.

Push out the gudgeon pin.

Reassembly is a reversal of the above.

IMPORTANT.—Attention must be given to the following points when assembling the piston to the connecting rod:

(1) That the piston is fitted the correct way round on the connecting rod. The crown of the piston is marked 'FRONT' to assist this, and the connecting rod is fitted with the gudgeon pin clamp screw on the camshaft side.

(2) That the gudgeon pin is positioned in the connecting rod so that its groove is in line with the clamp screw hole.

(3) That the clamp screw spring washer has sufficient tension.

(4) That the clamp screw will pass readily into its hole and screw freely into the threaded portion of the little-end, and also that it will hold firmly onto the spring washer.

Section A.23

FITTING GUDGEON PINS

A certain amount of selective assembly must be used when fitting new gudgeon pins. They must be a thumb-

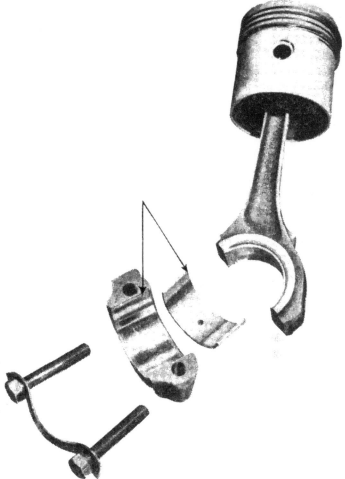

Fig. A.8

A piston and connecting rod assembly. Note the bearing locating tab

push fit for three-quarters of their travel, to be finally tapped in with a raw-hide mallet. This operation should be carried out with the piston and gudgeon pin cold.

Section A.24

REMOVING AND REPLACING PISTON RINGS

If no special piston ring expander is available, use a piece of thin steel such as a smoothly ground hacksaw blade or a disused ·020 in. (·50 mm.) feeler gauge.

Raise one end of the ring out of its groove. Insert the steel strip between the ring and the piston. Rotate the strip around the piston, applying slight upward pressure to the raised portion of the ring until it rests on the land above the ring grooves. It can then be eased off the piston.

Do not remove or replace the rings over the piston skirt, but always over the top of the piston.

Before fitting new rings, clean the grooves in the piston to remove any carbon deposit. Care must be taken not to remove any metal or sideplay between the ring and the groove will result, with consequent excessive oil consumption and loss of gastightness.

The cylinder bore glazing should be removed before fitting new rings to a worn bore.

When refitting the rings note that the second and third compression rings are tapered and marked with the letter 'T' (top) for correct reassembly.

New rings must be tested in the cylinder bore to ensure that the ends do not butt together. The best way to do this is to insert the piston approximately 1 in. (2·54 cm.) into the cylinder bore and push the ring down onto the top of the piston and hold it there in order to keep the ring square with the bore. The correct ring gap is ·008 to ·013 in. (·20 to ·33 mm.).

Section A.25

PISTON SIZES AND CYLINDER BORES

In addition to the standard pistons there is a range of four oversize pistons available for service purposes.

Oversize pistons are marked with the actual oversize dimensions enclosed in an ellipse. A piston stamped ·020 is only suitable for a bore ·020 in. (·508 mm.) larger

Piston marking	Suitable bore size	Metric equivalent
Standard	2·8757 in.	73·043 mm.
	2·8760 in.	73·050 mm.
Oversize		
+·010 in. (·254 mm.)	2·8857 in.	73·297 mm.
	2·8860 in.	73·304 mm.
+·020 in. (·508 mm.)	2·8957 in.	73·551 mm.
	2·8960 in.	73·558 mm.
+·030 in. (·762 mm.)	2·9057 in.	73·805 mm.
	2·9060 in.	73·812 mm.
+·040 in. (1·016 mm.)	2·9157 in.	74·059 mm.
	2·9160 in.	74·066 mm.

than the standard bore; similarly pistons with other markings are only suitable for the oversize bore indicated.

The piston markings indicate the actual bore size to which they must be fitted, the requisite running clearance being allowed for in the machining.

After reboring an engine, or whenever fitting pistons differing in size from those removed during dismantling, ensure that the size of the piston fitted is stamped clearly on the top of the cylinder block alongside the appropriate cylinder bore.

Pistons are supplied in the sizes indicated in the table on this page.

Section A.26

REMOVING AND REPLACING THE CAMSHAFT

Disconnect the battery.

Remove the inlet and exhaust manifold assembly (see Section A.6).

Remove the push-rods and take out the tappets (see Section A.13).

Remove the timing cover, timing chain and gears (see Section A.16 and Section A.17).

Disconnect the suction advance unit pipe from the distributor and take out the two bolts with flat washers securing the distributor to the housing. Do not slacken the clamping plate bolt or the ignition timing setting will be lost.

Withdraw the distributor.

Take out the bolt securing the distributor housing to the cylinder block. Using one of the tappet cover bolts as an extractor screwed into the tapped end of the distributor drive spindle, withdraw the spindle.

Remove the sump, oil pump and oil pump drive shaft (see Section A.28).

Disconnect the engine revolution indicator drive, remove the securing nuts and washers and withdraw the indicator drive gear.

Take out the three set screws and shakeproof washers which secure the camshaft locating plate to the cylinder block, and withdraw the camshaft.

If the camshaft bearing clearances are excessive new bearings should be fitted. To do this, the power unit should be removed. Drift out the old bearing towards the rear of the engine and press in a new one. Ensure that the oil holes in the bearing line up with the oil passages in the cylinder block. The front bearing must be reamed to give ·001 to ·002 in. (·025 to ·051 mm.) diametrical clearance.

Replacement of the camshaft is a reversal of the above procedure.

Note that the three set screws securing the camshaft thrust plate are not evenly spaced to ensure the correct alignment of the oil hole supplying the timing gear.

Section A.27

REFITTING THE DISTRIBUTOR DRIVE GEAR

Turn the engine until No. 1 piston is at T.D.C. on its compression stroke. When the valves on No. 4 cylinder are 'rocking' (i.e. exhaust just closing and inlet just opening), No. 1 piston is at the top of its compression stroke. If the engine is set so that the notch in the crankshaft pulley is in line with the long pointer on the timing chain cover, or the 'dimples' in the crankshaft and camshaft gears are in line, the piston is exactly at T.D.C.

Turn the engine back so that the notch in the pulley is 7° before the long pointer on the timing cover. This is the correct ignition setting giving a 7° advance. As a guide to the 7° position the two short pointers on the timing cover are 5° and 10° B.T.D.C. respectively.

Screw one of the tappet cover bolts into the threaded end of the distributor drive gear and, holding the drive

Fig. A.8

The arrow indicates the oil supply hole for the timing gear. The crankshaft and camshaft wheel keys should be in the positions shown when timing the engine

Fig. A.9

Replacing the distributor drive gear. Notice the slot angle. The large offset is uppermost

gear with the slot just below the horizontal and the large offset uppermost, enter the gear.

As the gear engages with the camshaft the slot will turn in an anti-clockwise direction until it is approximately in the one o'clock position.

Remove the bolt from the gear, insert the distributor housing and secure it with the special bolt and washer. Ensure that the correct bolt is used and that the head does not protrude above the face of the housing.

Refit the distributor, referring to Section B.7 for retiming instructions if the clamp plate has been released.

Section A.28

REMOVING AND REPLACING THE OIL PUMP

Remove the sump and oil sump strainer (see Section A.19).

Two bolts secure the oil pump cover and three studs secure the pump to the crankcase. Unscrew the stud nuts and remove the pump and drive shaft.

When refitting the pump use a new joint washer.

Section A.29

DISMANTLING AND REASSEMBLING THE OIL PUMP

Remove the oil pump from the engine.

The oil pump cover is attached to the body of the pump by two bolts and spring washers, and when these bolts are removed the oil pump cover, the outer rotor and the combined oil pump shaft and inner rotor may be extracted. The component parts are shown on the right-hand side of the illustration on page A.12.

Section A.30

REMOVING AND REPLACING THE FLYWHEEL
(Engine Out of Car)

Remove the clutch by unscrewing the six bolts and spring washers securing it to the flywheel. Release the bolts a turn at a time to avoid distortion of the cover flange. Two dowels locate the clutch cover on the flywheel.

To release the special flywheel bolts the engine sump and rear main bearing cap must also be removed.

Unlock and remove the six nuts and three lock plates which secure the flywheel to the crankshaft and remove the flywheel.

When replacing the flywheel, ensure that the 1 and 4 timing mark on the periphery of the flywheel is in line with and on the same side as the first and fourth throws of the crankshaft.

To assist correct location of the flywheel, the depression in the crankshaft flange face is stamped with a similar timing mark which should be in line with the one on the flywheel periphery.

Section A.31

REMOVING AND REPLACING THE CRANKSHAFT
(Engine Out of Car)

Remove the engine from the car (see Section A.18).

Take off the clutch and the flywheel (see Section A.30), the timing cover (see Section A.16), the timing wheels and chain (see Section A.17), the sump and the oil pump strainer (see Section A.19), and the rear engine mounting plate.

Remove the big-end bearing cap and then take off the main bearing caps (see Section A.21).

Mark each big-end bearing cap and bearing to ensure that it is reassembled to the correct journal, taking care, in the case of the bearings, that they are not damaged or distorted when marking. Punches should not be used for this purpose.

Lift the crankshaft out of the bearings.

Replacement of the crankshaft is a reversal of the above operations.

Before replacing the crankshaft, thoroughly clean out all oilways.

Note that each main bearing cap is stamped with a common number which is also stamped on the centre web of the crankcase near the main bearing.

Remember to fit the packing washers behind the crankshaft chain wheel (see Section A.17).

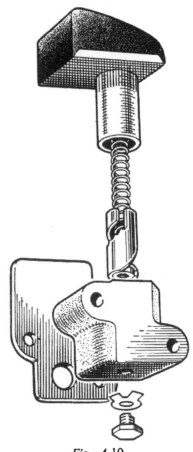

*Fig. A.*10

The chain tensioner components

Section A.32

TIMING CHAIN TENSIONER

Commencing at Engine No. 259, a timing chain tensioner, as illustrated in Fig. A.10, is fitted to the engine front mounting plate and secured by two bolts and a locking plate.

Also fitted are a modified camshaft locating plate and a camshaft timing gear in which the rubber tensioner ring is deleted. The timing chain now receives lubrication via the tensioner slipper and not as previously by an ejection of oil from the camshaft locating plate. The chain tensioner cannot be fitted to earlier vehicles.

Before removing the assembly from the engine, unlock the tab washer and remove the bottom plug from the tensioner body. Insert a $\frac{1}{8}$ in. Allen key to engage the cylinder and turn the key clockwise until the rubber slipper is fully retracted and locked behind the limit peg. Between a half and one full turn is all that is necessary.

Unlock and remove the bolts to release the chain tensioner assembly and the backplate.

A TRANSVERSE SECTION OF THE M.G. (Series MGA) ENGINE

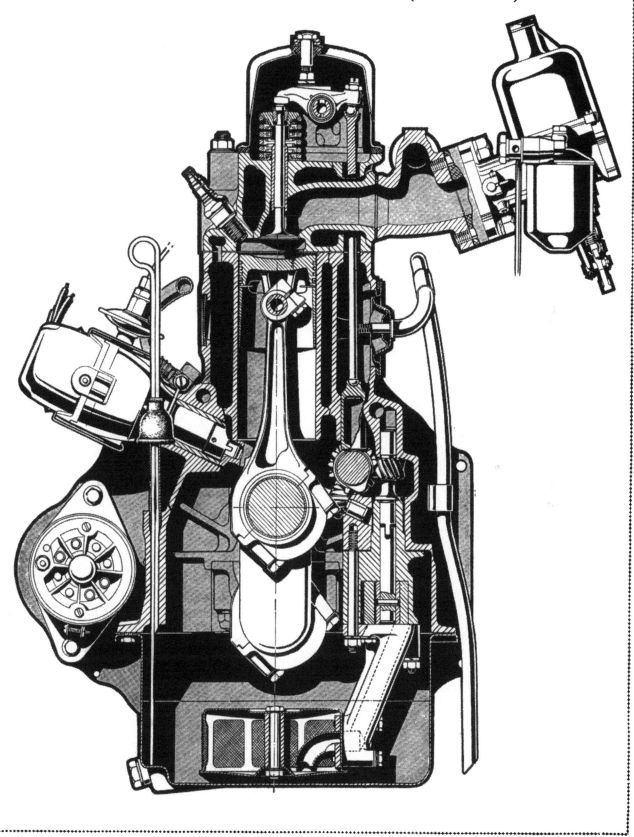

Withdraw the plunger and slipper assembly from the tensioner body and engage the lower end of the cylinder with the Allen key. Turn the key clockwise, holding the key and plunger securely until the cylinder and spring can be withdrawn from inside the plunger.

The components should be cleaned thoroughly in fuel and the ·125 in. (3·18 mm.) diameter inlet oil hole in the spigot and the ·040 in. (1·02 mm.) outlet oil hole in the slipper should be cleaned with compressed air before reassembling.

Check the bore in the tensioner body for ovality. If ovality is greater than ·003 in. (·0762 mm.) when measured on diameters near the mouth of the bore, then the complete chain tensioner must be renewed.

Inspect the slipper head for wear. If it is badly worn a new slipper head and cylinder assembly must be fitted to the existing body, provided the bore of the body is within the limits given above.

When the tensioner is in operation and the engine is running, oil from the lubrication system enters the spigot on the back face under pressure and lubricates the bearing surface through a hole in the slipper pad. The pad is held against the chain by spring and oil pressure.

Should the chain stretch with use, the slipper plunger rises and the limiting peg, bearing on the top of the helical slot, rotates the cylinder until the next recess in the lower edge of the slot comes into line with the limiting peg and prevents the plunger returning to its original position and allowing the chain to become slack again.

When reassembling, insert the spring in the plunger and place the cylinder on the other end of the spring.

Compress the spring until the cylinder enters the plunger bore, engaging the helical slot with the peg in the plunger. Hold the assembly compressed in this position and engage the Allen key. Turn the cylinder clockwise until the end of the cylinder is behind the peg and the spring is held compressed. Withdraw the key and insert the plunger assembly in the body. Replace the backplate and secure the assembly to the cylinder block.

When the timing chain is in position the tensioner is released for operation by inserting the key and turning it clockwise, allowing the slipper head to move forward under spring pressure only.

After refitting the tensioner check the slipper head for freedom of movement and ensure that it does not bind on the back plate when it is moved in the body.

Secure the bolts with the locking plate; replace the bottom plug and lock with a tab washer.

Section A.33

MODIFIED GEARBOX MOUNTING PLATE

Commencing at Engine No. 3289, the gearbox mounting plate has an oil escape recess and groove machined in the front face to relieve the air depression around the rear main bearing oil seal and prevent oil being drawn into the clutch housing.

The plate is interchangeable and retains the same part number.

Section A.34

MODIFIED TAPPETS AND PUSH-RODS

Commencing at Engine No. 5504, the ball ends of the push-rods and the seats in the tappets are increased in spherical diameter. The tappets and push-rods are interchangeable in sets and the new part numbers are :

Tappet 1H822
Push-rod 11G241

Section A.35

PISTON RINGS WITH CHROME PERIPHERY

At Engine No. 5682 the top compression ring on the piston is superseded by a piston ring with a chrome periphery to improve the life of the piston ring and to improve oil consumption.

The later piston ring is interchangeable with the old.

Section A.36

MODIFIED OIL FILTER

Commencing at Engine No. 15GB/U/H26661 to 15GB/U/H26700 then 26933 onwards, a new oil filter is fitted. The element may be removed from a later-type oil filter without disconnecting the oil pipe. Tecalemit and Purolator filters are used, and the elements, which are interchangeable, bear the B.M.C. Part No. 8G683 (Tecalemit and Purolator).

Section A.37

MODIFIED CRANKSHAFT

Originally the diameter of the oil return thread on the crankshaft was 2·139/2·1405 in. (54·33/54·37 mm.), but at Engine No. BP15GB/6615 this was reduced to 2·138/2·1385 in. (54·305/54·32 mm.).

For the correct functioning of the oil return thread it is imperative that it should be concentric with the bore of the housing and have between ·003 and ·006 in. (·075 and ·15 mm.) clearance measured from the crest of the thread to the housing. This may be checked with the aid of a long feeler gauge and a mandrel.

Section A.38

MODIFIED PISTON AND GUDGEON PIN

Commencing at Engine No 15GB/U/H38484, a new piston (Part No. 1H1114) with a modified gudgeon pin (Part No. 1H1113) is introduced, the gudgeon pin having a reduced internal diameter to give increased strength.

The later pistons and gudgeon pins are interchangeable with the originals but only as complete sets.

Section A.39

MODIFIED OIL PUMP AND STRAINER

On later engines a modified oil pump (Part No. 1H1191) and an oil strainer assembly of simplified construction (Part No. 1H1192) are fitted. The oil suction pipe position is now moved forward towards the front of the engine to eliminate any possibility of oil starvation during cornering or braking.

The new units are interchangeable as complete assemblies with the oil pumps and strainers previously used. The three oil pump to crankcase studs, however, have been lengthened to accommodate the greater thickness of the oil pump bottom cover, their part number now being 51K267.

These modifications were introduced at Engine No. 15GB/U/H46045 to 46100 and then 46342 onwards.

Section A.40

FITTING FLYWHEEL STARTER RINGS

To remove the old starter ring from the flywheel flange split the ring gear with a cold chisel, taking care not to damage the flywheel. Make certain that the bore of the new ring and its mating surface on the flywheel are free from burrs and are perfectly clean.

To fit the new ring it must be heated to a temperature of 300 to 400° C. (572 to 752° F.), indicated by a light-blue surface colour. If this temperature is exceeded the temper of the teeth will be affected. The use of a thermostatically controlled furnace is recommended. Place the heated ring on the flywheel with the lead of the ring teeth uppermost. The expansion will allow the ring to be fitted without force by pressing or tapping it lightly until the ring is hard against its register.

This operation should be followed by natural cooling, when the 'shrink fit' will be permanently established and no further treatment required.

Section A.41

FITTING VALVE SEAT INSERTS

Should the valve seatings become so badly worn or pitted that the normal workshop cutting and refacing tools cannot restore them to their original standard of efficiency, special valve inserts can be fitted.

The seating of the cylinder head must be machined to the dimensions given in Fig. A.11. Each insert should have an interference fit of ·0025 to ·0045 in. (·063 to ·11 mm.) and must be pressed and not driven into the cylinder head.

After fitting, grind or machine the new seating to the dimensions given in Fig. A.11. Normal valve grinding may be necessary to ensure efficient valve seating.

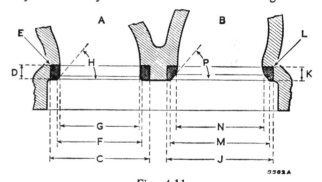

Fig. A.11

Valve seat machining dimensions

Exhaust (A)	Inlet (B)
C. 1·437 to 1·438 in. (36·5 to 36·52 mm.).	J. 1·530 to 1·531 in. (38·86 to 38·88 mm.).
D. ·186 to ·188 in. (4·72 to 4·77 mm.).	K. ·186 to ·188 in. (4·72 to 4·77 mm.).
E. Maximum radius ·015 in. (·38 mm.).	L. Maximum radius ·015 in. (·38 mm.).
F. 1·286 to 1·288 in. (32·20 to 32·71 mm.).	M. 1·487 to 1·507 in. (37·77 to 38·27 mm.).
G. 1·165 to 1·175 in. (29·59 to 29·84 mm.).	N. 1·302 to 1·322 in. (33·07 to 33·57 mm.).
H. 45°.	P. 45°.

Section A.42

FITTING CYLINDER LINERS

Should the condition of the cylinder bores be such that they cannot be cleaned up to accept standard

Engine type	Liner Part No.	Machine bores of cylinder block to this dimension before fitting liner	Outside diameter of liner	Interference fit of liner in cylinder block bore	Machine liner bore to this dimension after fitting
'B' 1500 c.c.	1H641	3·0165 to 3·017 in. (76·619 to 76·631 mm.)	3·0185 to 3·01925 in. (76·669 to 76·688 mm.)	·0015 to ·00275 in. (·038 to ·069 mm.)	2·8745 to 2·876 in. (73·01 to 73·04 mm.)

oversize pistons, dry cylinder liners can be fitted. This operation may be carried out by the use of specialized proprietary equipment or with a power press using pilot adaptors to the dimensions shown in Fig. A.12. The press must be capable of 3 tons (3048 kg.) pressure to fit new liners and 5 to 8 tons (5080 to 8128 kg.) to remove old liners.

Remove the engine from the vehicle as detailed in Section A.18. Dismantle the engine and remove the cylinder head studs. If liners have not previously been fitted the bores must be machined and honed to the dimensions given in the table below.

To remove worn liners

Place the cylinder block face downwards on suitable wooden supports on the bed of the press, making sure that there is sufficient space between the block and the bed of the press to allow the worn liner to pass down.

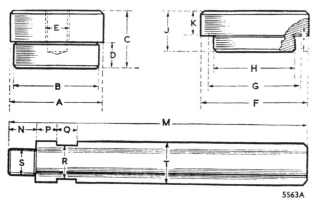

Fig. A.12

Cylinder liner pilots should be made to the above dimensions from 55-ton hardening and tempering steel and hardened in oil at a temperature of 550° C. (1,022° F.)

Pressing-out pilot

A. $2\frac{61}{64}$ $^{+.005}_{-.0}$ in. (75·8 $^{+.127}_{-.0}$ mm.).
B. $2·862$ $^{+·0}_{-·005}$ in. (72·69 $^{+·0}_{-·127}$ mm.).
C. $1\frac{3}{4}$ in. (44·45 mm.).
D. $\frac{3}{4}$ in. (19·05 mm.).
E. $\frac{3}{4}$ in. B.S.W. thread.

Pressing-in pilot

F. $3\frac{7}{16}$ in. (87·31 mm.).
G. $3\frac{1}{16}$ in. (77·39 mm.).
H. $2·850$ $^{+·0}_{-·005}$ in. (72·39 $^{+·0}_{-·127}$ mm.).
J. $1\frac{1}{4}$ in. (31·75 mm.).
K. $\frac{3}{4}$ in. (19·05 mm.).
L. $·015$ in. (·38 mm.).

Pilot extension

M. $14\frac{1}{2}$ in. (36·83 cm.).
N. $\frac{7}{8}$ in. (22·22 mm.).
P. $\frac{5}{8}$ in. (15·87 mm.).
Q. $\frac{5}{8}$ in. (15·87 mm.).
R. 1 in. (25·4 mm.) flats.
S. $\frac{3}{4}$ in. B.S.W. thread.
T. $1\frac{1}{4}$ in. (31·75 mm.).

Insert the pilot in the bottom of the liner and carefully press the liner from the bore.

To press in new liners

Thoroughly clean the inside of the bores and the outside of the liners. Stand the cylinder block upright on the bed of the press, insert the pilot guide in the top of the liner, and position the liner with its chamfered end in the top of the bore. Make certain that the liner is square with the top of the block and that the ram of the press is over the centre of the pilot. Press the liner into the bore.

Each liner must be machined to the dimensions given in the table on page A.21 after pressing into position.

Section A.43

MODIFIED PISTONS AND RINGS

New piston assemblies with compression and oil control rings of increased radial thickness (see 'GENERAL DATA') are introduced at Engine No. 15GB/U/H40824. The compression ring grooves are of reduced diameter. These changes have been made to reduce oil consumption.

The piston assemblies complete with gudgeon pins and rings are interchangeable in sets with those previously used. The new and old oil control rings are also interchangeable in sets, and the earlier-type compression rings may be used with the later-type pistons. It is not permissible, however, to fit the new-type compression rings to the old-type pistons.

Section A.44

MODIFIED POWER UNIT

A new power unit, having the type designation 15GD, is fitted from Car No. 61504 and to certain earlier cars.

The engine incorporates the various modifications made to the 15GB unit and has the starter motor placed higher on the gearbox mounting plate.

Modifications have also been made to the gearbox (see Section F.10), to the propeller shaft (see Section G.8), to the gearbox cover, and to the right-hand toeboard, so that neither the engine, the gearbox, nor the propeller shaft is interchangeable with those previously fitted.

The above changes involve alteration to the procedure for power unit removal and replacement.

Follow the instructions given in Section A.18, but note that the propeller shaft need not be disconnected from the rear axle flange. When the floorboards, toeboards, gearbox cover, and propeller shaft cover have been removed mark the propeller shaft and gearbox flanges and disconnect the propeller shaft from the gearbox.

A.22

M.G. 'MGA'. Issue 3. 34279

Section A.45

OIL COOLER KITS

Oil cooler kits are now fitted as standard equipment to all export cars from Car No. 102737 and are available as an optional extra for Home Trade cars. All bodies are now drilled to accept the oil cooler and its associated equipment.

The kits are available under Part No. 8G2282 for standard cars and Part No. 8G2325 for cars fitted with disc brakes all round.

The oil capacity of the cooler is approximately $\frac{3}{4}$ pint (·42 litre) and this quantity must be added to the sump when the cooler is fitted.

The fitting instructions, including the drilling of un-modified bodies, is as follows:

(1) Jack up and support the front of the car and remove the front off-side wheel.

(2) Remove the radiator case and grille to gain access to the horns and radiator duct panel.

(3) Disconnect and remove the horns complete with their brackets.

(4) Disconnect and remove the filter to crankcase pipe assembly.

(5) Mark out and drill two $\frac{11}{32}$ in. (8·73 mm.) holes in each front suspension member for repositioning the horns as shown in Fig. A.13 (c).

(6) Mark out and drill four $\frac{9}{32}$ in. (7·14 mm.) holes and cut two $1\frac{3}{8}$ in. (34·9 mm.) holes in the radiator duct panel to accept the oil cooler (Fig. A.13 [a]).

(7) Mark out and drill two $\frac{7}{32}$ in. (5·5 mm.) holes in the front off-side wheel arch for the fitting of the oil cooler pipe clips (Fig. A.13 [b]).

(8) Fit the packing-piece and oil cooler to the radiator duct panel and secure it with the four $\frac{1}{4}$ in. UNF. hexagon-headed screws, washers, and Aerotight nuts.

(9) Fit the two oil pipes to the cooler and secure the pipes to the wheel arch with the two clips, No. 10 UNF. screws, plain washers, spring washers, and nuts provided.

(10) Fit the two flexible pipes to the oil cooler pipes and then connect them to the oil filter and crankcase as shown in Fig. A.13.

(11) Refit, connect, and test the horns, replace the radiator grille and case, and refit the front road wheel. Remove the supports and jack.

(12) Add $\frac{3}{4}$ pint (·42 litre) of oil to the sump, run the engine, and check all pipe unions for leakage.

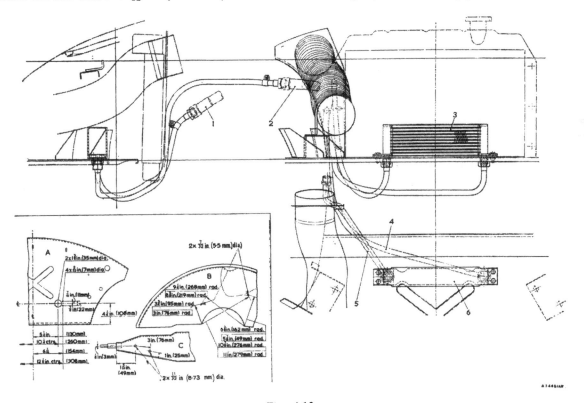

*Fig. A.*13

The general arrangement of the oil cooler, together with the installation dimensions

1. Flexible pipe—cooler to filter.	3. Oil cooler.	5. Oil pipe—cooler to block.
2. Flexible pipe—cooler to block.	4. Oil pipe—cooler to filter.	6. Oil cooler packing.

Notes

SECTION AA

THE ENGINE

(MGA 1600 [Mk. II])

This Section is a Supplement to Section A

Section No. AA.1   Piston sizes and cylinder bores.

THE M.G. (Series MGA 1600 Mk. II) ENGINE
(R.H. SIDE)

A5092W

THE M.G. (Series MGA 1600 Mk. II) ENGINE
(L.H. SIDE)

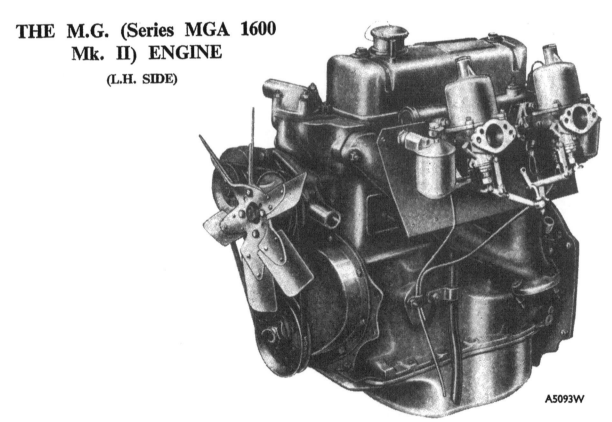

A5093W

Section AA.1

PISTON SIZES AND CYLINDER BORES

In production, pistons are fitted by selective assembly, and to facilitate this the pistons are stamped with identification figures on their crowns.

The number enclosed in a diamond, e.g. a piston stamped with a figure 2, is for use with a block having a similar stamp adjacent to the bore.

In addition to the standard pistons there is a range of four oversize pistons available for service purposes.

Oversize pistons are marked with the actual oversize dimensions enclosed in an ellipse. A piston stamped ·020 is only suitable for a bore ·020 in. (·508 mm.) larger than the standard bore; similarly, pistons with other markings are only suitable for the oversize bore indicated.

The piston markings indicate the actual bore size to which they must be fitted, the requisite running clearance being allowed for in the machining.

After reboring an engine, or whenever fitting pistons differing in size from those removed during dismantling, ensure that the size of the piston fitted is stamped clearly on the top of the cylinder block alongside the appropriate cylinder bore.

Pistons are supplied in the sizes indicated in the table below.

Piston marking	Suitable bore size	Metric equivalent
Standard	3·0011 in.	76·227 mm.
	3·0014 in.	76·235 mm.
Oversize		
+·010 in. (·254 mm.)	3·0111 in.	76·481 mm.
	3·0114 in.	76·489 mm.
+·020 in. (·508 mm.)	3·0211 in.	76·735 mm.
	3·0214 in.	76·743 mm.
+·030 in. (·762 mm.)	3·0311 in.	76·989 mm.
	3·0314 in.	76·997 mm.
+·040 in. (1·016 mm.)	3·0411 in.	77·243 mm.
	3·0414 in.	77·251 mm.

THE CYLINDER BLOCK AND CRANKCASE

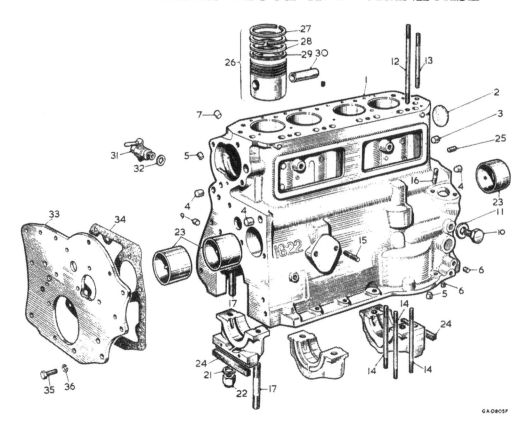

GA0805F

No.	Description	No.	Description
1.	Cylinder block assembly.	22.	Nut for main bearing stud.
2.	Plug—welch.	23.	Liner—camshaft bearing.
3.	Plug—oil gallery.	24.	Joint—front and rear main bearing cap.
4.	Plug—oil gallery.	25.	Dowel—gearbox mounting plate.
5.	Plug—taper—crankcase oil hole.	26.	Piston assembly (standard).
6.	Plug—oil relief valve vent hole.	27.	Ring—compression—top (standard).
9.	Plug—chain tensioner oil feed.	28.	Ring—compression—second and third (standard).
10.	Plug for oil hole (screwed).	29.	Ring—scraper (standard).
11.	Washer for screwed plug.	30.	Pin—gudgeon (standard).
12.	Stud—cylinder head (long).	31.	Tap—cylinder block drain.
13.	Stud—cylinder head (short).	32.	Washer for drain tap.
14.	Stud—oil pump (short).	33.	Plate—engine mounting.
15.	Stud—crankcase vent pipe clip.	34.	Joint washer for mounting plate.
16.	Stud—tachometer drive spindle housing.	35.	Screw—mounting plate to crankcase.
17.	Stud—main bearing.	36.	Washer—spring—for screw.
21.	Washer—spring—for main bearing stud.		

THE CYLINDER HEAD

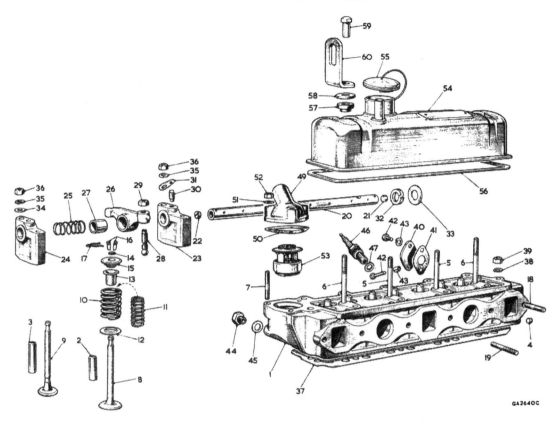

GA2640C

No.	Description	No.	Description
1.	Cylinder head with valve guides.	31.	Plate—locating screw (lock).
2.	Guide—inlet valve.	32.	Washer—spring—rocker shaft.
3.	Guide—exhaust valve.	33.	Washer—rocker shaft.
4.	Plug—oil hole.	34.	Washer—plain—for rocker bracket stud.
5.	Stud—rocker bracket (short).	35.	Washer—spring—for rocker bracket stud.
6.	Stud—rocker bracket (long).	36.	Nut—rocker bracket stud.
7.	Stud—water outlet elbow.	37.	Joint—cylinder head.
8.	Valve—inlet.	38.	Washer—plain—for cylinder head stud.
9.	Valve—exhaust.	39.	Nut for cylinder head stud.
10.	Spring—valve (outer).	40.	Plate—heater outlet elbow blanking.
11.	Spring—valve (inner).	41.	Joint washer for blanking plate.
12.	Collar—valve spring.	42.	Screw—plate to cylinder head.
13.	Shroud—valve guide.	43.	Washer—spring—for screw.
14.	Ring—valve packing.	44.	Plug—thermal transmitter boss.
15.	Cup—valve spring.	45.	Washer for plug.
16.	Cotter—valve (halves).	46.	Plug—sparking.
17.	Circlip for valve cotter.	47.	Gasket for plug (steel).
18.	Stud—exhaust manifold to cylinder head.	49.	Elbow—water outlet.
19.	Stud—inlet and exhaust manifold to cylinder head.	50.	Joint—elbow to cylinder head.
20.	Shaft—valve rocker (plugged).	51.	Washer—plain—for stud in cylinder head.
21.	Plug—valve rocker (plain).	52.	Nut for stud.
22.	Plug—valve rocker (screwed).	53.	Thermostat.
23.	Bracket—rocker shaft (tapped).	54.	Cover—valve rocker.
24.	Bracket—rocker shaft (plain).	55.	Cap and cable—oil filler.
25.	Spring—rocker spacing.	56.	Joint for cover.
26.	Rocker—valve—bushed.	57.	Bush—cover (rubber).
27.	Bush—valve rocker.	58.	Washer—cup—for nut.
28.	Screw—tappet adjusting.	59.	Nut—cover.
29.	Nut—tappet adjusting screw (lock).	60.	Bracket—engine sling.
30.	Screw—rocker shaft locating.		

Notes

B

SECTION B

THE IGNITION SYSTEM

Description.

GENERAL DESCRIPTION

The automatic advance device is housed in the distributor unit, and it consists of a centrifugally and vacuum-operated mechanism by means of which the ignition is advanced in proportion to the engine speed and load.

Like the rest of the electrical equipment, the ignition is wired on the 'positive earth' system.

Section B.1

LOCATING THE CAUSE OF UNEVEN FIRING

Start the engine and set it to run at a fairly fast idling speed.

Short-circuit each plug in turn by placing a hammer head or the blade of a screwdriver with a wooden or insulated handle between the terminal and the cylinder head. No difference in the engine performance will be noted when short-circuiting the plug in the defective cylinder. Shorting the other plugs will make uneven running more pronounced.

Having located the cylinder which is at fault, stop the engine and remove the cable from the terminal of the sparking plug. Restart the engine and hold the end of the cable about $\frac{3}{16}$ in. (4·8 mm.) from the cylinder head.

If the sparking is strong and regular, the fault probably lies in the sparking plug. Remove the plug, clean, and adjust the gap to the correct setting, or alternatively fit a new plug (see Section B.11).

If there is no spark or if it is weak and irregular, examine the cable from the sparking plug to the distributor. After a long period of service, the insulation may be cracked or perished, in which case the cable should be renewed.

Finally, examine the distributor moulded cap, wipe the inside and outside with a clean dry cloth, see that the carbon brush moves freely in its holder and examine the moulding closely for signs of breakdown. After long service it may become tracked, that is, a conducting path may have formed between two or more of the electrodes or between one of the electrodes and some part of the distributor in contact with the cap. Evidence of a tracked cap is shown by the presence of a thin black line in the places indicated. A replacement distributor cap must be fitted in place of one that has become tracked.

Section B.2

TESTING THE LOW-TENSION CIRCUIT

Spring back the securing clips on the distributor and remove the moulded cap and rotor. If the rotor is a tight fit, it can be levered off carefully with a screwdriver.

Check that the contacts are clean and free from pits, burns, oil or grease. Turn the engine and check that the contacts are opening and closing correctly and that the clearance is correct when the contacts are fully opened to between ·014 and ·016 in. (·36 and ·40 mm.).

Correct the gap if necessary.

Disconnect the cable at the contact breaker terminal of the coil and at the low-tension terminal of the distributor, and connect a test lamp between these terminals. If the lamp lights when the contacts close and goes out when the contacts open, the low-tension circuit is in order. Should the lamp fail to light, the contacts are dirty or there is a broken or loose connection in the low-tension wiring.

Section B.3

LOCATING A LOW-TENSION CIRCUIT FAULT

Having determined, by testing as previously described, that the fault lies in the low-tension circuit, switch on the ignition, and turn the engine until the contact breaker points are fully opened.

Refer to the wiring diagram and check the circuit with a voltmeter (0–20 volts) as follows.

NOTE.—If the circuit is in order, the reading on the voltmeter should be approximately 12 volts.

(1) **Battery to starter switch terminal.** Connect voltmeter to starter switch terminal and to earth. No reading indicates a faulty cable or loose connections.

(2) **Starter switch to control box terminal 'A' (brown lead).** Connect a voltmeter to the control box terminal 'A' and to earth. No reading indicates a faulty cable or loose connections.

(3) **Control box terminal 'A1'.** Connect a voltmeter to the control box terminal 'A1' and to earth. No reading indicates a fault in the series winding of the control box.

(4) **Control box terminal 'A1' to terminal on ignition switch (brown with blue lead).** Connect a voltmeter to the ignition switch terminal and to earth. No reading indicates a faulty cable or loose connections.

(5) **Ignition switch.** Connect a voltmeter to the second ignition switch terminal (white lead) and to earth. No reading indicates a fault in the ignition switch.

(6) **Ignition switch to fusebox terminal 'A3' (white lead).** Connect the voltmeter to the fuse unit terminal 'A3' and to earth. No reading indicates a faulty cable or loose connections.

(7) **Fuse unit terminal 'A3' to ignition coil terminal 'SW' (white lead).** Connect a voltmeter to the ignition coil terminal 'SW' and to earth. No reading indicates a faulty cable or loose connections.

(8) **Ignition coil.** Connect a voltmeter to the ignition terminal 'CB' (white with black lead) and to earth.

No reading indicates a fault in the primary winding of the coil and a new coil must be fitted.

(9) **Ignition coil to distributor (white with black lead).** Connect a voltmeter to the distributor low-tension terminal and to earth. No reading indicates a faulty cable or loose connections.

(10) **Contact breaker and capacitor.** Connect the voltmeter across the breaker points. No reading indicates a fault in the capacitor.

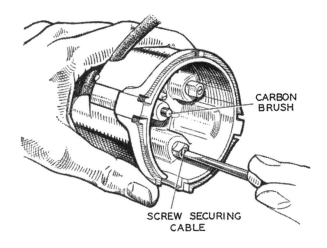

Fig. B.2

The method of connecting H.T. leads

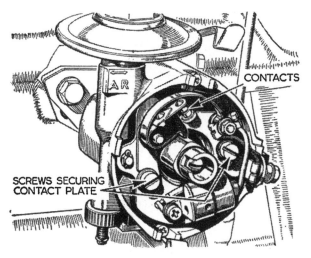

Fig. B.1

The distributor with the cover and rotor arm removed, showing the contact breaker mechanism

Section B.4

HIGH-TENSION CABLES

The high-tension cables must be examined carefully and any which have the insulation cracked, perished or damaged in any way must be renewed.

To fit the cables to the terminal of the ignition coil, thread the knurled moulded terminal nut over the lead, bare the end of the cable for about ¼ in. (6 mm.), thread the wire through the brass washer removed from the original cable and bend back the strands over the washer. Finally, screw the terminal into the coil.

To make the connections to the terminals in the distributor moulded cap, first remove the cap and slacken the screws on the inside of the moulding till they are clear of the cables. Fill the holes in the distributor cap with Silicone grease, then cut the new cables off to the required length, push them completely home and tighten the securing screws.

The cables from the distributor to the sparking plugs must be connected up in the correct firing order, which is 1, 3, 4, 2. Screw them firmly to the suppressors.

Section B.5

CONTACT BREAKER

After the first 500 miles (800 km.) and subsequently every 6,000 miles (9600 km.) check the contact breaker as follows:

(1) Turn the crankshaft until the contact breaker points are fully opened and check the gap with a gauge having a thickness of from ·014 to ·016 in. (·36 to ·40 mm.). If the gap is correct, the gauge should be a sliding fit. Do not alter the setting unless the gap varies considerably from the gauge thickness.

To adjust the setting, keep the engine in the position which gives maximum opening of the contacts and then slacken the two screws securing the fixed contact plate. Adjust the position of the plate until the gap is set to the thickness of the gauge and then tighten the two locking screws.

Remember that the cam only keeps the contact points fully open over a very small angle and that care must be taken to ensure that the points are in the fully open position.

(2) If the contacts are dirty or pitted, they must be cleaned by polishing them with a fine carborundum stone, and afterwards wiping them with a cloth moistened with petrol (gasoline). The moving contact can be removed from its mounting in order to assist cleaning. Check and adjust the contact breaker setting after cleaning the contacts.

(3) Check that the moving arm is free on its pivot. If it is sluggish, remove the arm and polish the pivot pin with a strip of fine cloth. Afterwards clean off all trace of emery dust and apply a spot of clean engine oil to the top of the pivot. The

contact breaker spring tension should be between 20 and 24 oz. (567 and 680 gm.) measured at the contacts.

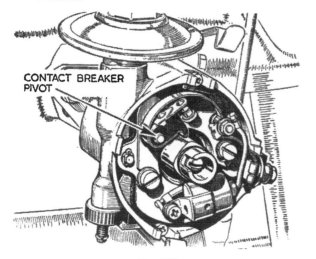

Fig. B.3

The pivot for the moving contact point

Section B.6

REMOVING AND REPLACING THE DISTRIBUTOR

The distributor can be removed and replaced without interfering with the ignition timing, provided the clamp plate pinch-bolt is not disturbed.

To facilitate the replacement of the distributor, turn the engine over until the rotor arm is pointing to the segment in the cover for No. 1 cylinder plug lead to provide a datum for replacement.

Remove the distributor cover, and disconnect the low-tension lead from the terminal on the distributor. Disconnect the suction advance pipe at the union on the distributor.

Extract the two bolts securing the distributor clamp plate to the distributor housing and withdraw the distributor.

To replace the distributor, insert it into the distributor housing until the driving dog rests on the distributor drive shaft. The rotor arm should then be rotated slowly until the driving dog lugs engage with the drive shaft slots, both of which are offset to ensure correct replacement. Turn the distributor body to align the clamping plate holes with those in the housing. The remainder of the assembly is now in the reverse order to that of removal.

NOTE.—Provided that the crankshaft has not been turned the rotor arm will be opposite the segment for No. 1 plug lead. The high-tension leads can then be replaced on their respective plug terminals in the order of firing, i.e. 1, 3, 4, 2, remembering that the distributor rotation is anti-clockwise when viewed from above.

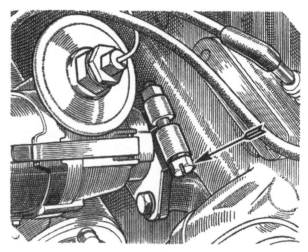

Fig. B.4

Showing the distributor clamp bolt

Section B.7

STATIC IGNITION TIMING

Before timing the ignition refer to 'GENERAL DATA' for the correct setting.

To set the distributor in the correct position for firing, the following procedure should be followed:

(1) Turn the crankshaft in the direction of rotation until No. 1 piston is at T.D.C. on its compression stroke. This can best be effected by turning the engine and observing the valves. When the valves are 'rocking' (i.e. exhaust just closing and inlet just opening) on No. 4 cylinder, No. 1 piston is approximately at T.D.C. on its compression stroke. If the crankshaft is now rotated until the notch in the

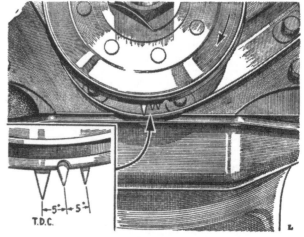

Fig. B.5

The notch in the pulley approaching the T.D.C. position for pistons 1 and 4. The inset shows the timing set at 5° B.T.C.

rear flange of the crankshaft pulley is in line with the long pointer on the timing cover or the dimples on the crankshaft and camshaft gears are in line, the piston is exactly at T.D.C. (see Fig. B.5).

(2) Turn the crankshaft back until the mark on the pulley is in the correct position (see 'GENERAL DATA'). As a guide to this position the two short pointers on the timing cover are 5° and 10° B.T.D.C.

(3) Set the contact breaker points to ·014 to ·016 in. (·36 to ·40 mm.) when in their position of maximum opening.

(4) Insert the distributor into its housing, and engage the drive dog lugs with the drive shaft slots (both of which are offset) by slowly rotating the rotor arm.

(5) Screw in the two bolts securing the distributor clamp plate to the distributor housing.

(6) Position the distributor so that the vacuum control unit side of the body is to the front and the unit is vertical.

(7) Rotate the distributor body anti-clockwise until the points are fully closed. Then slowly rotate it in a clockwise direction until the points just commence to open. Secure the distributor body in this position by tightening up the clamp plate pinch-bolt and nut. Finally, check that the rotor arm is opposite the correct segment for the cylinder which is at the top of its compression stroke.

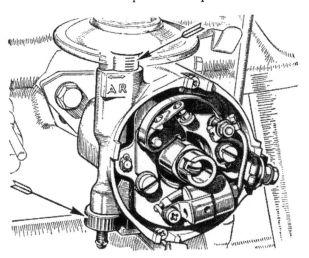

Fig. B.6

The adjusting nut provided to give the ignition point micrometer control

IMPORTANT.—To obtain an accurate setting an electrical method should be used to determine the actual position at which the points break, and the following method can be used.

With the low-tension lead connected to the distributor, turn on the ignition switch and connect a 12-volt lamp in parallel with the contact breaker points (i.e. one lead from the distributor low-tension terminal and the other to earth) and turn the distributor as detailed in paragraph (7) until the lamp lights, which indicates that the points have just opened.

NOTE.—If the distributor drive gear assembly has been removed from the engine it should be refitted in accordance with instructions given in Section A.27 before the above operation is carried out.

Should a stroboscopic lamp be used, care must be taken that with the engine running the speed is low enough to ensure that the centrifugal advance weights are not in operation. If the vacuum advance take-off is direct from the induction manifold the take-off must be disconnected before attempting the timing check, otherwise engine timing will be retarded.

Section B.8

DISMANTLING THE DISTRIBUTOR

The contact breaker plate may be removed as an assembly to give access to the centrifugal weights without completely dismantling the distributor. To do this the rotor arm must first be removed and then the low-tension terminal nuts slackened to enable the slotted connector to be withdrawn from between the head of the terminal bolt and the insulating washer. Next take out the spring clip securing the suction advance unit arm to the plate and release the plate assembly by extracting the two screws which secure it to the distributor body.

Before completely dismantling make careful note of the positions in which the various components are fitted, so that they may be replaced correctly.

(1) Remove the moulded cap.

(2) Lift the rotor off the top of the spindle.

(3) Remove the nut and washer from the moving contact anchor pin. Withdraw the insulating sleeve from the capacitor lead and low-tension lead connectors, noting the order in which they are fitted. Lift the moving contact from the pivot pin and remove the large insulating washer from the pivot pin and the small one from the anchor pin.

(4) Take out the two screws, each with a spring and flat washer, securing the fixed contact plate, and remove the plate.

(5) Take out the securing screw and remove the capacitor. Note that the earthing lead, which is attached to the same screw, passes under the capacitor to keep clear of the cams.

(6) Remove the spring clip retaining the suction advance unit arm to the contact breaker base-plate. Extract the two screws securing the base-

plate to the distributor body, noting that one also secures the earthing lead, and lift out the base-plate.

IMPORTANT.—Note the relative position of the rotor arm drive slot in the cam spindle and the offset drive dog at the driving end of the spindle, to ensure that the timing is not 180° out when the cam spindle is re-engaged with the centrifugal weights during assembly.

(7) Remove the cam retaining screw and spindle.

(8) Take out the centrifugal weights. These may be lifted out as two assemblies, each complete with a spring and toggle.

(9) To release the suction advance unit, remove the circlip, adjusting nut and spring. Withdraw the unit.

(10) To release the spindle from the body, drive out the parallel driving pin passing through the collar of the driving tongue member at the lower end of the spindle.

Section B.9

CAPACITOR

The best method of testing the capacitor is by substitution. Disconnect the original capacitor and connect a new one between the low-tension terminal of the distributor and earth.

Should a new capacitor be necessary, it is advisable to fit a complete capacitor and bracket, but should a capacitor only be available, use a hot iron to soften the solder securing the defective capacitor to the bracket. Care must be taken not to overheat the new capacitor when soldering it in position. The capacity of the capacitor is ·2 microfarad.

Section B.10

REASSEMBLING THE DISTRIBUTOR

Reassembly is a direct reversal of the dismantling procedure given in Section B.8, although careful attention must be given to the following points:

(1) As they are assembled, the components of the automatic advance mechanism, the distributor shaft, and the portion of the shaft on which the cam fits must be lubricated with thin, clean engine oil (Ref. F on page P.2).

(2) Turn the vacuum control adjusting nut until it is in the half-way position when replacing the control unit.

(3) When engaging the cam driving pin with the centrifugal weight, ensure that it is in the original position. When seen from above, the small offset of the driving dog must be on the right and the driving slot for the rotor arm must be downwards.

(4) Adjust the contact breaker to give a **maximum** opening of ·014 to ·016 in. (·36 to ·40 mm.).

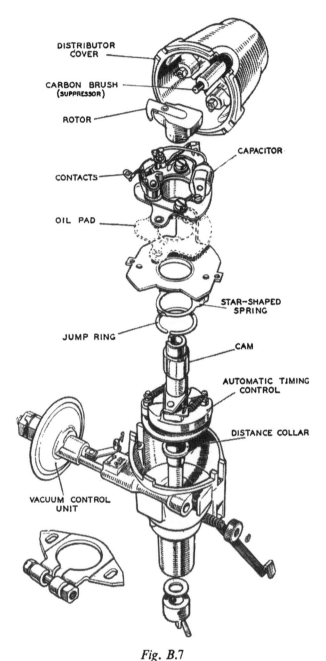

Fig. B.7

The component parts of the distributor

Section B.11

SPARKING PLUGS

Inspect, clean, adjust and renew sparking plugs at the recommended mileage intervals (see 'MAINTENANCE ATTENTION').

When sparking plugs are removed from the engine their gaskets should be removed with them and replaced on the plugs, which should be placed in a suitable holder. It is advisable to identify each plug with the number of the cylinder from which it was removed so that any faults revealed on examination can be traced back to the cylinder concerned.

When examining the plugs, place a new plug of the same type beside the others to afford a ready comparison of the relative condition of the used plugs.

Examine for signs of oil fouling. This will be indicated by a wet, shiny, black deposit on the insulator. This is caused by oil pumping due to worn cylinders and pistons, or gummed-up or broken rings. Under such conditions, oil from the cylinder walls is forced up past the rings on the suction stroke of the piston, and is eventually deposited on the plugs.

A permanent remedy for this cannot be effected, the only cure being the fitting of a new piston and rings, or, in extreme cases, a rebore may be necessary.

Next examine the plugs for signs of petrol (gasoline) fouling. This is indicated by a dry, fluffy, black deposit which is usually caused by over-rich carburation, although ignition system defects such as a run-down battery, faulty distributor, coil or condenser defects, or a broken or worn-out cable may be additional causes. If the plugs appear to be suitable for further use, proceed to clean and test them.

Fig. B.8

Showing plug gaskets in various conditions

First remove the plug gaskets and examine them for condition. Gaskets in different conditions are illustrated in Fig. B.8. The upper left gasket was obviously not properly compressed, owing to the plug not having been tightened down sufficiently. A large proportion of the heat of the plug is normally dissipated to the cylinder head through the copper gasket between the plug and the head. Plugs not screwed down tightly can thus easily become over-heated so that they operate out of their proper heat range, thus producing pre-ignition,

short plug life and 'pinking.' On the other hand, it is unnecessary and unwise to tighten up the plugs too much. What is required is a reasonably good seal between the plug and the cylinder head.

The lower left-hand gasket clearly indicates that the plug was pulled down too tightly or has been in service too long. Note its distorted condition and the evidence of blow-by, which is also a cause of plug overheating.

The upper right-hand gasket demonstrates a gasket in good condition, providing an adequate seal and a good path for heat dissipation.

For comparison a new gasket is shown at the lower right-hand corner of Fig. B.8. If gaskets are at all questionable they should be replaced by new ones.

If the plugs require cleaning it is preferable to make use of a proper plug cleaner of the type recommended by the plug manufacturers, and the makers' instructions for using the cleaner should be followed carefully.

Occasionally a blistered insulator or a badly burnt electrode may be noticed when examining the plugs.

If the plug is of the type normally recommended for the engine and it was correctly installed (down tightly on the gasket), this condition may have been brought about by a very lean mixture or an overheated engine. There is, however, a possibility that a plug of another type is required, but as a rule the recommended Champion plug should be adhered to (see **'GENERAL DATA'**).

After cleaning carefully, examine the plugs for cracked insulators and wear of the insulator nose due to excessive previous cleaning. In such cases the plugs have passed their useful life, and new plugs should be installed.

Examine the insulator for deposits underneath the side electrode which have possibly accumulated and which act as a 'hot spot' in service.

After cleaning the plugs in a special cleaner, blow all surplus abrasive out of the body recesses, and off the plug threads, by means of an air-blast. Next examine the threads for carbon. Any deposits can be removed and the threads cleaned with a wire brush. A wire buffing wheel may also be utilised, but reasonable care must be used in both methods in order not to injure the electrodes or the tip of the insulator. The thread section of the plug body is often neglected when cleaning the plugs, owing to the fact that it is not generally realised that, like the gaskets, the threads are a means of heat dissipation and that when they are coated with carbon it retards the flow of the heat from the plug, producing overheating. This simple procedure will also ensure absence of binding on the threads on replacement and also avoid unnecessary use of the plug spanner.

When replacing a plug, always screw it down by hand as far as possible and use the spanner for final tightening only. Whenever possible use a box spanner to avoid possible fracture of the insulator.

Examine the electrodes for correct gap by inserting a feeler between them. Avoid an incorrect reading in the case of badly pitted electrodes. See 'GENERAL DATA' for the correct clearance.

Remember that electrode corrosion and the development of oxides at the gap area vitally affects the sparking efficiency. The special cleaner can remove the oxides and deposits from the insulator, but the cleaner stream does not always reach this area with full effect owing to

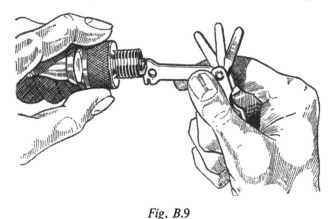

Fig. B.9

Adjusting the plug gap with the Champion setting tool

its location, and cannot necessarily deal with corrosion effectively as this sometimes requires too strong a blast for proper removal.

When plugs appear worthy of further use it is good practice to dress the gap area on both centre and side electrodes with a small file before resetting them to the correct gap. The intense heat, pressure, explosion shock, and electrical and chemical action to which the plugs are submitted during miles of service are so intense that the molecular structure of the metal points is eventually affected. Plugs then reach a worn-out condition and resetting the points can no longer serve a good purpose. When points are burnt badly, it is indicative that the plug has worn to such an extent that its further use is undesirable and wasteful.

Before replacing the plug in the engine, test it for correct functioning under air pressure in a plug tester, following out the instructions issued by the makers of the plug tester. Generally speaking, a plug may be considered satisfactory for further service if it sparks continuously under a pressure of 100 lb. per sq. in. (7 kg./cm.²) with the gap between the points set at ·022 in. (·56 mm.). It is essential that the plug points should be reset to the correct gap (see 'GENERAL DATA') before the plug is refitted to the engine.

While the plug is under pressure in the tester, it should be inspected for leakage by applying oil round the terminal. Leakage is indicated by the production of air

bubbles, the intensity of which will serve to indicate the degree of leakage. The leakage gases have a 'blow-torch' effect when the engine is running which rapidly raises the temperature of the plug, raising it above its designed heat range, thus producing overheating, pre-ignition, and rapid electrode destruction.

The top half of the insulator is frequently responsible for poor plug performance due to the following faults: splashes; accumulation of dirt and dust; cracked insulators, caused by a slipping spanner; overtightness of the terminals.

Examine for a cracked insulator at the shoulder and the terminal post and remove any accumulations of dirt and dust.

Section B.12

INTERFERENCE SUPPRESSORS

In order to reduce the interference with television reception to a minimum, all models are fitted with suppressors incorporated in the plug terminals of the high-tension leads.

Contrary to popular belief, these do not affect the ignition adversely.

Section B.13

DM2.P4 PRE-TILT DISTRIBUTOR

The DM2.P4 distributor fitted to later cars has a pre-tilted contact breaker unit. The moving contact breaker plate is balanced on two nylon studs and the angle through which the plate may be tilted is controlled by a stud riveted to the moving contact breaker

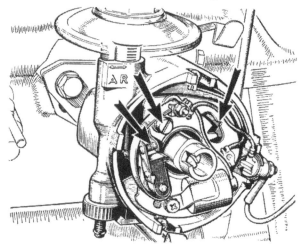

Fig. B.10.

The DM2.P4 distributor with the cover and rotor arm removed, showing the contact breaker mechanism

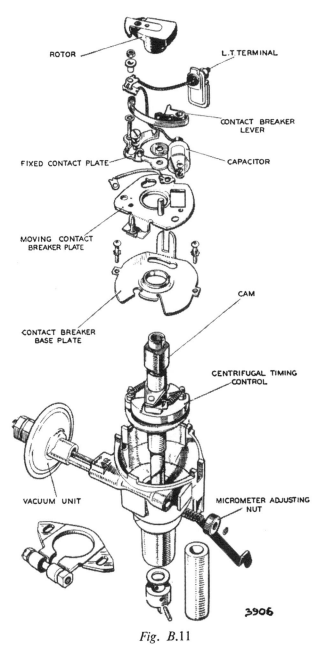

ROTOR

L.T. TERMINAL

CONTACT BREAKER LEVER

FIXED CONTACT PLATE

CAPACITOR

MOVING CONTACT BREAKER PLATE

CAM

CONTACT BREAKER BASE PLATE

CENTRIFUGAL TIMING CONTROL

VACUUM UNIT

MICROMETER ADJUSTING NUT

3906

Fig. B.11

The component parts of the DM2.P4 distributor

plate locating in a slot in the base plate. The plate carrying the fixed contact is secured by one screw only (centre arrow, Fig. B.10) on the new units.

To adjust the contact breaker gap (left-hand arrow, Fig. B.10), turn the engine by hand until the contacts show the maximum opening. This should measure ·014 to ·016 in. (·356 to ·406 mm.). If the setting is incorrect, slacken the securing screw and adjust the contact gap by inserting a screwdriver in the notched hole (right-hand arrow, Fig. B.10), and moving the plate

carrying the fixed contact. Turn clockwise to reduce the gap and anti-clockwise to increase the gap. Tighten the securing screw.

The base plate components are assembled with a special lubricant and no further lubrication is necessary during the normal service life of the distributor.

An improved version of the metalized capacitor is fitted and the eyelets on the cables connected to the contact breaker terminal post are squared and slotted to prevent them twisting round and short-circuiting against the distributor. A flexible actuating link connects the diaphragm in the vacuum unit with the moving contact breaker plate.

The new contact breaker base plates are not interchangeable with those of the previous type and, to avoid confusion, distributors incorporating the new base plate are issued under a new part number—1H811.

Section B.14

IGNITION VACUUM PIPE FUEL TRAP

At Engine No. 6625 a modified ignition vacuum pipe (Part No. 1H830) is fitted. A small trap which contains a fine-mesh gauze is incorporated in the pipe which will prevent fuel entering the vacuum control unit. The modified pipe may be fitted to earlier vehicles.

Section B.15

MODIFIED TYPE DM2 DISTRIBUTORS

A modified Type DM2 distributor incorporating a rolling weight centrifugal advance mechanism superseding the standard toggle type was introduced at Engine Nos. 16GC/U/H101 and 16GC/U/L101.

The rolling weight mechanism consists of a shaft and action plate with two action cams and two spring pillars riveted to the plate. Two rolling weights, each located by a boss on the under side of the cam foot, roll round the action plate cams and thus alter the position of the contact breaker cam relative to the distributor driving shaft when the distributor is rotating within the speed limits of centrifugal operation.

The rate of advance is controlled by springs anchored between the pillars on the action plate and two pillars on the cam foot. The maximum amount of advance is governed by the cam foot, which strikes one of the spring pillars when maximum advance has been reached.

The modified distributor is interchangeable with the previous model and the instructions for dismantling and assembly are given in Sections B.8 and B.10.

Notes

SECTION C

THE COOLING SYSTEM

C

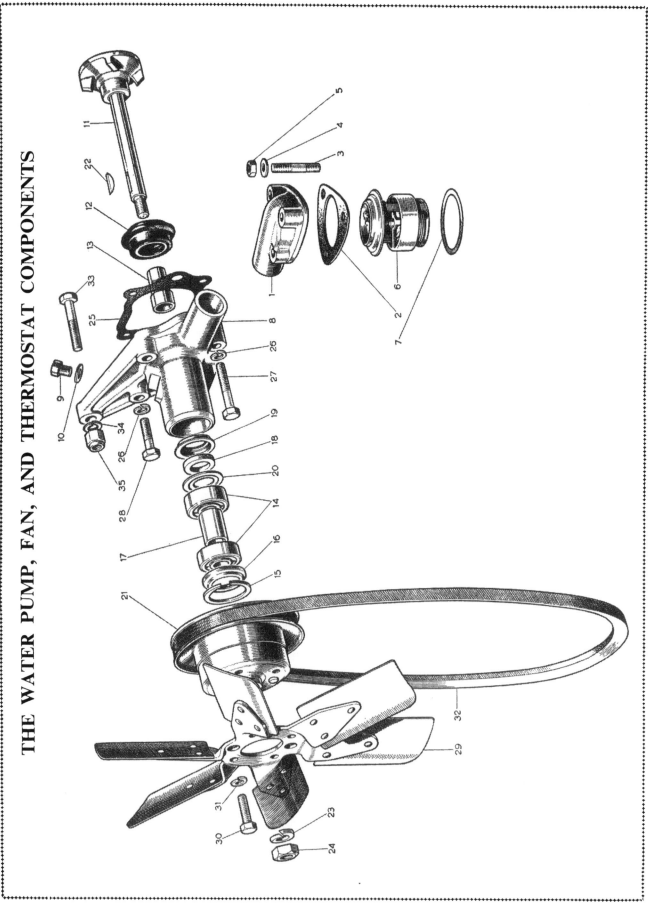

THE WATER PUMP, FAN, AND THERMOSTAT COMPONENTS

KEY TO THE WATER PUMP, FAN, AND THERMOSTAT

No.	Description
1.	Elbow—water outlet.
2.	Joint—elbow.
3.	Stud—elbow to cylinder.
4.	Washer—stud.
5.	Nut—stud.
6.	Thermostat.
7.	Joint—thermostat.
8.	Body.
9.	Plug.
10.	Washer—plug.
11.	Spindle with vane.
12.	Seal.
13.	Distance piece—gland.
14.	Bearing.
15.	Spring ring—bearing.
16.	Grease retainer—bearing.
17.	Distance piece—bearings.
18.	Washer—felt.

No.	Description
19.	Retainer—felt washer inner.
20.	Retainer—felt washer outer.
21.	Pulley and fan—water pump.
22.	Key—pulley.
23.	Spring washer.
24.	Nut—pulley to spindle.
25.	Joint—pump to block.
26.	Spring washer.
27.	Set screw—long—to block.
28.	Set screw—short—to block.
29.	Fan complete.
30.	Set screw—fan to pulley.
31.	Spring washer.
32.	Belt—wedge type—fan.
33.	Bolt—dynamo to water pump body.
34.	Spring washer.
35.	Nut—dynamo bolt.

DESCRIPTION

The cooling system is sealed, and the water circulation is assisted by a pump attached to the front of the engine and driven by a belt from the crankshaft. The water circulates from the base of the radiator and passes around the cylinders and cylinder head, reaching the header tank of the radiator core via the thermostat and the top water hose. From the header tank it passes down the radiator core to the base tank of the radiator. Air is drawn through the radiator by a fan attached to the water pump pulley.

The thermostat is set to open between 70° C. and 75° C. (158° F. and 167° F.).

IMPORTANT.—Never use a muff on the radiator grille to protect the cooling system in cold weather as this would seal the carburetter and heater unit air supply. The radiator must be protected by a blind such as the type available as an optional extra fitting.

Section C.1

REMOVING THE FILLER CAP

The cooling system is under appreciable pressure while the engine is hot after a run, and the radiator filler cap must be removed very carefully or left in position until the water has cooled.

If it is necessary to remove the filler cap when the engine is hot it is absolutely essential to remove it gradually, and the filler spout is provided with a specially shaped cam to enable this to be done easily.

Unscrew the cap slowly till the retaining tongues are felt to engage the small lobes on the end of the filler spout cam, and wait until the pressure in the radiator is fully released before finally removing the cap.

It is advisable to protect the hand against escaping steam while removing the cap.

Fig. C.1

Showing the radiator filler cap retaining cam

Fig. C.2

The engine drain tap on the right-hand side of the engine

Section C.2

DRAINING THE COOLING SYSTEM

Remove the radiator header tank filler cap.

Open the two drain taps. One is fitted on the left-hand side of the base of the radiator and the other at the rear of the cylinder block on the right-hand side.

NOTE.—If anti-freeze mixture is being used it should be drained into a suitable container and carefully preserved for replacement.

Section C.3

FILLING THE COOLING SYSTEM

Close the radiator and cylinder block drain taps.

Ensure that the water hose clips are tightened.

Fill up the system through the filler in the radiator header tank until approximately 1 in. (2·5 cm.) of water is visible in the filler neck.

When possible, rain-water should be used for filling the system.

Avoid overfilling to prevent loss of anti-freeze due to expansion.

Screw the filler cap firmly into position.

The cooling system is unsuitable for use with anti-freeze mixtures having an alcohol base owing to the high temperatures attained in the top tank. Only anti-freeze mixtures of the ethylene glycol or glycerine type should be employed.

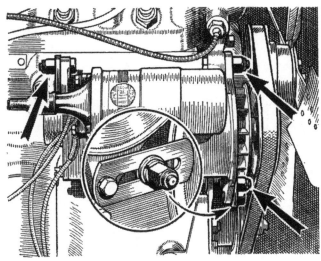

Fig. C.3

Dynamo mounting bolts which must be slackened for fan belt tension adjustment

Section C.4

REMOVING AND REPLACING THE RADIATOR

Drain the water from the cooling system as in Section C.2.

Release the clips on the top and bottom water hoses and detach the hoses from their connections.

Remove the three bolts securing each side of the radiator to the body and lift out the radiator.

Replace the radiator core by reversing the above procedure, noting that there is a packing strip between the radiator flanges and the body.

Close the drain taps and refill the cooling system with clean water and check for leaks.

Section C.5

DYNAMO AND FAN BELT ADJUSTMENT

The adjustment of the dynamo and fan belt tension is effected by slackening slightly the two bolts on which the dynamo pivots, releasing the bolt securing it to the slotted link, and the nut securing the slotted link to the engine. Raise the dynamo bodily until the belt tension is correct. Tighten up the bolts with the dynamo in this position.

NOTE.—A gentle hand pull only must be exerted on the dynamo, or the belt tension will be excessive and undue strain thrown on the dynamo bearings.

To check the tension for correctness rotate the fan blades. If the dynamo pulley slips inside the fan belt, the tension is insufficient. When the tension is correct, it should be possible to move the belt from side to side to the extent of 1 in. (2·5 cm.) at the centre of the longest belt run.

Section C.6

COLD WEATHER PRECAUTIONS

As the cooling system is sealed, relatively high temperatures are developed in the radiator upper tank. For this reason anti-freeze solutions having an alcohol base are unsuitable owing to their high evaporation rate producing rapid loss of coolant and a consequent interruption of the circulation of coolant.

Only anti-freeze of the ethylene glycol or glycerine type is suitable for use in the cooling system.

The correct quantities of anti-freeze for different degrees of frost resistance are given on page 7.

Before introducing anti-freeze mixture to the radiator it is advisable to clean out the cooling system thoroughly by swilling out the passages with a hose inserted in the filler cap, keeping the drain taps open. Only top up when the cooling system is at its normal running temperature, in order to avoid losing anti-freeze due to expansion.

Make sure that the cooling system is water-tight and examine all joints, replacing any defective rubber hose with new.

The capacity of the cooling system is 10 pints (5·7 litres, 12 U.S. pints).

Section C.7

REMOVING THE WATER PUMP

The water pump and fan assembly is attached to the front of the cylinder block by four bolts.

To remove the water pump it is first necessary to drain the water from the cooling system by opening the two drain taps as described in Section C.2, at the same time remembering to collect the water for re-use if it contains anti-freeze mixture.

Fig. C.4

The location of the radiator drain tap

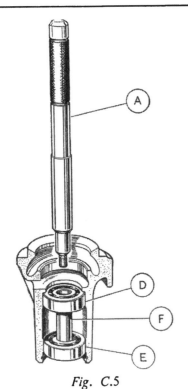

Fig. C.5

This cut-away view of the water pump shows the special driver about to enter the rear bearing for removal. The rear bearing felt and retainers are not shown here

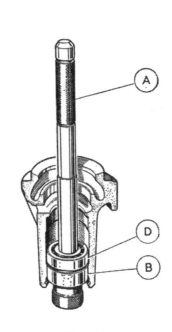

Fig. C.6

Here the dummy front bearing is in position and the driver is screwed into it. This is the method of removing the rear bearing

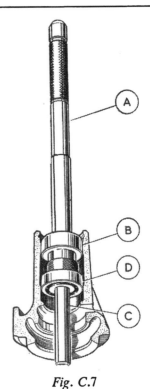

Fig. C.7

When reassembling the pump use the tool assembled as shown and thread the rear bearing onto the pilot with the felt ring and retainers. (Felt and retainers not shown here)

Release the clips on the top and bottom water hoses and detach the hoses from their connections.

Remove the three bolts at each side securing the radiator core to the body and lift out the radiator.

Disconnect the dynamo leads, remove the dynamo attachment bolts and take off the dynamo.

Take out the four bolts securing the fan and pump assembly to the front of the cylinder block and withdraw the assembly.

Replacement of the fan and pump assembly is a reversal of this procedure, but care must be taken to see that the joint gasket between the pump body and the cylinder block is in good condition. It is always advisable to fit a new gasket.

Section C.8

DISMANTLING THE WATER PUMP

When the fan and water pump assembly has been removed from the engine, as indicated in Section C.6, the water pump may be dismantled for attention in the following way:

Unscrew the four set bolts which attach the fan and belt pulley to the hub and remove the fan blades and pulley.

Unscrew the nut and spring washer from the end of the pump spindle and pull off the fan hub with a suitable extractor.

Remove the Woodruff key from the spindle and remove any burrs from the keyway. Withdraw the dished oil seal washer after removing its circlip.

Gently tap the pump spindle rearwards out of the pump body. Withdraw the sealing gland.

Should it be necessary to withdraw the ball races, the front one can be withdrawn with an extractor.

When the front bearing is removed it releases the distance tube between the bearings and gives access to the rear bearing.

When the rear bearing is extracted it permits the withdrawal of the felt washer along with its inner and outer retainers. The rear distance piece now remains in the pump body and may be removed if required.

Reassembly is a reversal of the dismantling procedure, but care must be taken to see that the seal assembly is in good condition before proceeding. If it shows signs of damage, a new seal must be fitted.

Renew the felt oil seal washer for the rear ball race if it shows signs of damage.

Repack with grease to Ref. C.

Section C.9

USING SPECIAL TOOL 18G60

This tool comprises three parts: the driver, dummy front bearing, and pilot (see the illustration on page Q.5). It is designed to remove and replace the water pump bearings without damage to the pump body.

After removing the water pump spindle and water seal, and the front bearing seal and retaining clip as detailed in Section C.8, support the pump body with the front end downwards.

Insert the driver (A) (Fig. C.5) into the rear bearing (D) and tap downwards until the front bearing (E) and the distance piece (F) are released.

Position the dummy front bearing (B) in the pump body and screw the driver into it (Fig. C.6). The rear bearing is now aligned with the housing bore and may be tapped out and the felt washer and retainers removed.

Replacing bearings

Support the body with the rear end downwards.

Assemble the driver, dummy front bearing, and pilot (c). Position the rear bearing, the felt ring inner retainer, the felt ring and outer retainer, in that order, on the pilot and press them into the pump body (Fig. C.7).

Position the distance tube and front bearing on the pilot and press the bearing into the pump body.

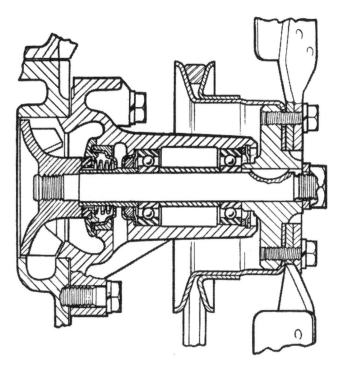

Fig. C.8

A section through the water pump showing the location of the components

Section C.10

MODIFIED WATER PUMP

A new water pump (Part No. 1H1149) which incorporates a one-piece bearing is fitted from engines numbered 15GB/U/H39365 to 15GB/U/H39400 inclusive, and 15GB/U/H39526 onwards. The pump is interchangeable with the original, but only as a complete unit.

Removing

Remove the radiator as detailed in Section C.4.

Remove the dynamo attachment bolts and take off the dynamo.

Unscrew the four bolts attaching the pump assembly to the front of the cylinder block and remove the fan and pump assembly.

Replacement of the fan and pump assembly is a reversal of the above procedure.

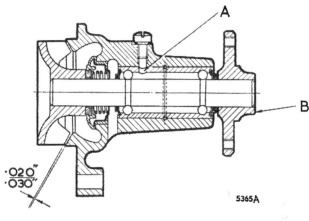

5365A

Fig. C.9

A section through the later-type water pump. When assembled, the hole in the bearing (A) must coincide with the lubricating hole in the water pump, and the face of the hub (B) must be flush with the end of the spindle

Dismantling

Unscrew the four set bolts which attach the fan and pulley to the hub and remove the fan and pulley.

Remove the fan hub with a suitable extractor.

Pull out the bearing locating wire through the hole in the top of the pump body.

Gently tap the pump bearing assembly rearwards out of the pump body. This will release the combined bearing and spindle assembly together with the seal and vane.

Remove the vane from the bearing assembly with a suitable extractor and remove the pump seal assembly.

Reassembly is a reversal of this procedure, but care must be taken to see that the seal assembly is in good condition. If there is any sign of damage the seal should be replaced by a new component. When the bearing assembly is fitted into the pump the hole in the bearing (A) must coincide with the lubricating hole in the water pump body. Should the interference fit of the fan hub have been impaired when the hub was withdrawn from the spindle, a new hub should be fitted.

SECTION D

THE FUEL SYSTEM

THE CARBURETTER COMPONENTS

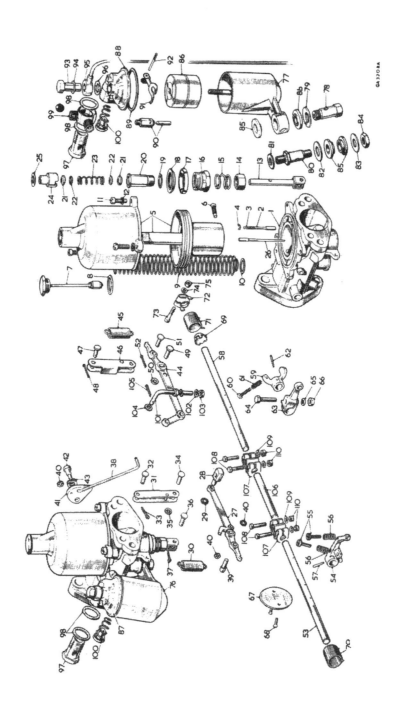

KEY TO THE CARBURETTER COMPONENTS

No.	Description
1.	Body—rear carburetter.
2.	Pin—piston lifting.
3.	Spring for pin.
4.	Circlip for pin.
5.	Chamber and piston assembly.
6.	Screw—needle locking.
7.	Cap and damper assembly.
8.	Washer—fibre—for cap.
9.	Spring—red—for piston.
10.	Washer—plain—for spring.
11.	Screw—chamber to body.
12.	Washer—spring—for screw.
13.	Jet assembly.
14.	Nut—adjusting.
15.	Spring for adjusting nut.
16.	Nut—gland sealing.
17.	Ring—sealing—aluminium.
18.	Ring—sealing—cork.
19.	Washer—bottom bearing—copper.
20.	Bearing—bottom.
21.	Washer—gland—cork.
22.	Washer—gland—brass.
23.	Spring—gland.
24.	Bearing—top.
25.	Washer—top bearing—copper.
26.	Needle—M6—standard jet.
27.	Lever—jet.
28.	Trunnion for lever.
29.	Washer—starlock—for trunnion.
30.	Spring—jet lever—return.
31.	Link assembly—jet lever.
32.	Pin—link to body.
33.	Split pin for pin.
34.	Pin—link to lever.
35.	Washer—starlock—for pin.
36.	Pin—lever to jet.
37.	Split pin for pin.

No.	Description
38.	Link tension.
39.	Swivel pin—link to lever.
40.	Washer—starlock—for link and swivel pin.
41.	Cam plate.
42.	Bolt—pivot—for cam plate.
43.	Washer—spring—for bolt.
44.	Lever—jet.
45.	Spring—jet lever—return.
46.	Link assembly—jet lever.
47.	Pin—link to body.
48.	Split pin for pin.
49.	Pin—link to lever.
50.	Washer—starlock—for pin.
51.	Pin—lever to jet.
52.	Split pin for pin.
53.	Spindle—throttle.
54.	Lever—stop—throttle.
55.	Screw—stop adjusting.
56.	Spring for screw.
57.	Pin—for throttle stop lever.
58.	Spindle—throttle.
59.	Lever—stop—throttle.
60.	Screw—stop adjusting.
61.	Spring for screw.
62.	Pin for throttle stop lever.
63.	Lever—throttle.
64.	Bolt for lever.
65.	Washer—spring—for bolt.
66.	Nut for bolt.
67.	Disc—throttle.
68.	Screw for disc.
69.	Plate—return spring anchor.
70.	Spring—return.
71.	Spring—return.
72.	Clip for return spring.
73.	Bolt for clip.
74.	Washer—plain—for bolt.

No.	Description
75.	Nut for bolt.
76.	Chamber—float.
77.	Chamber—float.
78.	Bolt—chamber to body.
79.	Washer—fibre—for bolt.
80.	Pillar—banjo.
81.	Washer for pillar.
82.	Washer—inner.
83.	Washer—outer.
84.	Nut—lock.
85.	Grommet—rubber.
86.	Float.
87.	Lid—float-chamber.
88.	Lid—float-chamber.
89.	Washer for lid.
90.	Needle and seat assembly.
91.	Lever—hinged.
92.	Pin for hinged lever.
93.	Cap nut for lid.
94.	Washer—aluminium—for nut.
95.	Banjo vent and drain pipe.
96.	Washer—fibre.
97.	Bolt—banjo.
98.	Washer—fibre—for bolt.
99.	Union—banjo—double.
100.	Filter.
101.	Rod—link.
102.	Washer—spring—for rod.
103.	Nut for rod.
104.	Washer—brass.
105.	Split pin for rod.
106.	Rod—throttle connecting.
107.	Coupling for rod.
108.	Bolt for coupling.
109.	Washer for bolt.
110.	Nut for bolt.

Section D.1

REMOVING THE FUEL TANK

Remove the hexagon drain plug and empty the tank.

Slacken the two clips on the filler neck hose and withdraw the filler extension.

Pull the hose from the tank. Take out the three screws and remove the tank filler neck seal and clamp plate.

Disconnect the fuel pipe at the union and the fuel gauge cable from the tank unit, each on the right-hand side of the tank.

Remove the two nuts from the bolts securing the rear of the tank to the anchorage brackets on the frame and remove the two bolts with spring washers which secure the front of the tank to the frame.

Withdraw the rear bolts and distance tubes.

Replacement is a reversal of the above instructions.

Section D.2

REMOVING THE FUEL PUMP

Raise the hood and remove the spare wheel.

Remove the hood stowage compartment floor. This is secured by five quick-release screws and each requires only a quarter turn anti-clockwise to release the cover.

Disconnect the inlet and outlet pipe unions.

Disconnect the earth lead and the supply lead from the terminals on the pump.

Remove the two set screws securing the fuel pump to the bracket on the frame cross-member.

Section D.3

CONSTRUCTION OF THE FUEL PUMP

The fuel pump is an S.U. Type HP high-pressure, 12-volt electric pump incorporating a radio suppressor. It is located close to the right-hand side of the fuel tank.

The pump consists of three main assemblies: the body, the magnet assembly and the contact breaker.

The body is composed of a hollow stamping or casting (8), into the bottom of which the filter (12) is screwed. The pump inlet union (29) is screwed in at an angle on one side. The outlet union (1) is screwed into the top and tightens down on the delivery valve cage (5), which is clamped between the two fibre washers (2 and 6). In the top of the delivery cage is the delivery valve, a thin brass disc (4) held in position by a spring clip (3). Inserted in the bottom of the cage is the suction valve (7), being a disc similar to (4) and held in position by a light spring on a seating machined in the body. Holes connect the space between the valves of the pumping chamber, which is a shallow depression on the forward face of the body. This space is closed by a diaphragm assembly (9) which is clamped at its outside edge between the magnet housing (27) and body (8) and at its centre between the retaining plate (11) and the steel armature (15). A bronze rod (16) is screwed through the centre of the armature, to which the diaphragm is attached, and it passes through the magnet core to the contact breaker, which is located at the other end. A volute spring (28) is interposed between the armature and the end plate of the coil to return the armature and diaphragm.

The magnet consists of a cast-iron pot have an iron core (17), on which is wound a coil of copper wire which energizes the magnet. Between the magnet housing and the armature are fitted 11 spherical-edged brass rollers (10). These locate the armature centrally within the magnet at all times, and allow absolute freedom of movement in a longitudinal direction. The contact breaker consists of a small bakelite moulding carrying two rockers (25 and 26) which are both hinged to the moulding at one end and are connected together at the top end by two small springs, arranged to give a 'throw over' action. A trunnion is fitted into the centre of the inner rocker, and the bronze push-rod (16) connected to the armature is screwed into this. The outer rocker (26) is fitted with a tungsten point, which makes contact with a further tungsten point on a spring blade (24). This spring blade is connected to one end of the coil, and the other end of the coil is connected to the terminal (20).

A short length of flexible wire is connected to the outer rocker and to the other terminal (23), which also serves to hold the bakelite moulding onto the magnet housing.

The rocker mechanism is insulated by fibre bushes. Two fibre bushes are fitted to one of the spindles of the 'throw over' mechanism in order to silence the operation of the contact breaker.

The body is die-cast in two pieces, the joint between them being sealed by a gasket.

Section D.4

ACTION OF THE FUEL PUMP

The action of the pump is as follows.

When the pump is at rest, the outer rocker lies in the outer position and the tungsten points are in contact. The current passes from the terminal through the coil back to the blade, through the points and to the earth return, thus energizing the magnet and attracting the armature. This comes forward, bringing the diaphragm with it and sucking fuel through the suction valve into the pumping chamber. When the armature has advanced nearly to the end of its stroke the 'throw over' mechanism operates, and the outer rocker flies back, separating the points and breaking the circuit. The spring (28) then pushes the armature and diaphragm back, forcing fuel through the delivery valve at a rate determined by the requirements of the engine. As soon as the armature gets near the end of this stroke the "throw over" mechanism again operates, the points again make contact, and the cycle of operations is repeated.

Section D.5

DISMANTLING AND REASSEMBLING THE FUEL PUMP

When a pump comes in for reconditioning the first thing to do is to determine whether it has been in contact with gum formation in the fuel, resulting in the parts

in contact with the fuel becoming coated with a substance similar to varnish. These deposits cause the eventual destruction of the neoprene diaphragm. The easiest way to identify this deposit is to smell the outlet union. If an unpleasant stale smell is noticed it indicates the presence of gum in the pump. The ordinary sharp, acrid smell of petrol (gasoline) denotes that no gum is present.

Assuming that trouble with gum formation is indicated, the whole of the parts coming into contact with fuel will have to be dismantled. Those made in brass or steel should be boiled in 20 per cent. caustic soda solution, given a dip in strong nitric acid and then washed in boiling water. Those made in aluminium should be well soaked in methylated spirits and cleaned.

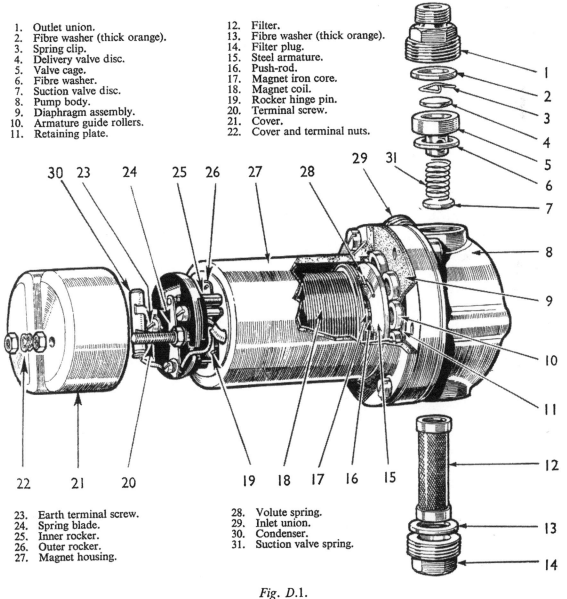

1. Outlet union.
2. Fibre washer (thick orange).
3. Spring clip.
4. Delivery valve disc.
5. Valve cage.
6. Fibre washer.
7. Suction valve disc.
8. Pump body.
9. Diaphragm assembly.
10. Armature guide rollers.
11. Retaining plate.
12. Filter.
13. Fibre washer (thick orange).
14. Filter plug.
15. Steel armature.
16. Push-rod.
17. Magnet iron core.
18. Magnet coil.
19. Rocker hinge pin.
20. Terminal screw.
21. Cover.
22. Cover and terminal nuts.
23. Earth terminal screw.
24. Spring blade.
25. Inner rocker.
26. Outer rocker.
27. Magnet housing.
28. Volute spring.
29. Inlet union.
30. Condenser.
31. Suction valve spring.

Fig. D.1.

The S.U. fuel pump.

To dismantle the pump

First undo the filter plug and remove the filter plug washer and the filter. The latter may be found to be clogged completely with gum. Next the inlet union and its washer should be removed, followed by the outlet union, outlet union washer, valve cage, valve cage washer and suction valve and spring. The valve cage should then be dismantled by removing the circlip retaining the delivery valve in place, and the valve itself can then be withdrawn.

Next undo the six screws holding the two main components of the pump together. All the components of the pump body—with the exception of the washer, but including the pump body itself—should now be cleaned to remove all trace of gum. New fibre washers should be used on replacement.

If there is no evidence of gum formation, proceed as follows:—First undo the six screws holding the two parts of the pump together. The action of the valves

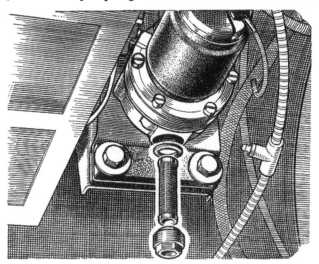

Fig. D.2.

The pump filter should be cleaned with a brush every 6,000 miles (10000 km.).

can then be checked by blowing and sucking in the inlet union, to check the suction valve; and the outlet union to check the delivery valve. In the former case it should be possible to blow freely but not to suck air back, and with the latter to suck and not blow.

Clean the filter in fuel with a brush and swill out the body of the pump.

Next unscrew the diaphragm assembly from its trunnion in the contact breaker. This is done by rotating the whole assembly in an anti-clockwise direction. Take care not to lose the brass rollers fitted behind the diaphragm. The easiest method is to hold the body in the left hand and to rotate the diaphragm.

Now remove the contact breaker cover by taking off

the nut which holds it in place on the terminal, and then undo the last nut on the terminal, which acts as a seating for the cover. Beneath this will be found a lead washer which is squeezed into the thread on the terminal. This should be cut away with a pocket knife, allowing the terminal to be pushed down a short way so that the tag on the coil end is free on the terminal.

Remove the 5 B.A. screw holding the contact blade in position, together with its spring washer and the contact blade.

Remove the two long 2 B.A. screws holding the bakelite pedestal in place, together with their spring washers. Take off the contact breaker assembly, using great care to get the coil end tag over the terminal without damaging the coil end.

Push out the hinge pin sideways and the pump is completely dismantled, since the rocker mechanism is supplied only as a complete assembly.

Do not disturb the core of the magnet; it can only be located correctly with special press tools.

To reassemble the pump

When reassembling, see that all parts are clean. The valves (4 and 7) should be fitted with the smooth side downwards. Care should be taken that the valve retaining clip (3) in the delivery valve cage (5) is correctly located in its groove. The thin, hard fibre washer (6) should be fitted under the valve cage and a thick one (2) above the valve cage and above the filter plug. The washer on the inlet union (29) is a thick fibre one.

The contact breaker should be assembled on its pedestal in such a manner that the rockers are free in their mountings, without appreciable side-play. Any excessive side-play on the outer rocker will allow the points to get out of line, while excessive tightness will make the action of the contact breaker sluggish. To obtain the required freedom in cases of tightness, it may be necessary to square up the outer rocker with a pair of thin-nosed pliers. **The hinge pin is case-hardened, and on no account should ordinary wire be used as a replacement. Always use the correct hardened pin.**

Should the spring contact breaker blade be removed, it must always be replaced bearing directly against the bakelite pedestal, i.e. underneath the tag.

When properly fitted the blade should rest against the ledge on the pedestal while the points are separated, and it should not be sufficiently stiff to prevent the outer rocker from coming right forward when the points are in contact. The points should make contact when the rocker is in its midway position. The simplest way to check this is to hold the blade in contact with the pedestal, taking care not to press on the overhanging portion, and see that you can get a ·030 in. (·76 mm.) feeler between the white rollers and the cast-iron body of the pump.

If necessary, the tip of the blade may be set to give the correct clearance.

Note.—The spring washer on the B.A. screw to which the earth connection is made should be fitted between the tag and the pedestal. The spring washer is not a reliable conductor, and the brass tag must bear directly against the head of the screw.

All four connections, namely, the two ends of the earthing tag and the two ends of the coil, should be soldered. The coil end leading to the terminal should be soldered to its tag and not to the retaining nut. In the case of the terminal screw which holds the bakelite cover in position, similar considerations apply, the assembly being : spring washer (1), wiring tag (2), lead washer (3), and recessed nut (4). (See Fig. D.5.) A lead washer has been found necessary at this point as some few cases of bad connection have been found. Under no circumstances must the spring washer be omitted, or the assembly shortened in any way. Any attempt to do so is likely to lead to breakage of the pedestal when the nut retaining the cover in position is tightened up.

The armature return spring should be fitted with its larger diameter towards the coil and its smaller diameter resting against the armature. This spring must not be stretched or otherwise interfered with, or the action of the pump will be affected.

Section D.6

RESETTING THE DIAPHRAGM FOR CONTACT BREAKER "THROW-OVER"

If the armature and centre rod have been unscrewed it will be necessary to reset as follows:—

1. **Swing to one side the spring blade which carries the contact points.**
2. Fit the impact washer in the recess of the armature.
3. Screw the armature into position.
4. Place the eleven guide rollers in position around the armature. **Do not use jointing compound on the diaphragm.**
5. Hold the magnet assembly in the left hand, in an approximately horizontal position.
6. Screw the armature inwards until the " throw-over " ceases to operate, and then screw it back gradually, a sixth of a turn (or one hole) at a time, and press the armature in after each part of a turn until it is found that when it is pushed in slowly and firmly the " throw-over " mechanism operates. **Unscrew the armature a further two-thirds of a turn** (four of the six holes). When a new diaphragm is fitted it is probable that considerable pressure will be required to push the armature right home.

7. Place the cast-iron body in position on the main body, taking care to see that the drain hole in the cast-iron body is at the bottom in line with the filter plug in the main body, and all the rollers are still in their correct positions.

If a roller drops out of position it will get trapped between the two ports, and this will cut a hole in the diaphragm.

Make sure that the cast-iron body is seating properly on the main body and insert the six securing screws. **Before tightening the screws down it is essential that the diaphragm should be stretched to its outermost position.**

Do this by inserting a match-stick behind one of the white fibre rollers on the outer rocker, thus

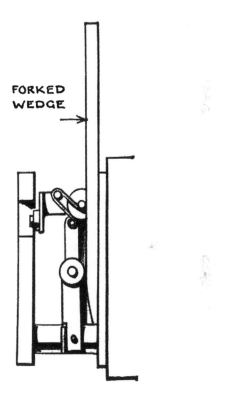

FORKED WEDGE

Fig. D.3.

The use of a forked wedge to keep the armature in the correct position for fitting the diaphragm.

holding the points in contact (after first re-positioning the spring blade into its normal position). If a current is then passed through the pump the magnet will be energised and will pull the armature and diaphragm forward, and while it is in this position the six screws should be tightened. Although the diaphragm-stretching operation can be effected by the matchstick method, a special tool for the purpose is available

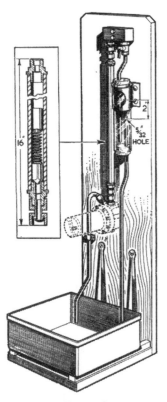

Fig. D.4.

Checking rig.

from the S.U. Carburetter Co. or their Distributors. The tool is a steel wedge, to be inserted under the trunnion in the centre of the inner rocker in order to stretch the diaphragm to its outermost position before tightening the six flange screws.

8. Finally, check that when the spring blade is in its normal position the clearance hole in it is so positioned around the locking screw that each contact point, according to the operation of the outer rocker, **wipes over the centre line of the other contact point,** and that this action is not to one side of the centre on either contact. The width of the gap at the points is approximately ·030 in. (·76 mm.).

9. The pump should now be placed on test, using a cut-away cover to enable the contact breaker action to be observed and at the same time to prevent the rocker hinge pin from falling out.

A test rig of the type illustrated in Fig. D.4 is advised; either petrol (gasoline) or paraffin (kerosene) may be used for testing purposes. Test figures are given in "General Data." The use of a glass tube and rubber connections between the sump and the test tank is advised. When the pump is switched on it should prime itself promptly, and the paraffin, which is normally used for testing, should rise in the glass container until it

flows over the top of the pipe having the $\frac{5}{32}$ in. (4 mm.) hole drilled in it 2 in. (5 cm.) below the top of the pipe. If the output of the pump is not up to normal, the $\frac{5}{32}$ in. (4 mm.) diameter hole will be able to deal with all the paraffin pumped and the liquid will not flow over the top of the pipe. If a time-test is used, one pint (·57 litre) of fuel per minute should be pumped.

These test rigs can be obtained complete from the Service Parts Department of the S.U. Carburetter Company or their Distributors.

This, therefore, constitutes a simple form of flow-meter which establishes in a simple manner whether the pump is giving a sufficient output or not. If there is any air leak in the pump or in its connections, bubbles will be seen coming out of the pipe projecting downwards into the flow-meter. Bubbles will certainly come through here for a short while after starting up, but they should cease after the pump has been running for a minute or so. The tap should then be turned right off and the pump should stand without repeating its action for at least fifteen seconds. If it repeats within this time, the suction valve is not seating correctly.

The tap should then be turned on slowly to see if the pump idles satisfactorily, and that the outer rocker comes forward till it makes contact with the pedestal, and while it is in this position the tip of the blade should be pressed inwards to reduce the stroke of the pump gradually. However much this stroke is reduced, the pump should go on pumping normally until it fails altogether owing to there being no gap left. If, instead of pumping, it buzzes, it usually indicates excessive flexibility in the diaphragm. This, of course, is not likely to be experienced with a new diaphragm. Then, with the tap turned on fully, the pump should be tested on 9 volts (or on 4½ volts if it is a 6-volt pump) and it should work satisfactorily under these conditions, although probably with a reduced output.

It is as well to let the pump run for ten minutes or so before carrying out these various tests. The cover, which is red for 12-volt, should then be fitted and held in place with two ordinary brass nuts fitted on the end of the terminal. The type of pump can always be identified by the colour of the sleeving on the coil ends, this being red for low pressure and brown for high pressure (both 12-volt.).

Note.—There are three important points which are repeatedly overlooked by operators. These seriously affect the functioning of the pump; they are:—

1. To keep the contact breaker blade out of contact while obtaining the correct diaphragm setting.
2. To press firmly and steadily on the armature, instead of jerking it, while obtaining the setting.
3. Omission to stretch the diaphragm to the limit of its stroke while tightening up the body screws.

Section D.7

TRACING FUEL PUMP TROUBLES

Should the pump cease to function, first disconnect the fuel delivery pipe from the pump. If the pump then works the most likely cause of the trouble is a sticking needle in the float-chamber of the carburetter. Should the pump not work, disconnect the lead from the terminal and strike it against the body of the pump after switching on the ignition. If a spark occurs it indicates that the necessary current is available at the terminals, and that the trouble arises with the pump mechanism. If no spark can be detected, then it is an indication that the current supply has failed and that attention should be given to the wiring and battery. If current is present, further investigation should be carried out by removing the bakelite cover which is retained by the terminal nut. Touch the terminal with the lead. If the pump does not operate and the contact points are in contact yet no spark can be struck off the terminal, it is very probable that the contact points are dirty and require cleaning. These may be cleaned by inserting a piece of card between them, pinching them together and sliding the card backwards and forwards.

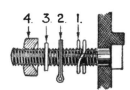

Fig. D.5.

The correct sequence of assembly of the connecting components on the terminal screw.

It is possible that there may be an obstruction in the suction pipe, which should be cleared by blowing air through it, or that some irregularity in the pump itself is preventing the correct movement. This may be due either to the diaphragm having stiffened, or to foreign matter in the roller assembly which supports the diaphragm, in which case the diaphragm should be removed and the whole assembly cleaned and reassembled in accordance with the instructions on page D.6.

On the other hand, if the points are not making contact, see that the tips of the inner rocker (25) are in contact with the magnet housing. If they are not, it is an indication that the armature has failed to return to the end of its normal travel.

To cure this, loosen the six screws, which attach the magnet housing to the pump body, and make sure that the diaphragm is not sticking to the face of the magnet housing by carefully passing a penknife between the two. The hinge pin (19) should then be removed and the six retaining screws tightened up again. The tips of the inner rockers will probably now be found to be making contact with the face of the magnet housing, but if they are not, it will be necessary to remove and dismantle the whole magnet assembly in order to ascertain if an accumulation of foreign matter has caused a jam. Remember that whenever the magnet housing is removed care should be taken to see that the guide rollers (10) do not drop out.

Pump noisy

If the pump becomes noisy and works rapidly, it is usually an indication that there is an air leak on the suction side of the pump. Check the level of the fuel in the tank and see that it is not too low.

The simplest way to test for air leakage is to disconnect the fuel pipe from the carburetter and place its end in a glass jar (approximately 1 pint or half a litre) and allow the pump to deliver fuel into it. If air bubbles appear when the end of the pipe has become submerged in the fuel, it is a clear indication of an air leak on the suction side of the pump in the fuel feed pipe between the tank and the pump which should be found and cured. Check all the unions and joints, making sure that the filter union and inlet unions are all quite air-tight.

Failure to deliver fuel

Should the pump continue beating without delivering fuel, it is probable that some dirt has become lodged under one of the valves, in which case they should be dismantled by unscrewing the top or delivery union and lifting out the valve cage, when they can be cleaned and reassembled. When replacing it see that the thin, hard, red fibre washer is *below* the valve cage and the thick, orange one above.

If the pump struggles to operate and becomes very hot, it is probable that the filter has become clogged or there is an obstruction on the suction side. The filter is readily removed for cleaning by unscrewing its retaining plug at the bottom of the pump.

Section D.8

FUEL PUMP MAINTENANCE

Apart from keeping the contacts clean and removing the filter at regular intervals for cleaning, there is no maintenance required on the fuel pump.

The filter can be removed by unscrewing the hexagon plug at the bottom of the pump, when it can be cleaned in fuel with a stiff brush. Never use rag to clean a filter.

D.9

Fig. D.6.

When connecting the mixture control wire, give twist as indicated to ensure correct operation of the lock.

Section D.9

THE CARBURETTERS

The S.U. carburetters are of the controllable jet type drawing air through oil-wetted air cleaners.

A damper is provided in each unit, consisting of a plunger and non-return valve attached to the oil cap nut. The damper operates in the hollow piston rod which is partly filled with oil. Its function is to give a slightly enriched mixture on acceleration by controlling the rise of the piston, and also to prevent flutter.

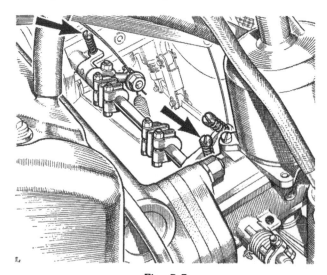

Fig. D.7.

The slow-running adjustment screws are indicated by the arrows.

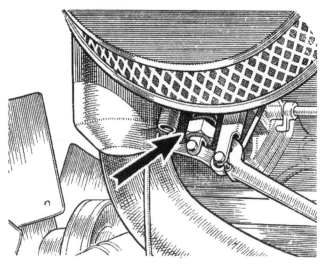

Fig. D.8.

The arrow indicates the jet adjusting nut.

Section D.10

CARBURETTER ADJUSTMENTS

Slow running is governed by the setting of the jet adjusting nuts and the throttle stop screws, all of which must be correctly set and synchronised if satisfactory results are to be obtained.

The two throttles are interconnected by a coupling shaft and spring coupling clips enabling them to be set and correctly synchronised when adjustments are being made.

The mixture control levers are also connected, between the carburetters, by a short adjustable link.

Before blaming the carburetter settings for bad slow running, make certain that the trouble is not caused by badly adjusted contact points, faulty plugs, incorrect valve clearance or faulty valves and springs.

Adjusting the jets

Run the engine until it attains its normal running temperature.

Slacken off the pinch-bolt of one of the spring coupling clips locating the carburetter inter-connecting shaft to the carburetter throttle spindles so that each carburetter can be operated independently.

Disconnect the mixture control cable and the connecting link between the two jet adjusting levers.

Unscrew both throttle lever setting screws until the throttles are completely closed. Turn the adjusting screw of the rear carburetter in a clockwise direction approximately one turn to set the throttle for fast idling; lift the piston of the front carburetter $\frac{1}{2}$ in. (13 mm.) to leave the carburetter out of action.

With the engine running, set the jet adjusting nut of the rear carburetter so that a mixture strength is obtained

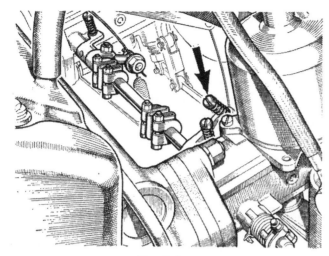

Fig. D.9

The screw indicated by the arrow is used to adjust the amount of throttle opening when the mixture control is in use

which will give the best running speed for this particular throttle opening, taking care to see that the jet head is in firm contact with the adjusting nut the whole time.

The correctness or otherwise of this setting can be checked by raising the suction piston about $\frac{1}{32}$ in. (·8 mm.) with the piston-lifting pin. This should cause a very slight momentary increase in the speed of the engine without impairing the evenness of the running. If the engine stops, the mixture is too weak. If the speed increases and continues to increase when the piston is raised as much as $\frac{1}{4}$ in. (6 mm.) the mixture is too rich.

When the setting of the mixture is correct for the rear carburetter, unscrew the throttle adjusting screw until the throttle is fully closed and lift the piston $\frac{1}{2}$ in. (13 mm.) to put it out of action. Repeat the adjustment operations on the front carburetter.

When both carburetters are correctly adjusted for mixture, set the throttle adjustment screw of each to give the required slow running. Adjust the link between the mixture levers so that each lever is moved the same amount when the mixture control is used.

Slow-running and synchronization

Turn the throttle adjustment screw of each carburetter to give a fast idling speed, taking care to turn each screw the same amount. Now unscrew each throttle lever adjustment screw an equal amount, a fraction of a turn at a time until the desired slow-running speed is obtained.

Accuracy of synchronization can be checked by listening at each carburetter air intake in turn through a length of rubber tubing and noticing if the noise produced by the incoming air is the same at both. Any variation in the intensity of the sound indicates that one throttle is set more widely open than the other.

When the same intensity of sound is given by both carburetters the coupling shaft clip should be tightened to ensure that the throttles work in unison.

Since the delivery characteristics when both carburetters are working together vary somewhat from those existing when each is working separately, it will be necessary to check again for correctness of mixture strength by lifting each piston in turn as indicated in **'Adjusting the jets'**, and adjusting as necessary.

Section D.11

REMOVING THE CARBURETTERS

Release the spring clips and detach the breather hose from the air cleaner and rocker cover.

Disconnect the fuel supply pipe at the rear carburetter banjo union.

Remove the split pin and flat washer and release the mixture cable and clevis pin from the mixture control linkage and release the mixture outer cable abutment from its bracket.

Detach the throttle return spring and release the throttle cable.

Unscrew the union nut and disconnect the ignition vacuum control pipe from the rear carburetter.

Remove the nut and flat washer to release the vent pipe from the top of each float-chamber.

Remove the four nuts securing the carburetter flanges and withdraw the carburetters and air cleaners as an assembly.

Replacement is a reversal of the above instructions.

Section D.12

CENTRING THE JET

First remove the clevis pin at the base of the jet which attaches the jet head to the jet operating lever; withdraw the jet completely, and remove the adjusting nut and the adjusting nut spring. Replace the adjusting nut without its spring and screw it up to the highest position. Slide the jet into position until the jet head is against the base of the adjusting nut.

When this has been done remove the dashpot piston and test for free piston movement by lifting it with a finger. If it is not perfectly free, slacken the jet holding screw and manipulate the lower part of the assembly including the projecting part of the bottom half jet bearing, adjusting nut and jet head. Make sure that the assembly is now slightly loose. The piston should rise and fall quite freely as the needle is now able to move the jet into the required central position. Tighten the jet holding screw and check the position again.

If it is still not free, slacken the jet holding screw and repeat the operation. When the piston is completely free-moving, remove the adjusting nut and replace its spring. Replace the nut, screwing it to its original position.

Experience shows that a large percentage of the carburetters returned for correction have had jets removed and incorrectly centred on replacement.

Section D.13

SOURCES OF CARBURETTER TROUBLE

Piston sticking

The piston assembly comprises the suction disc and the piston forming the choke, into which is inserted the hardened and ground piston rod which engages in a bearing in the centre of the suction chamber and in which is, in turn, inserted the jet needle. The piston rod running in the bearing is the only part which is in actual contact with any other part, the suction disc, piston, and needle all having suitable clearances to prevent sticking. If sticking does occur the whole assembly should be cleaned carefully and the piston rod lubricated with a spot of thin oil. No oil must be applied to any other part except the piston rod. A sticking piston can be ascertained by removing the dashpot damper, inserting a finger in the air intake and lifting the piston, which should come up quite freely and fall back smartly onto its seating when released. On no account should the piston return spring be stretched or its tension altered in an attempt to improve its rate of return.

Float-chamber flooding

This is indicated by the fuel dripping from the drain pipe, and is generally caused by grit between the float-chamber needle and its guide. This is cured by removing the float-chamber, washing the valve and float-chamber components and reassembling.

Float needle sticking

If the engine stops, apparently through lack of fuel, when there is plenty in the tank and the pump is working properly, the probable cause is a sticking float needle. An easy test for this is to disconnect the pipe from the

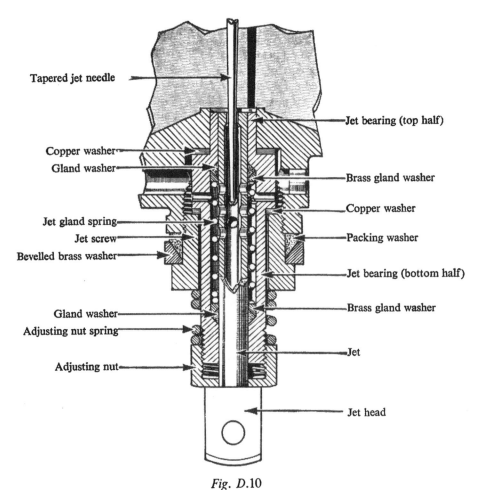

Fig. D.10

An enlarged view of the jet assembly, showing the component parts

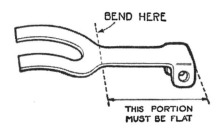

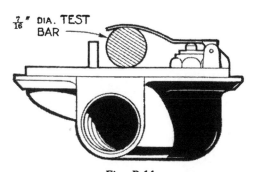

Fig. D.11

Showing the place where the float lever should be set and the method of checking the correct adjustment of the lever

electric pump to the carburetters and switch the ignition on and off quickly while the end of the pipe is directed onto a pad of cloth or into a container.

If fuel is delivered, starvation has almost certainly been caused by the float needle sticking to its seating, and the float-chamber lid should therefore be removed, the needle and seating cleaned and refitted. At the same time it will be advisable to clean out the entire fuel feed system, as this trouble is caused by foreign matter in the fuel and unless this is removed it is likely to recur. It is of no use whatever renewing any of the component parts of either carburetter, and the only cure is to make

sure that the fuel tank and pipe lines are entirely free from any kind of foreign matter or sticky substance capable of causing this trouble.

Section D.14

AIR CLEANER

Every 3,000 miles (4800 km.) or more frequently in dusty conditions the air cleaner should be serviced as follows.

Unscrew and remove the two bolts, remove the outer cover and withdraw the element from the body of each cleaner. Wash the element thoroughly in petrol (gasoline), drain and dry. Wet the element with S.A.E. 20 engine oil and allow to drain before replacing.

Reassemble the front element with the corrugations clear of the breather spigot in the filter case.

Section D.15

ACCELERATOR RETURN SPRING

On later cars, commencing with Car No. 24954, an additional accelerator return spring is introduced as a safety precaution and the original return spring is replaced by a new spring.

The new parts may be fitted to earlier cars as a set as follows.

Remove the original accelerator return spring and fit the new spring (Part No. AHH5621).

Remove the accelerator cable and replace the anchor pin with the new anchor pin (Part No. AHH5626), or fit the new accelerator cable (Part No. AHH5625).

Fit the anchor bracket (Part No. AHH5623) to one of the accelerator cable guide screws and fit the auxiliary return spring (Part No. AHH5624).

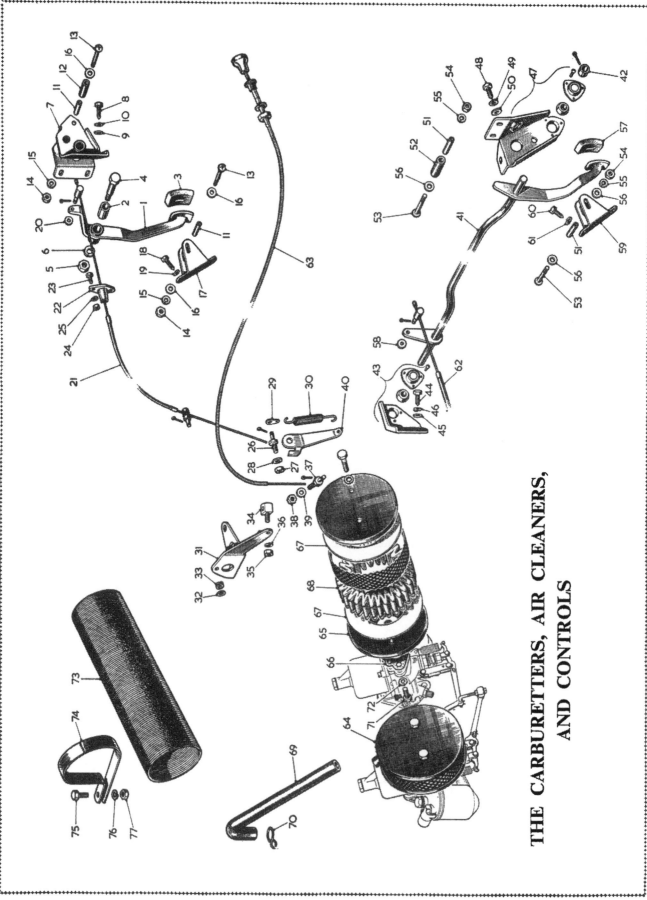

THE CARBURETTERS, AIR CLEANERS, AND CONTROLS

KEY TO THE CARBURETTERS, AIR CLEANERS AND CONTROLS

No.	Description	No.	Description
1.	Pedal—accelerator assembly.	37.	Clamp—choke control cable.
2.	Bush.	38.	Nut—cable to clamp.
3.	Pad—rubber pedal.	39.	Washer—plain—cable to clamp.
4.	Fulcrum pin—accelerator pedal.	40.	Anchor bracket—accelerator spring.
5.	Nut—fulcrum pin.	41.	Cross-shaft assembly—accelerator.
6.	Washer—spring—fulcrum pin.	42.	Locating collar—accelerator shaft.
7.	Bracket—accelerator pedal mounting.	43.	Bracket assembly—R/H—accelerator pedal shaft.
8.	Screw—bracket to goalpost inner upright.	44.	Screw—bracket to goalpost—R/H.
9.	Washer—plain—pedal bracket.	45.	Washer—plain.
10.	Washer—spring—pedal bracket.	46.	Washer—spring.
11.	Distance-tube—bracket/adjusting stop.	47.	Bracket assembly—L/H—accelerator pedal shaft.
12.	Stop—accelerator pedal bracket.	48.	Screw—bracket to goalpost—L/H.
13.	Screw—distance-tube/adjustable stop.	49.	Washer—spring.
14.	Nut—tube and stop screw.	50.	Washer—plain.
15.	Washer—spring—tube and stop screw.	51.	Distance-tube.
16.	Washer—plain—bracket/adjusting stop.	52.	Stop—accelerator pedal.
17.	Adjustable stop—accelerator pedal.	53.	Screw—distance-tube/adjustable stop.
18.	Screw—adjustable stop to goalpost.	54.	Nut—distance-tube/adjustable stop.
19.	Washer—spring—stop to goalpost screw.	55.	Washer—spring.
20.	Washer—plain—cable to pedal.	56.	Washer—plain—adjustable stop screw.
21.	Cable—accelerator.	57.	Pad—rubber—accelerator pedal.
22.	Guide—cable.	58.	Washer—plain—cable to shaft assembly.
23.	Screw—guide to toe-board plate.	59.	Adjustable stop—accelerator pedal.
24.	Nut—guide screw.	60.	Screw—adjustable stop to toe-board.
25.	Washer—spring—guide screw.	61.	Washer—spring.
26.	Clamp—accelerator cable.	62.	Cable—accelerator.
27.	Nut—cable to clamp.	64.	Cleaner—front air.
28.	Washer—plain—cable to clamp.	65.	Cleaner—rear air.
29.	Bracket—accelerator return spring.	66.	Gasket.
30.	Return spring—accelerator.	67.	Insert pad.
31.	Abutment bracket—accelerator/choke cables.	68.	Element.
32.	Washer—plain—cable to abutment.	69.	Pipe—air filter.
33.	Washer—double coil spring—cable to abutment.	70.	Clips—air pipe.
34.	Abutment—choke control.	71.	Set screw—cleaner to carburetters.
35.	Nut—choke abutment.	72.	Washer—cleaner set screw.
36.	Washer—spring—choke abutment.		

Notes

SECTION E

THE CLUTCH

THE CLUTCH COMPONENTS

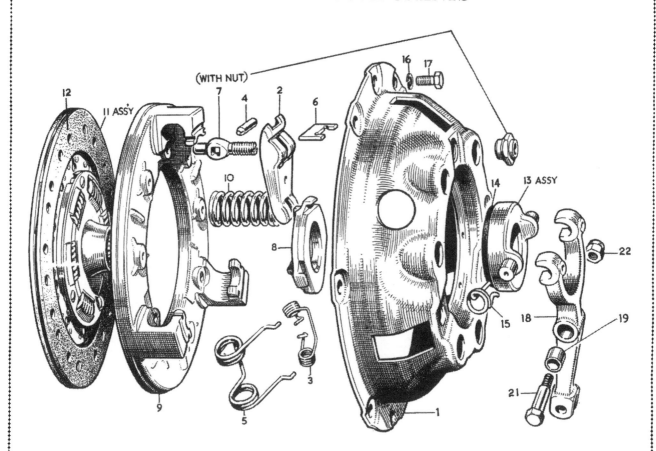

KEY TO THE CLUTCH COMPONENTS

No.	Description	No.	Description	No.	Description
1.	Cover—clutch.	8.	Plate—bearing thrust.	15.	Retainer.
2.	Lever—release.	9.	Plate—pressure.	16.	Washer—spring—cover screw.
3.	Retainer—lever.	10.	Spring—pressure plate.	17.	Screw—cover to flywheel.
4.	Pin—lever.	11.	Plate assembly—driven.	18.	Lever assembly—withdrawal.
5.	Spring—anti-rattle.	12.	Lining.	19.	Bush.
6.	Strut.	13.	Ring assembly—thrust.	21.	Bolt—lever.
7.	Eyebolt with nut.	14.	Ring—carbon.	22.	Nut—bolt.

DESCRIPTION

The clutch is of the single-plate dry-disc type operated hydraulically.

The driven plate assembly

This consists of a splined hub and flexible steel driven plate (C), to the outer diameter of which are fixed the annular friction facings. This plate is attached to the splined hub by a spring mounting which provides a torsional cushion.

Withdrawal bearing assembly

This comprises the graphite release bearing (G) mounted in a cup attached to the throw-out fork and a release plate (H) attached to the inner ends of the release levers (J) by means of the retainer springs (I). Release is accomplished by moving the release bearing forward into contact with the release plate and thus applying pressure to the release levers.

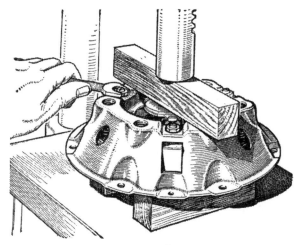

Fig. E.2.

Compressing the springs with wood blocks and press.

Cover assembly

Each release lever is pivoted on a floating pin (N), which remains stationary in the lever and rolls across a short flat portion of the enlarged hole in the eye-bolts (M) (see Fig. E.1). The outer ends of the eye-bolts extend through holes in the clutch cover and are fitted with adjusting nuts (L) by means of which each lever is located in its correct position. The outer or shorter ends of the release levers engage the pressure plate lugs by means of struts (O) which provide knife-edge contact between the outer ends of the levers and the pressure plate lugs, eliminating friction at this point. Thus the pressure plate (P) is pulled away from the driven plate (C), compressing the six thrust coil springs (E) which are assembled between the pressure plate and the clutch cover (D).

When the foot pressure is removed from the clutch pedal the clutch springs force the pressure plate forward against the driven plate, gradually and smoothly applying the power of the engine to the rear wheels.

Hydraulic operation

A master twin-bore cylinder bolted to the chassis frame contains two pistons operated by the clutch and brake pedals. For clutch withdrawal, fluid pressure is transmitted to the slave cylinder bolted to the clutch housing, moving the piston, push-rod and clutch lever and disengaging the clutch.

The master cylinder

The assembly of the clutch master cylinder is similar to that of the brake master cylinder (Section M) except that no combination inlet and outlet valve is fitted and therefore no pressure is maintained in the clutch line when the pedal is released. (See the illustration on page M.4.)

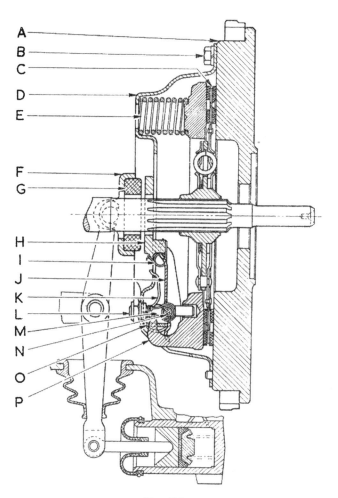

A
B
C
D
E
F
G
H
I
J
K
L
M
N
O
P

Fig. E.1.

A section through the clutch.

E.3

The slave cylinder

This is bolted to the clutch housing and normally requires no maintenance; its assembly is similar to that of the front wheel brake cylinder, and the push-rod is not adjustable.

If the system is drained of fluid it will be necessary to bleed the cylinder after reassembly and refilling.

Section E.1

ADJUSTMENT

It is essential that there should be a clearance between the master cylinder push-rod and the piston when the clutch pedal is released. This clearance, $\frac{1}{32}$ in. (·8 mm.), is adjusted by slackening the locknut and rotating the push-rod in the appropriate direction. Ensure that the pedal is not obstructed by the toeboard or by the floor covering. The free movement at the pedal pad must be sufficient to allow the piston fully to return in the cylinder and still retain the $\frac{1}{32}$ in. (·8 mm.) clearance at the push-rod.

Excessive movement may indicate lack of fluid or the need for bleeding; whenever the system is drained, bleeding will be required after refilling.

Section E.2

REMOVING THE CLUTCH

Remove the power unit as detailed in Section A.18.

Remove the clutch housing bolts and withdraw the gearbox from the engine, taking care to support the gearbox until the first motion shaft is clear of the driven plate and release lever plate.

Loosen each of the hexagon bolts securing the clutch to the flywheel by slackening them a turn at a time until spring pressure is released. The clutch cover can now be disengaged from the flywheel dowels and the whole assembly lifted from the flywheel.

Section E.3

DISMANTLING THE CLUTCH

Two methods are possible in dismantling the clutch: (*a*) Using the clutch gauging fixture, and (*b*) Using a press and blocks of wood.

Using the clutch gauging fixture (Fig. E.6)

Consult the code card to determine the correct spacers for the particular clutch. Place the spacers on the base-plate in the positions indicated on the code card and place the clutch on the spacers. Screw the actuator into the central hole in the base-plate and press the handle

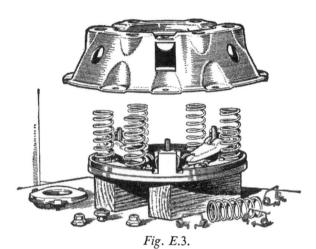

Fig. E.3.

Showing the component parts of the clutch assembly.

to clamp the clutch. Screw the set bolts firmly into the base-plate. The clutch can now be compressed or released as required.

Compress the clutch with the actuator and remove the adjusting nuts gradually to relieve the load of the thrust springs. Lift the cover off the clutch and carry out whatever additional dismantling may be necessary.

Using a press and wood blocks (Fig. E.2)

Place the cover on the bed of a press with the pressure plate resting on wood blocks so arranged that the cover is left free to move downwards. Place a block or bar across the top of the cover, resting it on the spring bosses.

Apply pressure to the cover with the spindle of the press and, holding it under compression, remove the three adjusting nuts. The pressure from the press may now be released gradually until the clutch springs are fully extended.

While stripping down the cover-plate assembly, the parts should be marked so that they may be reassembled in the same relative position to each other, to ensure that the correct balance is maintained. When a new pressure plate is fitted it is essential that the complete cover and pressure plate assembly be accurately balanced, for which reason it is not a practical proposition to fit new pressure plates unless balancing facilities are available.

All parts are available for inspection when the cover is lifted off.

To remove the release levers, grasp the lever and eyebolt between the thumb and fingers so that the inner end of the lever and the threaded end of the eyebolt are as near together as possible, keeping the eyebolt pin seated in its socket in the lever. The strut can then be lifted over the ridge on the end of the lever, making it possible to lift the eyebolt off the pressure plate. It is advisable to replace any parts which show signs of wear.

E.4

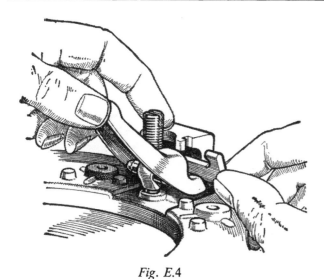

Fig. E.4

To assemble the levers hold the threaded end of the eyebolt and lever close together as shown and insert the struts in the slots of the pressure plate lugs sufficiently to permit the plain end of the eyebolt to be inserted in the hole in the pressure plate

Section E.4

ASSEMBLING THE CLUTCH

Lay the pressure plate on the wood block on the bed of the press (or on the base-plate of the special tool) and place the springs on it in a vertical position, seating them on their small locating bosses. Thoroughly clean all parts and renew any which show appreciable wear.

Assemble the release levers, eyebolts and eyebolt pins, holding the threaded end of the eyebolt and the inner end of the lever as close together as possible. With the other hand insert the strut in the slots of the pressure plate lug just sufficiently to allow the plain end of the eyebolt to be inserted in the hole in the pressure plate. Move the strut upwards into the slots in the pressure plate lugs, over the ridge on the short end of the lever, and drop it into the grooves formed in the lever.

Lay the cover over the parts, taking care that the anti-rattle springs are in position as shown in Fig. E.1 and that the springs are directly under the seats in the cover. Also make sure, if using the original parts, that the eyebolts, eyebolt nuts, pressure plate lugs and cover are fitted in their correct relative positions, as marked when dismantling, to ensure correct balance being maintained.

Compress the springs either by the actuator, if the special tool is being used, or by the use of a wooden block across the cover and a press. Take care to guide the eyebolts and the pressure plate lugs through the correct holes in the cover. Make sure also that the thrust springs remain correctly in their seats.

Replace the eyebolt nuts on the eyebolts and release the pressure compressing the cover assembly.

Section E.5

ADJUSTING THE RELEASE LEVERS

Satisfactory operation of the clutch is dependent upon accurate adjustment of the release levers, so that the pressure plate face is maintained parallel to the flywheel face. This cannot be accomplished by setting the levers parallel to the face of the release bearings after the clutch has been assembled to the flywheel, because of the variations in the thickness of the driven plate.

For accurate adjustment either a gauge plate or the universal gauging fixture must be used.

Using the gauging fixture

After carrying out any necessary servicing, reassemble the parts on the clutch pressure plate, place the cover on it and the whole assembly on the base-plate of the gauging fixture. It is essential that the correct spacers be used, as indicated on the code card.

Bolt the cover to the base-plate and screw the adjusting nuts onto the bolts until the tops of the nuts are flush with the tops of the bolts. Screw the actuator into the base-plate and work the handle a dozen times to settle the mechanism. Remove the actuator. Screw the pillar firmly into the base-plate and place the appropriate adaptor (see code card) on the pillar with the recessed side downwards; place the gauge finger in position.

Turn the adjusting nuts until the finger just touches each release lever, pressing downwards on the finger assembly to ensure that it is bearing squarely on the adaptor. Remove the finger and the pillar and replace the actuator; operate the actuator several times. Re-check with the finger assembly and make any necessary further adjustments.

Lock the adjusting nuts.

Using a gauge plate

Place the gauge on the flywheel in the position normally occupied by the driven plate, and mount the cover assembly on the flywheel in the same position as before dismantling. Tighten the holding screws a turn or two at a time when pulling against the spring pressure, otherwise the cover may be distorted. Before the cover is tightened down be sure that the gauge plate is properly centred and the three flat machined lugs are directly under the levers. The release lever plate must be detached from the levers before the levers are adjusted.

After the cover assembly has been mounted, a short straight-edge should then be laid across the centre boss of the special gauge plate and one nut adjusted until the

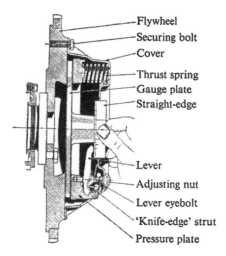

Flywheel
Securing bolt
Cover
Thrust spring
Gauge plate
Straight-edge

Lever
Adjusting nut
Lever eyebolt
'Knife-edge' strut
Pressure plate

Fig. E.5

Setting the release levers by means of the special gauge plate, Part No. 68886, and a short straight-edge

lever just makes contact with the straight-edge. The other levers can then be set in turn by the same method. If carefully done, this setting will be within the permissible tolerance of ·005 in. (·12 mm.).

The adjusting nut is then locked in position by punching the protruding flange of the nut into the slot of the eyebolt, thereby definitely locking it in position. When carrying out this operation take care not to upset the adjustments previously made.

After adjustment is completed, loosen the holding screws a turn or two at a time until the spring pressure is released, which will allow the clutch assembly and the gauge plate to be removed.

Section E.6

REFITTING THE CLUTCH

Position the driven plate assembly on the flywheel, taking care to place the larger-chamfered spline end of the driven plate hub away from the flywheel.

Centralize the driven plate by means of the special alignment bar, Part No. 18G276, which fits the splined bore of the driven plate hub and the pilot bearing in the flywheel. As an alternative a spare first motion shaft can be used.

Locate the cover assembly on the flywheel dowels and secure with the bolts, tightening them a turn at a time by diagonal selection. Do not remove the clutch alignment bar until all the bolts are securely tightened.

Remove the clutch alignment bar and refit the gearbox. The weight of the gearbox must be supported during refitting in order to avoid strain on the shaft and distortion or displacement of the release plate or driven plate assembly.

Section E.7

SERVICING THE CLUTCH

As the clutch facings wear, the pressure plate moves closer to the flywheel face, and the outer or shorter ends of the release levers follow. This causes the inner or longer ends of the levers to travel farther towards the gearbox. As the release bearing moves rearwards it must result in pushing the piston in the clutch slave cylinder inwards. The piston then forces the excess fluid back into the master cylinder via the compensating orifice.

Provided that the minimum $\frac{1}{32}$ in. (·8 mm.) free movement is maintained between the clutch pedal push-rod and the master cylinder piston, this automatic compensation for wear will always take place.

Should there be no free movement at this point the master cylinder piston will not be allowed to return fully to its stop and therefore the compensating orifice will be cut off.

Excessive pedal movement causes coil binding of the springs and imposes an undue load on the bearing and on the crankshaft, causing excessive and rapid bearing wear. It therefore follows that the required pedal travel is the sum of the two movements:

(1) *The free movement*, or travel necessary to take up the clearance between the master cylinder push-rod and the master cylinder piston, provided to ensure that the clutch is fully engaged when the foot is removed from the pedal. (See Section E.1.)

(2) *The effective movement*, or travel necessary to release the clutch, i.e. the amount of effective pedal movement necessary to move the release plate the distance required to free the clutch completely.

If any difficulty is experienced in freeing the clutch when the correct release movement is provided, on no account should efforts be made to improve matters by attempting to increase the effective pedal travel. The actual cause of the trouble must be ascertained and rectified.

To obtain a clean release, the release lever plate should move a distance of $\frac{5}{16}$ in. (8 mm.) towards the flywheel.

Spring pressure

A tolerance of not more than 10 to 15 lb. (4·5 to 6·8 kg.) pressure is allowable on the compression load of the operating springs when at their assembled height, and all clutch springs are tested for this before assembly.

The clutch operating springs are not affected by high clutch temperatures, as the pressure plate absorbs heat rapidly, the springs have only line contact, and a draught is continually passing under them when the engine is running.

Tolerances

Wear on the working faces of the driven plate is about ·001 in. (·02 mm.) per 1,000 miles (1600 km.) under normal running conditions. The accuracy of the alignment of the face of the driven plate must be within ·015 in. (·38 mm.).

Driven plates

It is important that neither oil nor grease should contact the clutch facings.

Lubrication of the splines of the driven plate is provided at assembly only, when CS881 graphite grease or zinc-based " Keenol " is used.

It is essential to install a complete driven plate assembly when renewal of the friction surfaces is required. If the facings have worn to such an extent as to warrant renewal, then slight wear will have taken place on the splines, and also on the torque reaction springs and their seatings. The question of balance and concentricity is also involved. Under no circumstances is it satisfactory to repair or rectify faults in clutch driven plate centres, and we do not countenance this as manufacturers.

Condition of clutch facings in service

It is natural to assume that a rough surface will give a higher frictional value against slipping than a polished one, but this is not necessarily correct. A roughened surface consists of small hills and dales, only the " high spots " of which make contact. As the amount of useful friction for the purpose of taking up the drive is dependent upon the area in actual contact, it is obvious that a perfectly smooth face is required to transmit the maximum amount of power for a given surface area.

Since non-metallic facings of the moulded asbestos type have been introduced in service the polished surface is common, but it must not be confused with the glazed surface which is sometimes encountered due to conditions to be discussed subsequently. The ideally smooth or polished condition will therefore provide proper surface contact, but a glazed surface entirely alters the frictional value of the facing, and will result in excessive clutch slip. These two conditions might be simply illustrated by comparison between a piece of smoothly finished wood and one with a varnished surface; in the former the contact is made directly by the original material, whereas in the latter instance a film of dry varnish is interposed between the contact surfaces and actual contact is made by the varnish.

If the clutch has been in use for some little time under satisfactory conditions, the surface of the facings assumes a high polish through which the grain of the material can be seen clearly. This polished facing is of light colour when in perfect condition.

Should oil in small quantities gain access to the clutch

Fig. E.6.

Using the actuator to compress the clutch springs for dismantling or setting the assembly.

and find its way onto the facings, it will be burnt off as a result of the heat generated by the slipping occurring under normal starting conditions. The burning of this small quantity of lubricant has the effect of gradually darkening the facings, but provided the polish of the facing remains such that the grain of the material can be distinguished clearly it has little effect on clutch performance.

Should increased quantities of oil obtain access to the facing, then one of two conditions, or a combination of these, may arise, depending upon the nature of the oil.

1. The oil may burn off and leave a carbon deposit on the surface of the facings, which assume a high glaze, producing further slip. This is a very definite, though very thin, deposit, and in general it hides the grain of the material.

2. The oil may partially burn and leave a resinous deposit on the facings. This has a tendency to produce a fierce clutch, and may also cause excessive " spinning " due to the tendency of the face of the linings to adhere to the surface of the flywheel or pressure plate.

3. There may be a combination of conditions (1) and (2) which produces a tendency to " judder " on such engagement.

Still greater quantities of oil produce a dark and soaked appearance of the facings, and the result will be further slip, accompanied by fierceness or " juddering."

If the conditions enumerated above are experienced, the clutch driven plate should be replaced by a new one. **The cause of the presence of the oil must be traced and removed.** It is, of course, necessary for the clutch and flywheel to be cleaned out thoroughly before assembly.

Where the graphite release bearing ring is badly worn in service, a complete replacement assembly should be

Fig. E.7.
Checking the setting of the release levers.

fitted, returning the old assembly for salvage of the metal cup. These graphite rings are inserted into their metal cup by heating the metal cup to a cherry red, then forcing the graphite ring into position. Immediately the ring is forced into position, the whole should be quenched in oil. Alignment of the thrust pad in relation to its face and the trunnions should be within ·005 in. (·12 mm.).

In almost every case of rapid wear on the splines of the clutch driven plate, misalignment is responsible.

Looseness of the driven plate on the splined shaft results in noticeable backlash in the clutch. Misalignment also puts undue stress on the driven member, and may result in the hub breaking loose from the plate, with consequent total failure of the clutch.

It may also be responsible for a fierce chattering or dragging of the clutch, which makes gear changing difficult. In cases of persistent difficulty it is advisable to check the flywheel for truth with a dial indicator. The dial reading should not vary more than ·003 in. (·07 mm.) anywhere on the flywheel face.

Section E.8

CLUTCH WITHDRAWAL LEVER PIVOT BOLT

On later types a modified clutch withdrawal lever pivot bolt (Part No. 11G3196) is fitted. The bolt is increased in diameter and has a shoulder to provide an abutment for the self-locking nut (Part No. LNZ.206) which supersedes the nut and spring washer previously fitted.

A larger bearing bush for the withdrawal lever is needed and so a modified lever (Part No. 11G3193) complete with bush is fitted. The bosses on the front cover (Part No. 11G3197) are modified to take the larger diameter bolt.

The modified parts as a whole can be fitted to earlier vehicles.

SECTION F

THE GEARBOX

General description.

THE GEARBOX COMPONENTS

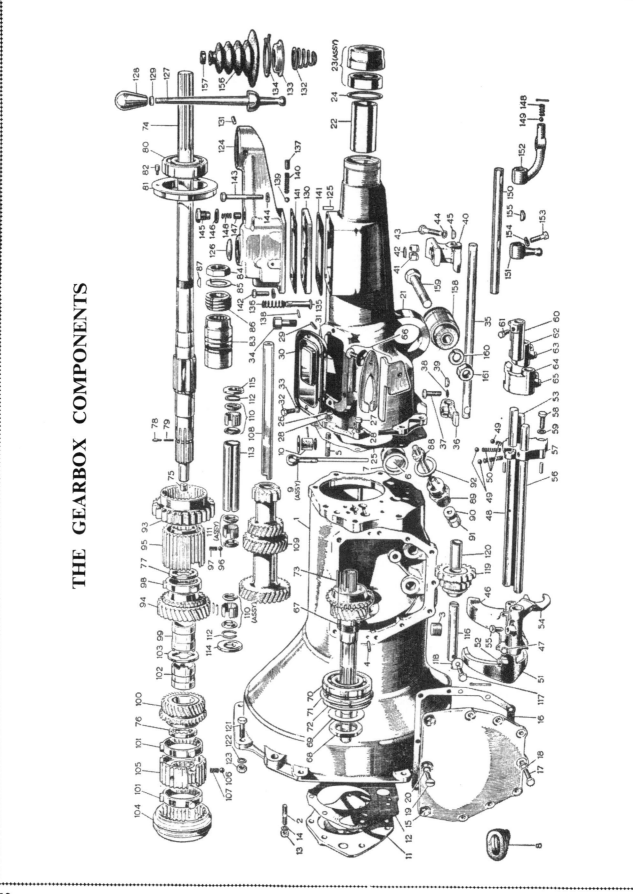

KEY TO THE GEARBOX COMPONENTS

No.	Description
1.	Casing—gearbox.
2.	Stud—front cover.
3.	Plug—drain.
4.	Dowel—side cover to gearbox.
5.	Stud—gearbox extension.
6.	Plug—blanking.
7.	Joint washer—blanking plug.
8.	Dust cover—clutch withdrawal lever.
9.	Dipstick.
10.	Felt.
11.	Cover—front.
12.	Joint—front cover.
13.	Nut—front cover studs.
14.	Spring washer—front cover stud.
15.	Cover—side.
16.	Joint—side cover.
17.	Set screw—side cover.
18.	Spring washer—side cover screw.
19.	Countersunk screw—side cover.
20.	Shakeproof washer—countersunk screw.
21.	Extension—gearbox.
22.	Bush.
23.	Oil seal.
24.	Joint washer—oil seal.
25.	Joint extension to gearbox.
26.	Nut—gearbox extension stud.
27.	Set screw—gearbox extension.
28.	Spring washer—stud and set screw.
29.	Plug—taper—gearbox extension.
30.	Cover—extension side.
31.	Joint—extension side cover.
32.	Set screw—extension side cover.
33.	Spring washer.
34.	Breather assembly.
35.	Shaft—remote control.
36.	Lever—selector—front.
37.	Set screw—front cover.
38.	Spring washer—set screw.
39.	Key—selector lever.
40.	Lever—selector—rear.
41.	Bush—rear selector lever.
42.	Circlip—lever bush.
43.	Set screw—rear lever.
44.	Spring washer—set screw.
45.	Key—selector lever.
46.	Fork—1st and 2nd speed.
47.	Screw—fork locating.
48.	Shaft—1st and 2nd speed fork.
49.	Ball—shaft.
50.	Spring—ball.
51.	Fork—3rd and 4th speed.
52.	Screw—fork locating.
53.	Shaft—3rd and 4th speed fork.
54.	Fork—reverse.
55.	Screw—fork locating.
56.	Shaft—reverse fork.
57.	Block—shaft locating.
58.	Set screw—block to casing.
59.	Spring washer—block screw.
60.	Selector—1st and 2nd gear.
61.	Screw—selector locating.
62.	Selector—3rd and 4th gear.
63.	Screw—selector locating.
64.	Selector—reverse gear.
65.	Screw—reverse gear selector.
66.	Interlock arm complete.
67.	Shaft—1st pinion.
68.	Nut—shaft.
69.	Lock washer.
70.	Bearing—ball—shaft.
71.	Spring ring—bearing.
72.	Shim—bearing.
73.	Rollers—needle—shaft.
74.	Shaft—3rd motion.
75.	Restrictor—oil.
76.	Washer—thrust—front.
77.	Washer—thrust—rear.
78.	Peg—thrust washer—front.
79.	Spring—peg.
80.	Bearing—rear—3rd motion shaft.
81.	Housing—bearing.
82.	Peg—locating.
83.	Distance-piece — speedometer gear.
84.	Nut—shaft and speedometer gear.
85.	Lock washer.
86.	Gear—speedometer drive.
87.	Key—gear.
88.	Pinion—speedometer drive.
89.	Bush—pinion.
90.	Oil seal—pinion.
91.	Ring—oil seal retaining.
92.	Joint—bush to rear cover.
93.	Gear—1st speed.
94.	Gear—2nd speed.
95.	Synchroniser—2nd speed.
96.	Ball—synchroniser.
97.	Spring—ball.
98.	Baulk ring—2nd speed gear.
99.	Bush—2nd speed gear.
100.	Gear—3rd speed.
101.	Baulk ring—3rd and 4th gear.
102.	Bush—3rd speed gear.
103.	Ring—interlocking—2nd and 3rd bushes.
104.	Coupling—sliding—3rd and 4th speed.
105.	Synchroniser—3rd and 4th speed.
106.	Ball—synchroniser.
107.	Spring—ball.
108.	Layshaft.
109.	Gear unit—layshaft.
110.	Bearing—needle roller—layshaft —outer.
111.	Bearing—needle roller—layshaft —inner.
112.	Spring ring—needle rollers.
113.	Distance-piece—bearing.
114.	Washer—thrust—front.
115.	Washer—thrust—rear.
116.	Shaft—reverse.
117.	Screw—locking—shaft.
118.	Lock washer—screw.
119.	Gear—reverse.
120.	Bush.
121.	Bolt—gearbox to mounting plate.
122.	Washer—spring.
123.	Nut—mounting plate bolt.
124.	Tower—remote control.
125.	Dowel—remote control tower.
126.	Core plug—tower.
127.	Lever—change speed.
128.	Knob—change speed lever.
129.	Locknut—change speed lever knob.
130.	Stop plate.
131.	Snug—change speed ball.
132.	Spring—change speed lever.
133.	Cover—ball spring.
134.	Circlip—ball spring cover.
135.	Plunger—reverse selector.
136.	Spring—reverse plunger.
137.	Plug—reverse plunger.
138.	Dowel—reverse plunger.
139.	Ball—reverse plunger.
140.	Spring—reverse plunger detent.
141.	Gasket—control tower.
142.	Bolt—short—tower.
143.	Bolt—long—tower.
144.	Spring washer.
145.	Plug—ball retaining—box cover.
146.	Washer—plug.
147.	Plunger.
148.	Spring—plunger.
149.	Ball—selector lever.
150.	Shaft—remote control.
151.	Lever—front—selector.
152.	Lever—rear—selector.
153.	Set screw—front and rear lever.
154.	Spring washer.
155.	Key.
156.	Draught excluder—rubber—gear lever.
157.	Ring—lever draught excluder.
158.	Flexible bush—rear engine mounting.
159.	Bolt—rear mounting bush.
160.	Washer—spring.
161.	Nut—rear mounting bush bolt.

GENERAL DESCRIPTION

The gearbox has four forward speeds and one reverse. Top gear is obtained by direct drive, third and second by gears in constant mesh, and first and reverse by sliding spur gears.

A sliding joint of the reverse spline type is fitted to the rear end of the third motion shaft and is lubricated from the gearbox.

Section F.1

REMOVING THE GEARBOX

Remove the power unit as in Section A.18.

Remove the starter motor and unscrew the bolts and nuts securing the bell housing and exhaust pipe support brackets and withdraw the gearbox and rear extension from the engine. Take care to keep the gearbox flange parallel with the crankcase face until the first motion shaft is clear of the clutch.

Section F.2

DISMANTLING THE GEARBOX

Extract the dipstick, drain plug and speedometer drive.

Unscrew the nuts and remove the gear lever remote control tower and joint washer.

Unscrew and remove the six bolts and the rear extension cover and joint washer. Remove the interlock arm and bracket.

Remove the one nut and seven set screws securing the gearbox extension to the gearbox. Pull the extension from the gearbox, at the same time manœuvring the remote control shaft selector lever from the selectors.

Unscrew the three countersunk screws and the seven hexagon-headed set screws holding the gearbox cover; remove the cover and overshoot stop.

Cut the locking wire and unscrew the three change speed fork set screws.

Unscrew the two set screws and remove the shifter shaft locating block with shifter shafts from the gearbox; note the two dowels in the block; take care to catch the three selector balls and springs.

Withdraw the forks from the box in the following order—reverse, top and third, and first and second.

Unscrew the clutch lever pivot nut; screw out the pivot bolt and remove the lever with the thrust bearing.

Unscrew the nuts and remove the gearbox front cover; note the bearing shims between the cover and the bearing. Tap out the layshaft, allowing the gear cluster to rest in the bottom of the box.

Unscrew the retaining set screw and remove the reverse shaft and gear.

Withdraw the mainshaft assembly to the rear.

Withdraw the first motion shaft complete with 18 spigot needle rollers, using tool No. 68894 if necessary.

Lift out the layshaft gear cluster and the two thrust washers.

Rear extension

Release the front and rear selector levers from the remote control shaft by removing the clamping screws and sliding the levers from the rod. Extract the keys from the shaft and withdraw the remote control shaft from the rear extension.

Section F.3

DISMANTLING THE THIRD MOTION SHAFT

Remove the following items in this order: baulk ring; synchromesh sleeve and hub; second baulk ring. If the synchromesh sleeve is removed from the hub take care not to lose the three locating balls and springs which will be released in consequence.

Press down the third speed gear cone thrust washer plunger; rotate the thrust washer to align its splines with those on the shaft and remove the washer.

Withdraw the third speed gear and its splined bush.

Withdraw the bush interlocking washer to release the second speed gear with its bush and baulk ring.

Remove the thrust washer from the splines on the shaft and withdraw the first and second speed hub and gear; if necessary slide the gear from the hub, taking care not to lose the three balls and springs.

Tap up the locking tab and unscrew the rear retaining nut; withdraw the washer, speedometer drive gear and key and the distance sleeve from the shaft.

Press the rear bearing and housing from the shaft.

Section F.4

ASSEMBLING THE THIRD MOTION SHAFT

Assemble from the front end.

1. Locate the rear thrust washer on the front end of the splines, ground face to the front.
2. Push the longer brass bush up to the splines with the dog towards the front.

    Note.—This bush must be fitted so that the oil hole is in line with the one in the shaft and the cut-away portion of the third speed splined bush will be over the locating peg hole when the dogs of the two bushes are engaged with the bush interlocking washer.

3. Fit the second speed baulk ring and gear onto the bush with the plain side of the gear towards the front.
4. Slide on the bush interlocking ring and the shorter

splined bush, locating the dogs of both bushes in the interlocking ring.

(5) Insert the spring and locating peg into the hole in the shaft.

(6) Fit the third speed gear onto the bush with the cone towards the front.

(7) Thread on the front thrust washer, machined face towards the gear, while holding down the locating peg with a thin punch through the hole in the gear cone, and push the washer over it; turn the washer to allow the locating peg to engage in one of the splines.

(8) Fit the three springs and balls to the third speed synchronizer and push on the synchronizer sleeve (striking dog).

(9) Push on the top and third gear synchromesh assembly hub with its two baulk rings. The plain side of the hub faces the rear.

Assemble the following items from the rear:

(1) Insert the three balls and springs in the second gear hub and push the synchronizer sleeve (striking dog) into position on the hub.

(2) Fit the first speed gear and synchromesh hub assembly, and the baulk ring, to the splines on the shaft.

(3) Press the rear bearing into its housing and fit it to the shaft, outer flange of the housing to the rear.

(4) Push on the distance sleeve, speedometer drive gear and key, lock washer and nut.

Section F.5

LAYSHAFT GEAR

The assembly sequence of the layshaft bearings is as follows: a circlip at the rear, a needle race, a single long distance tube, a circlip, a needle race, a circlip, a needle race, a circlip, two races being fitted at the front end and one at the rear.

When assembling, fit a circlip to the innermost groove in the gear, pushing it in from the front, or large gear, end.

Hold the layshaft vertically in the vice, stepped end downwards.

Smear the shaft with grease and assemble a roller bearing on the shaft against the vice jaws and then slide the gear cluster over the shaft and the bearing with the large gear downwards.

Remove the shaft from the vice and push the bearing into the gear against the circlip. Fit a retaining circlip and follow with the end roller bearing assembly and retaining circlip.

Slide the distance tube into the other end of the gear, followed by the other end bearing and circlip. Withdraw the shaft from the gear.

Fig. F.1

The arrow indicates the third speed thrust washer and locating peg. Note the hole in the gear cone

Section F.6

ASSEMBLING THE FIRST MOTION SHAFT

Fit the bearing to the shaft with the spring ring away from the gear. Replace the lock washer and tighten the retaining nut; bend over the locking tab. Fit the shaft to the housing. Do not fit the front end cover until the layshaft has been refitted.

Section F.7

ASSEMBLING THE REAR EXTENSION

Locate the remote control shaft in the rear extension.

Fit the front and rear selector levers to the remote control shaft; note that they are secured and located by keys and set screws.

Fit the rear extension to the gearbox, locating the control shaft front selector lever in the shifter rod selectors.

Replace the interlock arm on the rear extension side cover flange and refit the cover.

Section F.8

ASSEMBLING THE GEARBOX

Place the layshaft gear in the box complete with end thrust washers but do not fit the shaft.

Assemble and replace the first motion shaft, and replace the 18 needle-roller bearings.

Insert the third motion shaft from the rear; use the gasket fitted between the box and rear extension to position the dowel and bearing housing. Push home the shaft, the rear bearing and housing, and enter the spigot in the needle-roller race of the first motion shaft.

Fit the layshaft and thrust washers. Line up the cutaway portion of the front end with the layshaft locating groove in the front cover.

Fit the reverse gear and shaft; tighten and lock the set screw.

Refit the front end cover, replacing the bearing shims that were removed on dismantling.

Refit the clutch lever and fork.

Fit the selectors to the shifter shaft rear ends.

Bolt the shifter shaft locating block to the rear face of the gearbox; replace the balls and springs and insert the shifter shafts.

Position the gear change forks in the box in the following sequence: reverse, first and second, third and top. Push the shifter shafts into the box and through the forks; insert, tighten, and wire up the set screws.

Position the selectors on the rear ends of the shifter shafts; insert, tighten, and wire up the set screws.

Refit the gearbox rear extension.

Locate the change speed gate in the gearbox and fit the side cover, using a new joint as necessary.

Screw in the speedometer drive gear assembly, plugs and breather.

The remote control assembly is fitted to the gearbox, and the gearbox filled with oil, after the power unit is installed in the chassis.

Section F.9

MODIFIED GEARBOX FRONT END COVER

Commencing at Engine No. 7981, and a few earlier gearboxes, a modified gearbox front end cover is introduced. The new cover is fitted with an oil seal to prevent the possibility of oil leaking into the clutch housing. There is also a venting duct in the cover necessitating modified fork rods.

The parts may not be fitted to earlier cars.

The new part numbers are :

Gearbox front end cover	..	..	1H3137
Gearbox cover oil seal	..	..	1H3138
Reverse fork rod	..	..	11G3137

First and second fork rod	..	..	11G3079
Third and fourth fork rod	..	..	11G3140

Refitting the front cover

It is essential that the front cover should be concentric with the first motion shaft in order to avoid oil leaks. This is effected as follows.

Mount the cover, less oil seal, onto the gearbox, and push right home on the studs. Ensure that the cover is free to move in all directions on the studs. If not, the points at which the holes bind on the studs must be relieved until the cover is free to 'float'. Remove the cover and refit the oil seal, using Service tool 18G134 with adaptor 18G134Q.

Fit Service tool 18G598 to the bore of the front cover, and push it in until it is tight. Lightly oil the seal, and carefully fit the front cover, retaining the centralizer 18G598 firmly in position. Fit all spring washers and nuts and tighten them finger-tight only. Using a suitable socket spanner, tighten all nuts, by diametric selection, one half-turn at a time until the nuts are fully tightened. Remove the centralizer.

Section F.10

MODIFIED GEARBOX

Coincident with the introduction of the 15GD series power unit (see Section A.44) the following changes have been incorporated in the gearbox.

The main gearbox casing has been modified to accommodate the new high position of the starter motor on the engine. The gearbox extension has also been changed to suit the new gearbox third motion shaft. The propeller shaft (see Section G.8) is now bolted to a flange which is splined to the gearbox third motion shaft and secured by a nut and spring washer. This arrangement supersedes that of the splined sliding joint for the propeller shaft on the third motion shaft.

To remove the gearbox remove the power unit as in Section A.44. Detach the gearbox from the engine as in Section F.1.

The new gearbox is not interchangeable with that previously fitted.

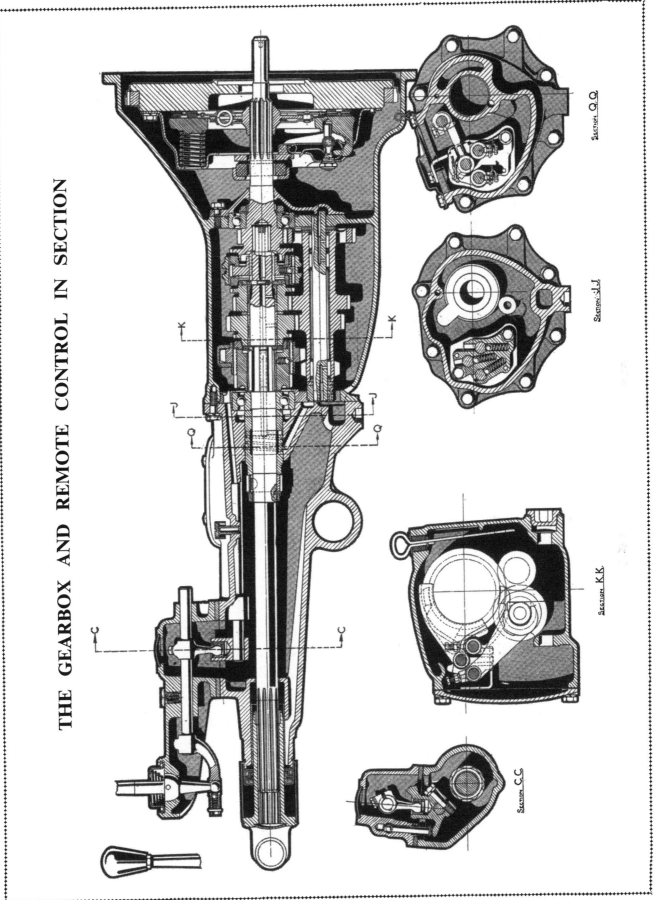

THE GEARBOX AND REMOTE CONTROL IN SECTION

Section Q.Q.

Section J.J.

Section K.K.

Section C.C.

F

F.7

Notes

SECTION FF

THE GEARBOX

(MGA 1600 and MGA 1600 [Mk. II])

Section FF.1   Modified gearbox assemblies.

Section FF.1

MODIFIED GEARBOX ASSEMBLIES

Three modifications to the gearbox have been introduced to prevent automatic disengagement of third gear. If trouble of this nature is experienced remove the gearbox from the car (Section F.1) and check the following points:

(1) Follow the gearbox dismantling procedure given in Section F.2 as far as removing the shift shaft locating block from the gearbox casing. Remove the third and fourth gear shifting rod from the locating block, being careful to catch the ball and spring that will be released. The free length of this spring should be 1·187 in. (30·16 mm.) and the poundage should be between 18 and 20 lb. (8·16 and 9·07 kg.) when the spring is compressed to ·75 in. (19·05 mm.). As these springs adopt a permanent set in service without necessarily affecting the poundage, it is advisable to ensure that the spring is in order by checking its poundage.

(2) Check the depth of the bore in the fork rod locating block, using a depth gauge micrometer. This dimension should be 2·094 in. (53·18 mm.) (see [A], Fig. FF.1). On unmodified gearboxes the depth was 2·157 in. (54·77 mm.). In such cases fit a packing washer ·063 in. thick (1·59 mm.) in the bottom of the bore.

(3) Check the depth of the detent notches in the third and fourth speed selector fork rod. Give particular

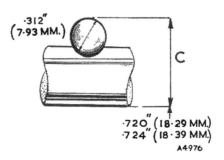

Fig. FF.2

The third and fourth speed fork rod with detent ball placed in a notch for measurement of distance (C)

attention to the third speed notch nearest the centre of the rod. It is not intended that the smaller central (neutral position) notch should be included in this check. Measure the diameter of the detent ball, using a micrometer (Fig. FF.2). Place the ball in each of the deeper notches in turn and measure the distance (C) (Fig. FF.2). If this measurement is greater than ·724 in. (18·39 mm.) a new fork rod providing dimension (C) in both the third and fourth speed notches should be selected and fitted. The depth of these two detent notches was increased by ·018 in. (·46 mm.) on later gearboxes to give the dimension shown in Fig. FF.2.

(4) Following the procedure given in Section FF.2, remove the third motion shaft (mainshaft) assembly from the gearbox. Remove the top and third gear synchromesh sleeve and hub with its baulk rings. Press down the third speed gear thrust washer locating plunger. Turn the thrust washer to align its splines with those on the shaft and remove the washer. Withdraw the third speed gear and its bronze bush. Withdraw the bush interlocking washer to release the second speed gear with its bronze bush and baulk ring. Check with a micrometer the outside diameter of the bronze bushes. This should be 1·3115 to 1·312 in. (33·308 to 33·321 mm.). Ensure that this dimension is constant throughout the length of each bush. If the bushes are worn fit new phosphor-bronze bushes (Part Nos. 11G3028 and 11G3029). These were reintroduced at Gearbox No. 24001 to replace the sintered bronze bushes used previously.

Reassemble the third motion shaft (mainshaft), following the instructions given in Section F.4, and immerse the bronze bushes in warm oil to facilitate fitting.

Reassemble the gearbox, using the method given in Sections F.4 and F.5.

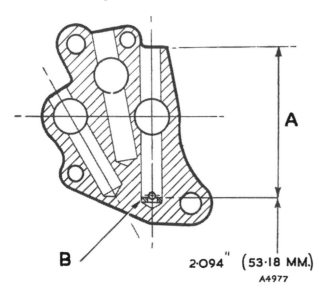

Fig. FF.1

The fork locating block in section, showing a washer (B) ·063 *in.* (1·59 mm.) *thick fitted to provide the revised bore depth* (A) *of* 2·094 *in.* (53·18 mm.)

SECTION G

THE PROPELLER SHAFT

General description.

GENERAL DESCRIPTION

The propeller shaft and universal joints are of the Hardy Spicer type with needle-roller bearings.

A single shaft connects the rear axle and the gearbox. To accommodate fore and aft movement of the axle, a sliding joint of the reverse-spline type is fitted between the gearbox and the front universal joint flange. Each joint consists of a centre spider, four needle-roller bearing assemblies and two yokes.

Section G.1

LUBRICATING THE UNIVERSAL JOINTS

A lubricator is fitted to each front and rear spider and should be charged fully after overhauling and subsequently given three or four strokes with the gun every 3,000 miles (5000 km.). The correct lubricant is shown at Ref. C (page P.2).

If a large amount of lubricant exudes from the oil seal the joint should be dismantled and new oil seals fitted.

The sliding joint is automatically lubricated from the gearbox.

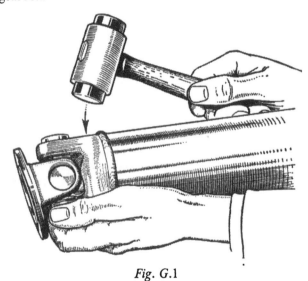

Fig. G.1

Where to apply light blows to the yoke after removing the retaining circlip

Section G.2

TESTING FOR WEAR

Wear on the thrust faces is ascertained by testing the lift in the joint, either by hand or with the aid of a length of wood suitably pivoted.

Any circumferential movement of the shaft relative to the flange yokes indicates wear in the needle-roller bearings, or in the splined shaft in the case of the forward joint.

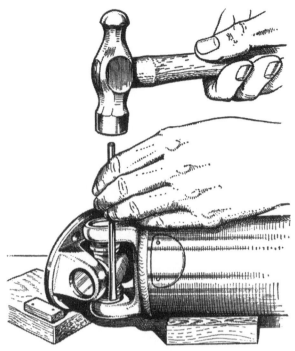

Fig. G.2

When dismantling a universal joint the bearings may be tapped out with a small-diameter rod from the inside as shown. Take care not to damage the roller races

Section G.3

REMOVING THE PROPELLER SHAFT

Before removing the bolts and nuts securing the propeller shaft universal joint flange to the rear axle flange, carefully mark the flanges to assist in refitting them in their original positions. **This is important.**

Remove the bolts securing the propeller shaft to the rear axle flange. The shaft can now be removed from the car downwards and rearwards.

Section G.4

DISMANTLING THE PROPELLER SHAFT

Remove the enamel and dirt from the snap-rings and bearing races. Remove all the snap-rings by pinching their ears together with a pair of thin-nosed pliers and prising them out with a screwdriver.

If a ring does not slide out of its groove readily, tap the end of the bearing race slightly to relieve the pressure against the ring. Remove the lubricator from the journal and, holding the joint in one hand, tap the radius of the yoke lightly with a copper hammer. The bearing should begin to emerge; turn the joint over and finally remove with the fingers. If necessary, tap the bearing race from inside with a small-diameter bar, taking care not to damage the bearing face, or grip the needle bearing race in a vice and tap the flange yoke clear.

Be sure to hold the bearing in a vertical position, and when free remove the race from the bottom side to avoid dropping the needle rollers.

Repeat this operation for the opposite bearing.

Rest the two exposed trunnions on wood or lead blocks to protect their ground surfaces, and tap the top lug of the flange yoke to remove the bearing race.

Turn the yoke over and repeat the operation.

Section G.5

TO EXAMINE AND CHECK FOR WEAR

The parts most likely to show signs of wear after long usage are the bearing races and the spider journals. Should looseness, load markings, or distortion be observed, the affected part must be renewed complete; no oversized journals or races are provided.

It is essential that the bearing races are a light drive fit in the yoke trunnions. In the event of wear taking place in the yoke cross-holes, rendering them oval, the yokes must be renewed. In case of wear in the cross-holes in the fixed yoke, which is part of the tubular shaft assembly, it should be replaced by a complete tubular shaft assembly.

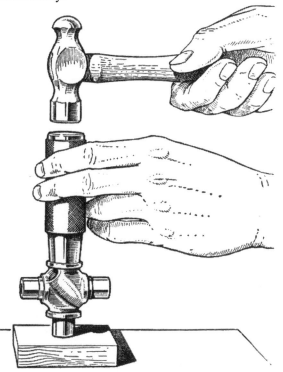

Fig. G.3

When replacing the gasket retainer, use should be made of a hollow drift to tap it into place without damage

Section G.6

REASSEMBLING THE SHAFT

See that all the drilled holes in the journals are thoroughly cleaned out and free of grease or oil.

Assemble the needle rollers in the bearing races and fill with lubricant to Ref. C (page P.2). Should difficulty be experienced in retaining the rollers under control, smear the walls of the races with lubricant to Ref. C (page P.2) to retain the needle rollers in position while reassembling.

Insert the spider in the flange yoke, ensuring that **the lubricator boss is fitted away from the yoke.** Using a soft-nosed drift, about $\frac{1}{32}$ in. ($\cdot$8 mm.) smaller in diameter than the hole in the yoke, tap the bearing into position. Repeat this operation for the other three bearings. Replace the circlips and be sure that these are firmly located in their grooves. If the joint appears to bind, tap lightly with a wooden mallet; this will relieve any pressure of the bearings on the end of the journals.

It is always advisable to replace the cork gasket and the gasket retainers on the spider journals by means of the tubular drift shown in Fig. G.3. The spider journal shoulders should be shellacked prior to fitting the retainers, to ensure a good oil seal.

Section G.7

REPLACING THE PROPELLER SHAFT

Wipe the faces of the flanges clean, and place the propeller shaft in position on the car. Ensure that the flange registers engage correctly, that the components are replaced in exactly the same relation as before removal and that the joint faces bed down evenly all round. Insert the bolts and tighten the self-locking nuts.

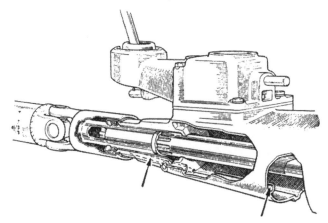

Fig. G.4

The sliding joint, showing the lubrication channels for the sliding joint bush

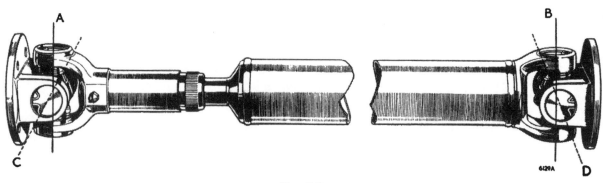

Fig. G.5

When the splined shaft is assembled to the drive shaft it is essential to see that the forked yokes on both shafts have their axes parallel to each other. In other words, the yoke (A) must be in alignment with the yoke (B), and the flange yoke (C) must be in alignment with the flange yoke (D)

Section G.8

MODIFIED PROPELLER SHAFT

Coincident with the introduction of the modified power unit (15GD series) described in Section A.44, a propeller shaft incorporating a splined sliding joint at its front end was fitted.

In addition to the nipples fitted to each universal joint, a nipple is also provided on the sleeve yoke for the lubrication of the sliding joint splines. This nipple must receive attention every 3,000 miles (5000 km.) with the gun filled with lubricant to Ref. C (page P.2).

Instructions given for the servicing of the earlier propeller shaft apply, in the main, to the modified propeller shaft. The following points, however, should be noted:

(1) In order to remove the propeller shaft it is necessary to remove the four nuts and bolts securing the front universal joint flange to the gearbox flange, as well as those securing the rear flange to the rear axle flange.

(2) Check the sliding splines for wear by attempting to turn the splined sleeve yoke in relation to the splined shaft. If excessive circumferential movement is present a reconditioned propeller shaft assembly will be required.

(3) When fitting new universal joints it will be found helpful to separate the two parts of the propeller shaft at the sliding joint.

(4) Before refitting the splined sleeve yoke to the shaft push the threaded dust cover, the metal washer, and the felt washer over the splines onto the splined shaft. When assembling the joint ensure that the trunnions of the front and rear universal joints are in line (see Fig. G.5). This can be checked by observing that the arrows marked on the splined sleeve yoke and the splined shaft are in line.

(5) Fit the propeller shaft to the car with the sliding joint at the gearbox end.

SECTION H

THE REAR AXLE

General description.

Lubrication.

THE REAR AXLE COMPONENTS

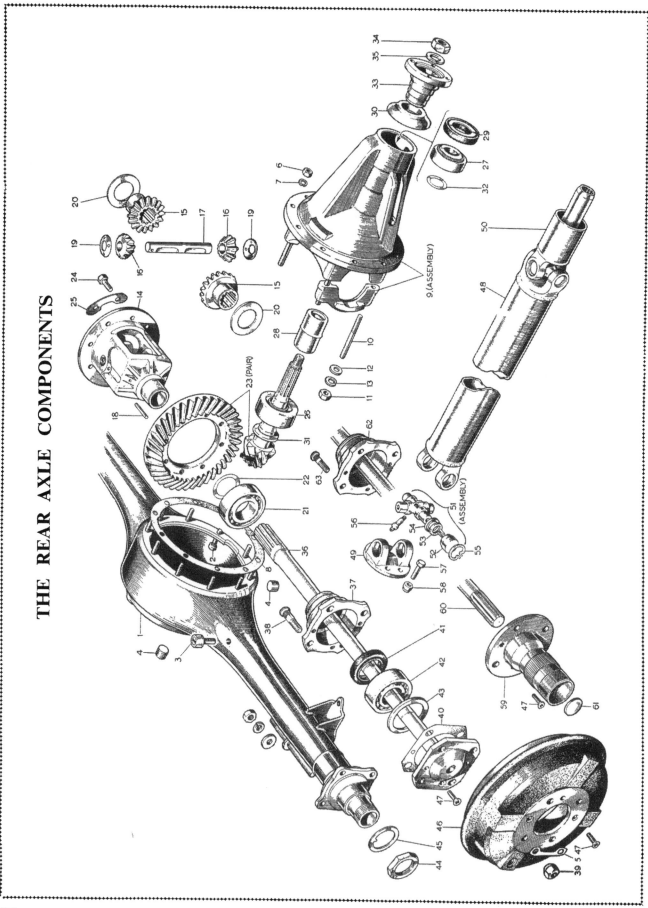

KEY TO THE REAR AXLE COMPONENTS

No.	Description	No.	Description	No.	Description
1.	Casing—rear axle. DW.	22.	Washer—packing—bearing (·002 in.).	42.	Bearing—rear hub.
2.	Bolts—differential carrier.	23.	Crown wheel and bevel pinion 10/43.	43.	Spacer—bearing. DW.
3.	Breather assembly.	24.	Bolt—crown wheel to case.	44.	Locknut.
4.	Plug—oil—drain and filler.	25.	Washer—lock—crown wheel bolt.	45.	Washer—tab—locknut.
5.	Washer—tab—drum retaining.	26.	Bearing—bevel pinion—rear.	46.	Drum—brake.
6.	Nut—differential carrier bolt.	27.	Bearing—bevel pinion—front.	47.	Screw—drum hub—axle shaft. DW.
7.	Washer—spring.	28.	Spacer—bearing.	48.	Tubular shaft assembly.
8.	Joint—carrier to case.	29.	Oil seal—bearing—front.	49.	Yoke—flange.
9.	Carrier.	30.	Dust cover—oil seal.	50.	Yoke—sleeve assembly.
10.	Stud.	31.	Washer—bevel pinion (·112 in.).	51.	Journal and needle—kit set.
11.	Nut.	32.	Shim—front bevel pinion bearing (·004 in.).	52.	Bearing assembly—needle.
12.	Washer—plain.	33.	Flange—universal joint.	53.	Gasket.
13.	Washer—spring.	34.	Nut—flange.	54.	Retainer.
14.	Case—differential.	35.	Washer—spring—flange nut.	55.	Circlip.
15.	Wheel—differential.	36.	Shaft—rear axle. DW.	56.	Lubricator—journal.
16.	Pinion—differential.	37.	Hub assembly—rear. DW.	57.	Bolt—shaft flange yoke—rear.
17.	Pin—pinion.	38.	Stud—wheel. DW.	58.	Nut—bolt.
18.	Peg—pin locating.	39.	Nut—wheel stud. DW.	59.	Hub extension R/H—rear. WW.
19.	Washer—pinion—thrust.	40.	Gasket—shaft to hub housing.	60.	Shaft—axle. WW.
20.	Washer—wheel—thrust.	41.	Seal—oil—rear hub.	61.	Welch plug—hub extension. WW.
21.	Bearing—differential.				

H.3

GENERAL DESCRIPTION

The rear axle is of the three-quarter-floating type, incorporating hypoid final reduction gears. The axle shafts, pinion and differential assemblies can be withdrawn without removing the axle from the vehicle.

The rear axle wheel bearing outer races are located in the hubs; the inner races are mounted on the axle tube and secured by nuts and lock washers. Wheel studs in the hubs pass through the brake-drums and axle shaft driving flanges.

The differential and pinion shaft bearings are pre-loaded, the amount of pre-load being adjustable by shims. The position of the pinion in relation to the crown wheel is determined by a spacing washer. The backlash between the gears is adjustable by shims.

Suspension is by semi-elliptic leaf springs, rubber-mounted, and the shackles are fitted with rubber bushes of the flexing type.

LUBRICATION

The axle is filled or topped up with oil through the filler plug in the rear cover by means of an oil gun with a special adaptor.

It is of the utmost importance that only hypoid oils of the approved grades and manufacture be used if satisfactory service is to be obtained from the hypoid gears.

Inspect the oil level every 3,000 miles (5000 km.) and top up as necessary to the level of the filler opening with oil to Ref. B.

Every 6,000 miles (10000 km.) drain off the old oil and refill with new. The capacity of the axle is 2¾ pints (3·3 U.S. pints; 1·56 litres).

The hub bearings are lubricated from the axle and no provision is made for any other attention.

Section H.1

REMOVING AND REPLACING A BRAKE-DRUM AND AXLE SHAFT

Jack up the car and place blocks under the spring as close as possible to the axle.

Remove the wheel.

Release the hand brake.

Unscrew and remove the two countersunk Phillips screws locating the drum and tap it from the hub. It may be necessary to slacken off the brake adjustment slightly if the shoes hold the drum.

Unscrew the countersunk Phillips locating screw in the axle shaft driving flange.

Withdraw the axle shaft by gripping the flange or

carefully prising it with a screwdriver. If the latter method is used the paper washer may be damaged and must be renewed when reassembling.

To replace the shaft and drum, reverse the above sequence of operations but note that in some models the flange locating screw is shorter than the drum locating screws, and make sure that the bearing spacer is in position.

Fig. H.1.

Withdrawing an axle shaft. Note the bearing spacer which here is being replaced in the hub.

Section H.2

REMOVING AND REPLACING A HUB

Remove the drum and axle shaft as detailed in Section H.1.

Remove the bearing spacer.

Knock back the tab of the locking washer and unscrew the nut with a suitable spanner.

Tilt the lock washer to disengage the key from the slot in the threaded portion of the axle casing; remove the washer.

The hub can then be withdrawn with a suitable puller such as special tools Nos. 18G.304 and 18G.304B. The bearing and oil seal will be withdrawn with the hub.

The bearing is not adjustable and is replaced in one straightforward operation.

When reassembling it is essential that the outer face of the bearing spacer should protrude from ·001 in. (·025 mm.) to ·004 in. (·091 mm.) beyond the outer face of the hub and the paper washer, when the bearing is pressed into position. This ensures that the bearing is gripped between the abutment shoulder in the hub and the driving flange of the axle shaft.

Fig. H.2

Using the special service extractor on the rear hub

Section H.3

RENEWING THE BEVEL PINION OIL SEAL

Mark the propeller shaft and the pinion driving flanges so that they may be replaced in the same relative positions. Disconnect the propeller shaft.

Knock back the lock washer and unscrew the nut in the centre of the driving flange. Remove the nut and washer and withdraw the flange and pressed-on end cover from the pinion shaft.

Extract the oil seal from the casing.

Press a new oil seal into the casing with the edge of the sealing ring facing inwards.

Replace the driving flange end cover, taking care not to damage the edge of the oil seal. Tighten the nut with a torque wrench to a reading of 140 lb. ft. (19·34 kg. m.).

Reconnect the propeller shaft, taking care to fit the two flanges with the locating marks in alignment.

Section H.4

REMOVING THE DIFFERENTIAL PINIONS

Drain the oil from the axle casing.

Remove the axle shafts as detailed in Section H.1.

Mark the propeller shaft and pinion shaft driving flanges so that they may be replaced in the same relative positions; unscrew the self-locking nuts and disconnect the joint.

Unscrew the 10 nuts securing the bevel pinion and gear carrier to the axle casing; withdraw the gear carrier complete with the pinion shaft and differential assembly.

Make sure that the differential bearing housing caps are marked so that they can be replaced in their original positions, then remove the four nuts and spring washers. Withdraw the bearing caps and differential assembly.

Tap out the dowel pin locating the differential pinion shaft. The diameter of the pin is $\frac{3}{16}$ in. (4·8 mm.) and it must be tapped out from the crown wheel side as the hole into which it fits has a slightly smaller diameter at the crown wheel end to prevent the pin from passing right through. It may be necessary to clean out the metal peened over the entry hole with a $\frac{3}{16}$ in. (4·8 mm.) drill in order to facilitate removal of the dowel pin. Drive out the differential pinion shaft. The pinions and thrust washers can then be removed from the cage.

Section H.5

REPLACING THE DIFFERENTIAL PINIONS

Examine the pinions and thrust washers and renew as required.

Replace the pinions, thrust washers and pinion shaft in the differential cage and insert the dowel pin. Peen over the entry hole.

Reassembly is now a reversal of the instructions given in Section H.4. Refill the axle with fresh oil to Ref. B (page P.2).

If it proves necessary to fit any new parts other than those detailed in Sections H.2, H.3, or H.5 the axle assembly must be set up as in Section H.7.

Section H.6

DISMANTLING THE CROWN WHEEL AND PINION

Remove the differential assembly as detailed in Section H.4.

Remove the differential bearings from the differential cage, using special tool 18G47C with adaptors 18G47T. Note that the word 'THRUST' is stamped on the thrust face of each bearing and that shims are fitted between the inner ring of each bearing and the differential cage.

Knock back the tabs of the locking washers, unscrew the nuts from the bolts securing the crown wheel to the differential cage, and remove the crown wheel.

Knock back the tab of the locking washer and unscrew the pinion nut; remove the driving flange and the pressed end cover.

Drive the pinion shaft towards the rear; it will carry with it the inner race and the rollers of the rear bearing, leaving the outer race and the complete front bearing in position.

The inner race of the front bearing may be removed with the fingers and the outer races of both bearings withdrawn with special tool 18G264, using adaptors 18G264E and 18G264F.

Slide off the pinion sleeve and shims; withdraw the rear bearing inner race from the pinion shaft with special tool 18G285, noting the spacing washer against the pinion head.

Assembly and adjustment procedure are detailed in Section H.7.

Section H.7

ASSEMBLING AND SETTING THE CROWN WHEEL AND PINION

Apart from the fitting of components as detailed in Sections H.2, H.3, and H.5 it is not permissible to fit any new parts (e.g. crown wheel and pinion, pinion bearings, differential bearings, etc.) to the axle assembly without working through the procedure given in this Section. Furthermore, if a new crown wheel or a new pinion is needed, **a mated pair—crown wheel and pinion—must be fitted.**

Fitting a new crown wheel and pinion involves four distinct operations:

(1) Setting the position of the pinion.
(2) Adjusting the pinion bearing preload.
(3) Setting the crown wheel position.
(4) Adjusting the backlash between the gears.

The following special service tools are required to enable these operations to be carried out correctly:

Bevel pinion and differential setting gauge.
Bevel pinion inner race remover and replacer.
Bevel pinion outer race remover and replacer.
Bevel pinion preload gauge.

1. SETTING THE PINION POSITION

(1) Fit the bearing outer races to the gear carrier, using the special pinion race replacing tool.

(2) Smooth off the pinion head with an oil-stone, but do not erase any markings that may be etched on the pinion head.

(3) Assemble the pinion and rear bearings with a washer of known thickness behind the pinion head.

(4) Position the pinion in the gear carrier without the shims, bearing spacer, and oil seal.

(5) Fit the inner ring of the front bearing and the universal joint driving flange and tighten the nut gradually until a bearing preload of 10 to 12 lb. in. (·12 to ·14 kg. m.) is obtained.

(6) Remove the keep disc from the base of the magnet. Adjust the dial indicator to zero on the machined step 'B' of the setting block.

(7) Clean the pinion head and place the magnet and dial indicator in position (Fig. H.4). Move the indicator arm until the foot of the gauge rests on the centre of the differential bearing bore at one side and tighten the knurled locking screw. Obtain the maximum depth reading and note any variation from the zero setting. Repeat the check in the opposite bearing bore. Add the two variations together and divide by two to obtain a mean reading.

(8) Take into consideration any variation in pinion head thickness. This will be shown as an un-bracketed figure etched on the pinion head and will always be minus (−). If no unbracketed figure is shown the pinion head is of nominal thickness. Using the mean clock gauge reading obtained and the unbracketed pinion head figure (if any), the following calculation can be made:

(a) **If the clock reading is minus** add the clock reading to the pinion head marking, the resulting sum being minus. **Reduce** the washer thickness by this amount.

Example:

Clock reading	..	..	−·002 in
Pinion marking	..	..	−·005 in
Variation from nominal	..		−·007 in.

Reduce the washer thickness by this amount.

Fig. H.3

Setting the gauge to zero on the special block for determination of the pinion position. The arrow indicates the extension to the contact foot.

Table of washer and shim thicknesses	
Pinion head washer thickness	·112 to ·126 in. in steps of ·002 in.
Pinion bearing preload shims	·004 to ·012 in. in steps of ·002 in., plus ·020 in. and ·030 in.
Crown wheel bearing shims	·002 in., ·004 in. and ·006 in.
Pinion bearing preload ..	10 to 12 lb. in. without oil seal; 13 to 15 lb. in. with oil seal
Crown wheel bearing pinch	·002 in. each side

Fig. H.4

The gauge in position on the pinion with the dial indicating a variation from the standard setting

(b) **If the clock reading is plus and numerically less** than the pinion marking **reduce** the washer thickness by the difference.

Example:

Pinion marking	.. ..	—·005 in.
Clock reading	.. ..	+·003 in.
Variation from nominal	..	—·002 in.

Reduce the washer thickness by this amount.

(c) **If the clock reading is plus and numerically greater** than the pinion marking **increase** the washer thickness by the difference.

Example:

Clock reading	.. ..	+·008 in.
Pinion marking	.. ..	—·003 in.
Variation from nominal	..	+·005 in.

Increase the washer thickness by this amount.

The only cases where no alterations are required to the washer thickness are when the clock reading is **plus** and **numerically equal** to the unbracketed pinion marking, or when the clock reading is zero and there is no unbracketed marking on the pinion head.

(9) Allowance should then finally be made for the mounting distance marked on the pinion head in a rectangular bracket as follows.

If the marking is a **plus** figure **reduce** the washer thickness by an equal amount.

If the marking is a **minus** figure **increase** the washer thickness by an equal amount.

A tolerance of ·001 in. is allowed in the thickness of the washer finally fitted.

2. ADJUSTING PINION BEARING PRELOAD

Fit the appropriate washer to the pinion head.

Assemble the pinion shaft, bearings, distance tube, and shims to the gear carrier; fit the oil seal and driving flange. Shims to a thickness of ·008 to ·011 in. (·2 to ·28 mm.) should be used as a starting-point for adjustment of the bearing preload.

Tighten the driving flange nut gradually with a torque wrench to 140 lb. ft. (19·34 kg. m.) and check the preload on the bearings during tightening to ensure that it does not exceed 13 to 15 lb. in. (·15 to ·173 kg. m.), i.e. 3 lb. in. (·034 kg. m.) greater than the recommended figure since the oil seal is now fitted. If the preload is too great, more shims must be added. If the preload is too small when the nut is tightened correctly, the shim thickness must be reduced.

3. SETTING THE CROWN WHEEL POSITION

(1) Before fitting the crown wheel and differential assembly to the gear carrier it is necessary to calculate the shim thickness required behind each bearing to give the required pinch. To facilitate the calculation, machining variations are indicated by stamped numbers on the carrier adjacent to the bearing bores. The dimensions to be considered are shown in Fig. H.5, (A) being the distance from the centre-line to the bearing register of the carrier on the left-hand side, and (B) the distance from the centre-line to the bearing register of the carrier on the right-hand side. The (C) dimension is from the bearing register on one side of the cage to the register on the other side, while the (D) dimension is from the rear face of the crown wheel to the bearing register on the opposite side. Any variation from nominal on the (A) dimension will be found stamped on the carrier adjacent to the bearing bore, and similarly with the (B) dimension. The variations from nominal on (C) and (D) dimensions are stamped on the machined face of the differential cage.

It is possible to calculate the shim thickness required on the **left-hand side** by the use of the following formula:

$$A + D - C + \cdot 007 \text{ in.}$$

Substituting the actual variations shown, this formula gives the shim thickness required to compensate for the variations in machining plus the

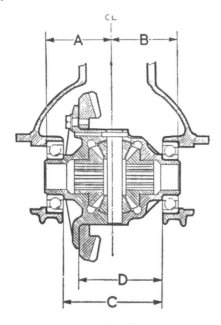

Fig. H.5

The dimensions referred to in the instructions for differential setting

Fig. H.6

To measure variations in bearing thickness, first zero the gauge on the appropriate portion of the gauge block. Here the gauge is seen set for an 'A' type axle. Use the position marked 'B' for the 'MGA' axles

extra ·002 in. (·05 mm.) to give the necessary bearing pinch. In addition, allowance must be made for variations in bearing thickness in the following manner.

Rest the bearing, with the inner race over the recess and the outer ring thrust face downwards, on the small surface plate of tool 18G191B. Drop the magnet onto the surface plate and set the clock gauge to zero on the small gauge block on the step marked 'B'. (See Fig. H.6.) This is the thickness of the standard bearing. Swing over the indicator until it rests on the plain surface of the inner race and, holding the inner race down against the balls, take a reading (Fig. H.7). Normally the bearing will be standard to −·003 in., though in some cases the tolerance may be from standard to −·005 in. A negative variation shown by this test indicates the additional thickness of shimming to be added to that side of the differential.

The formula for the **right-hand side** is:

$$B - D + \cdot 006 \text{ in.}$$

and here again final allowance must be made for variation in bearing thickness.

Fig. H.7

With the gauge set to zero, place the bearing on the surface plate with the outer ring thrust face down, and take a reading while the indicator foot contacts the inner ring

(2) When a framed number is marked on the back of the crown wheel, e.g. $\boxed{+2}$, it must be taken into account before assembling the shims and bearings to the differential cage. This mark assists in relating the crown wheel with the pinion.

If, for example, the mark is $+2$, then shims to the value of ·002 in. (·05 mm.) must be transferred from the left-hand side (the crown wheel side) to the right-hand side. If the marking is -2, then shims to the value of ·002 in. (·05 mm.) must be moved from the right-hand side to the left-hand side.

4. ADJUSTING THE BACKLASH

(1) Assemble the bearings and shims as calculated to the differential cage.

(2) Bolt the crown wheel to the differential case, but do not knock over the locking tabs. Tighten the nuts to a torque wrench reading of 60 lb. ft. (8·3 kg. m.).

Fit the shims and differential bearings with the thrust faces outwards.

Mount the assembly on two 'V' blocks and check the amount of run-out of the crown wheel, as it is rotated, by means of a suitably mounted dial indicator.

The maximum permissible run-out is ·002 in. (·05 mm.) and any greater irregularity must be corrected. Detach the crown wheel and examine the joint faces on the flange of the differential case and crown wheel for any particles of dirt.

When the parts are thoroughly cleaned it is unlikely that the crown wheel will not run true.

Tighten the bolts to the correct torque wrench reading and knock over the locking tabs.

(3) Fit the differential to the gear carrier. Replace the bearing caps and tighten the nuts to a torque wrench reading of 65 lb. ft. (8·99 kg. m.). Bolt the special tool surface plate to the gear carrier flange and mount the clock gauge on the magnet bracket in such a way that an accurate backlash figure may be obtained. (See Fig. H.8.). The minimum backlash allowed in any circumstances is ·005 in. (·127 mm.) and the maximum is ·011 in. (·280 mm.). The correct figure for the backlash to be used with

Fig. H.8

Measuring the crown wheel backlash

H

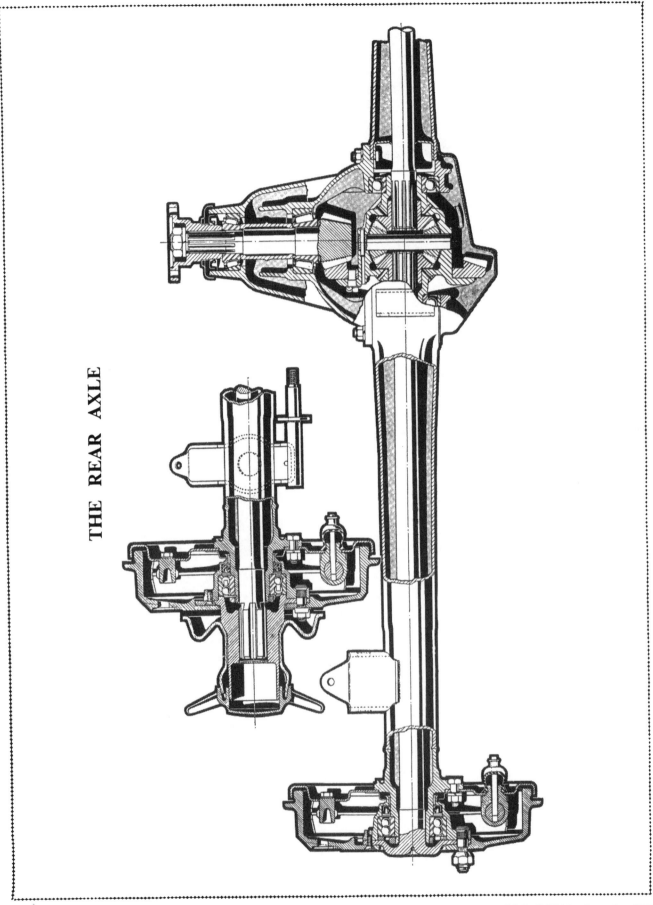

THE REAR AXLE

any particular crown wheel and pinion is etched on the rear face of the crown wheel concerned and must be adhered to strictly.

NOTE.—To ensure adequate clearance when fitting a crown wheel and pinion to earlier axles it may be found necessary to use a pair of gears of which the crown wheel is unmarked.

A movement of ·002 in. (·05 mm.) shim thickness from one side of the differential to the other will produce a variation in backlash of approximately ·002 in. (·05 mm.). Thus it should be possible to set up the differential, even though the backlash is incorrect, by removing the bearings on one occasion only.

Great care must be taken to ensure absolute cleanliness during the above operations, as any discrepancies resulting from dirty assembly would affect the setting position of the crown wheel or pinion.

Section H.8

REMOVING AND REFITTING THE AXLE

Raise the rear of the car.

Mark the propeller shaft coupling flanges so that they may be replaced in the original relative positions. Remove the four bolts and self-locking nuts and release the rear end of the propeller shaft from the axle. Remove the nuts and spring and flat washers securing each end of each check strap to the anchor pins and remove the check straps.

Remove the split pin and clevis pin securing the brake cables to each brake operating lever. Remove the small nut and Phillips recessed-head screw securing the hand brake cable clip to the axle casing. Remove the self-locking nut and large flat washer securing the brake balance lever to the pivot on the axle casing.

Remove the nut and spring washer securing the lower end of each damper link to the rear spring clamp plate.

Unscrew the brake fluid supply pipe union and release the flexible pipe from the battery box support bracket.

Release the exhaust pipe from the exhaust manifold and the three supporting brackets and remove the exhaust pipe assembly.

Remove the nut and spring washer from the spring front anchor pin.

Support the axle casing and remove the rear shackle plates, brackets and rubbers. Lower the axle support until the axle and spring assembly rests on the road wheels. Withdraw the front anchor pins and roll the assembly from beneath the car.

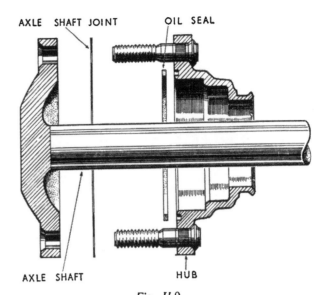

Fig. H.9

The position of the additional hub oil seal fitted on later models

Uncouple the propeller shaft at the rear flange by unscrewing the four self-locking coupling nuts and bolts. Support the tail end of the propeller shaft.

Remove the rear shackle nuts and bolts.

Remove the spring front anchorage bolts after removing the retaining nuts and spring washers.

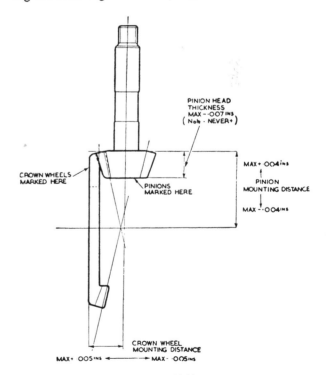

Fig. H.10

Crown wheel and pinion markings

The axle is now free to be withdrawn on the stand rearwards from the car.

Replacement is the reverse of the above sequence of operations.

Section H.9

MODIFIED REAR AXLE HUB BEARING NUTS

The left-hand hub bearing nut on the rear axle now has a left-hand thread and is turned clockwise to unscrew. The right-hand hub nut is unchanged and retains the right-hand thread.

This change is introduced at:

Car No. 10917 on cars fitted with disc wheels.
Car No. 11450 on cars fitted with wire wheels.

Section H.10

REAR HUB OIL SEAL

On the rear axles fitted to later cars a modified rear hub is introduced with an additional oil seal. The hub assembly has a groove machined in the face and a rubber oil sealing ring is fitted in the groove, between the hub and the axle shaft to hub joint (see Fig. H.9).

The oil seal may only be fitted to earlier axles together with the later-type hub.

This change is introduced at Axle No. 5225 on cars fitted with wire wheels and Axle No. 3725 on cars fitted with disc wheels.

Section H.11

BEVEL PINION AND DIFFERENTIAL BEARING SETTING GAUGE

In a recent modification to the bevel pinion and differential bearing setting gauge (Part No. 18G191B and illustrated in Fig. H.3) a stepped gauge block and a small dial gauge with a $\frac{1}{2}$ in. (12·7 mm.) extension to the contact foot replace the original cut-out block and large dial gauge.

When in use this new block should be placed on the surface plate, together with the dial gauge and magnet assembly (with the keep disc removed), and the gauge set to zero on the appropriate step for the 'B' type axle with the later-type hub.

SECTION HH

THE REAR AXLE
(MGA 1600)

For details of the rear axle fitted to the 'MGA 1600' refer to Section H

Notes

SECTION HHH

THE REAR AXLE
(MGA 1600 fitted with Dunlop disc brakes)

Section HHH.1

REMOVING AND REPLACING A HUB EXTENSION AND AXLE SHAFT

Jack up the car and place blocks under the spring as close as possible to the axle.

Remove the wheel.

Release the hand brake.

Unscrew and remove the four self-locking nuts securing the hub extension driving flange to the hub.

Withdraw the hub extension and axle shaft by gripping the driving flange or the winged hub nut, which may be temporarily refitted for this purpose.

Remove the welch plug and apply pressure to the end of the axle shaft with a hand press to remove the hub extension from the spline on the shaft.

To replace the shaft and driving flange reverse the above sequence of operations. If the welch plug has been distorted on removal a new one should be fitted.

Fig. HHH.2
Unscrew and remove the four self-locking nuts to withdraw the hub extension

Section HHH.2

REMOVING AND REPLACING A HUB

Remove the hub extension and axle shaft as detailed in Section HHH.1. Remove the wheel brake unit by the method described in Section MMM.7.

Knock back the tab of the hub nut locking washer, unscrew the nut, using spanner 18G152, and pull off the washer. The left-hand hub bearing nut has a left-hand thread and is turned in a clockwise direction to unscrew.

The hub and brake disc assembly can then be withdrawn, using rear hub remover 18G304 together with adaptors 18G304B and thrust pad 18G304J. The bearing and oil seal will be withdrawn with the hub.

The bearing is not adjustable and is replaced in one straightforward operation. Replace the hub and drift it into position with replacer 18G134 and adaptor 18G134P. The remainder is a reversal of the above sequence of operations.

Fig. HHH.1
Using hub remover 18G304 with adaptors 18G304B and thrust pad 18G304J

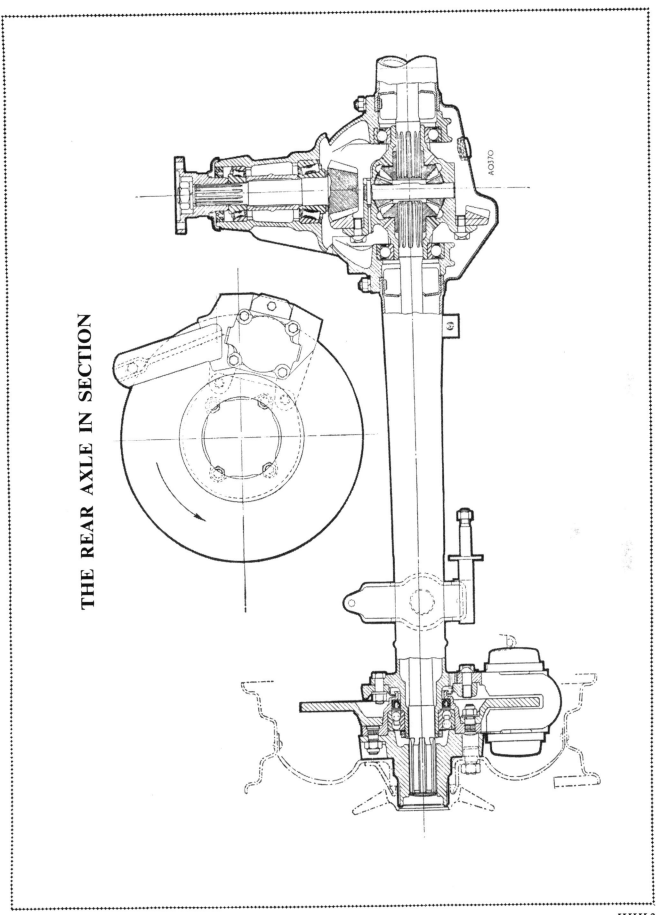

THE REAR AXLE IN SECTION

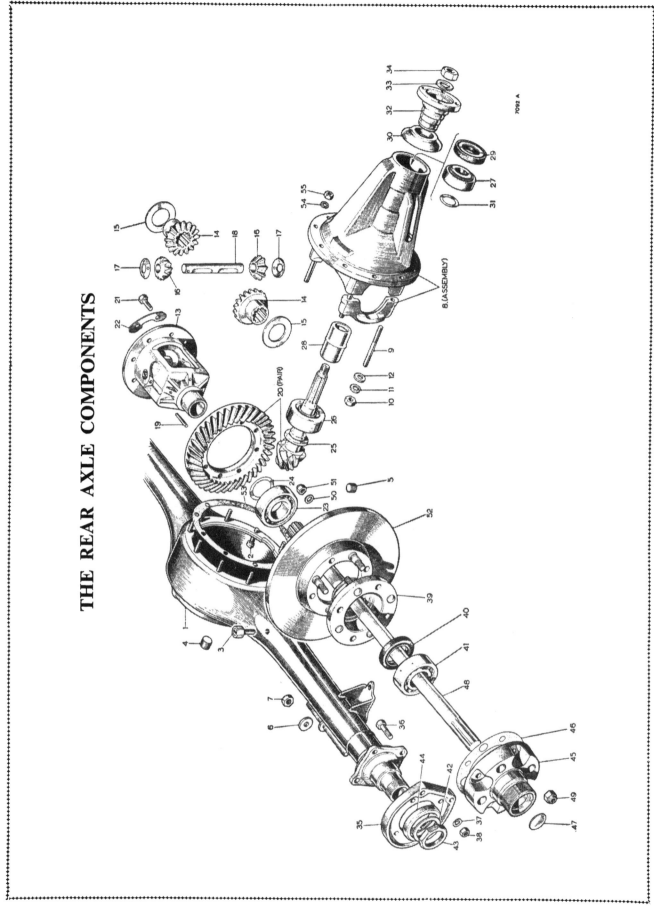

THE REAR AXLE COMPONENTS

KEY TO THE REAR AXLE COMPONENTS

No.	Description	No.	Description
1.	Axle centre case.	20.	Crown wheel and pinion.
2.	Serrated bolt.	21.	Bolt.
3.	Breather.	22.	Lock washer.
4.	Filler plug.	23.	Differential bearing.
5.	Drain plug.	24.	Packing washer.
6.	Plain washer.	25.	Pinion head washer.
7.	Nut.	26.	Pinion bearing.
8.	Gear carrier.	27.	Pinion bearing (outer).
9.	Stud.	28.	Pinion bearing spacer.
10.	Nut.	29.	Oil seal.
11.	Spring washer.	30.	Dust cover.
12.	Washer.	31.	Pinion bearing shim.
13.	Differential case.	32.	Flange.
14.	Differential wheel.	33.	Spring washer.
15.	Thrust washer.	34.	Nut .
16.	Differential pinion.	35.	Adaptor plate.
17.	Thrust washer.	36.	Bolt.
18.	Pinion centre.	37.	Spring washer.
19.	Peg.		

No.	Description
38.	Nut.
39.	Wheel bearing housing.
40.	Oil seal.
41.	Rear wheel bearing.
42.	Tab washer.
43.	Locknut.
44.	Dust cover.
45.	Hub extension.
46.	Joint washer.
47.	Welch plug.
48.	Axle shaft.
49.	Pinnacle nut.
50.	Plain washer.
51.	Pinnacle nut.
52.	Rear brake disc.
53.	Differential joint.
54.	Spring washer.
55.	Nut.

Notes

I

SECTION I

THE REAR ROAD SPRINGS

General Description.

I

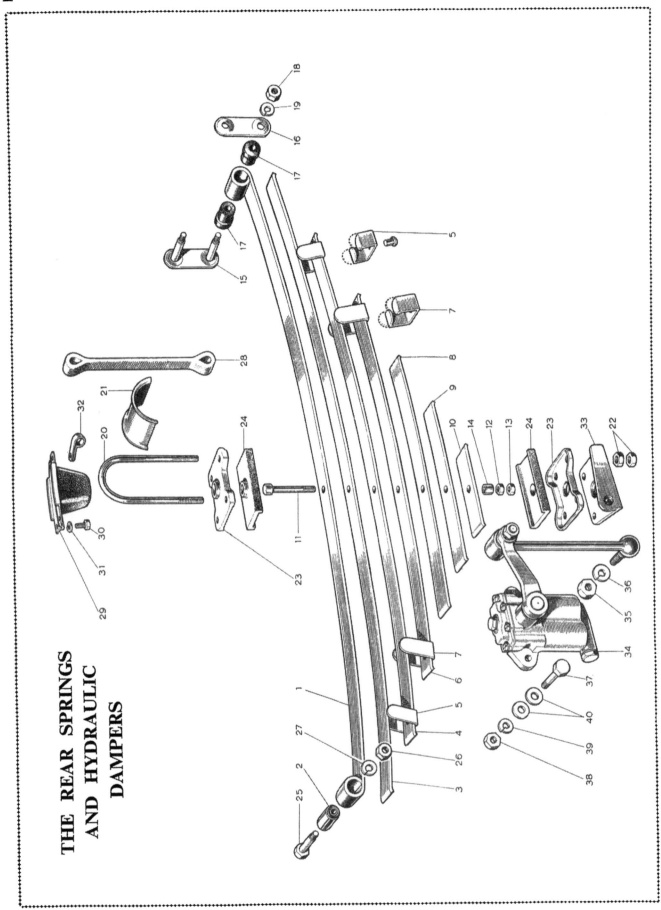

THE REAR SPRINGS
AND HYDRAULIC
DAMPERS

I.2

146

KEY TO THE REAR SPRINGS AND HYDRAULIC DAMPERS

No.	Description
1.	Leaf—main.
2.	Bush.
3.	Leaf—second.
4.	Leaf—third.
5.	Clip.
6.	Leaf—fourth.
7.	Clip.
8.	Leaf—fifth.
9.	Leaf—sixth.
10.	Bottom plate.
11.	Bolt—locating.
12.	Nut—locating bolt.
13.	Locknut—locating bolt.
14.	Distance-piece—bolt.

No.	Description
15.	Shackle plate and pins.
16.	Shackle plate—inner.
17.	Bush—rubber.
18.	Nut—shackle plate.
19.	Washer—spring.
20.	" U " clip—rear spring.
21.	Plate—top " U " clip.
22.	Nut—" U " clip.
23.	Plate—spring locating.
24.	Pad—spring seating.
25.	Bolt—spring front end.
26.	Nut—front end bolt.
27.	Washer—spring.
28.	Rebound strap.

No.	Description
29.	Bump rubber.
30.	Screw—bump rubber to frame.
31.	Washer—spring.
32.	Clip—tail-lamp harness (on bump screw).
33.	Bracket—shock absorber arm to rear spring—L/H.
34.	Shock absorber—rear—L/H.
35.	Nut—arm to bracket.
36.	Washer—spring.
37.	Bolt—shock absorber to frame.
38.	Nut—frame bolt.
39.	Washer—spring.
40.	Washer—plain.

GENERAL DESCRIPTION

The semi-elliptic leaf springs provided for the rear suspension are secured beneath the rear axle by " U " bolts.

The front ends of the springs are anchored in flexing rubber bushes and the rear ends are mounted in similar bushes in swinging shackles.

Rubber pads are fitted between the spring and the axle.

Section I.1

REMOVAL OF REAR ROAD SPRINGS

Raise the rear of the car and support the chassis with a sling attached to the rear bumper bolts, or channelled out or well-padded wood blocks forward of the rear springs. Support the axle on a suitable stand.

Remove the " U " clip locknuts and nuts and drive up the clips to release the hydraulic damper anchor plate, also removing the spring clamp plates and rubbers.

Remove rear shackles and front anchor pin and the spring.

Section I.2

DISMANTLING AND REASSEMBLING THE SPRING

Remove the locating plates and rubber pads.

Remove the locknut, nut and distance-piece from the spring centre bolt, this will release the three bottom leaves. The remaining leaves are parted by prising open the clips on Nos. 3 and 4 leaves.

Clean each leaf, and examine for cracks or breakage. Check the centre bolt for wear or distortion. This bolt forms the location for the spring on its axle pad and should be in good condition.

Important.—When fitting new leaves it is important that they are of the correct length and thickness, and have the same curvature as the remaining leaves.

It is advisable, even when no leaves are broken, to fit replacement springs when the originals have lost their camber due to settling.

Reassembling

The springs should be assembled clean, dry and free from any lubricant unless they are liberally coated with Shell Ensis 260 Fluid.

Place the leaves together in their correct order, locating them with the centre bolt.

The dowel head of the bolt must be on top of the spring. Replace the distance-piece and clamp the leaves together.

Knock down the spring clips to close firmly round the main leaf.

Before replacing the shackle bolts, bushes and shackle plates they must be inspected for wear and, if necessary, replaced by new components.

Before tightening the spring bolts it is absolutely essential that the normal working load be applied to the springs so that the flexing rubber bushes are deflected to an equal extent in both directions during service. Failure to take this precaution will inevitably lead to early deterioration of the bushes.

Section I.3

MAINTENANCE OF THE REAR SPRINGS

As the rear springs are mounted in rubber, spraying with oil should be strictly avoided.

The only attention required is an occasional tightening of the spring seat bolts to make sure they are quite tight.

I.4

SECTION J

THE STEERING GEAR

General description.

Maintenance.

J

THE STEERING GEAR COMPONENTS

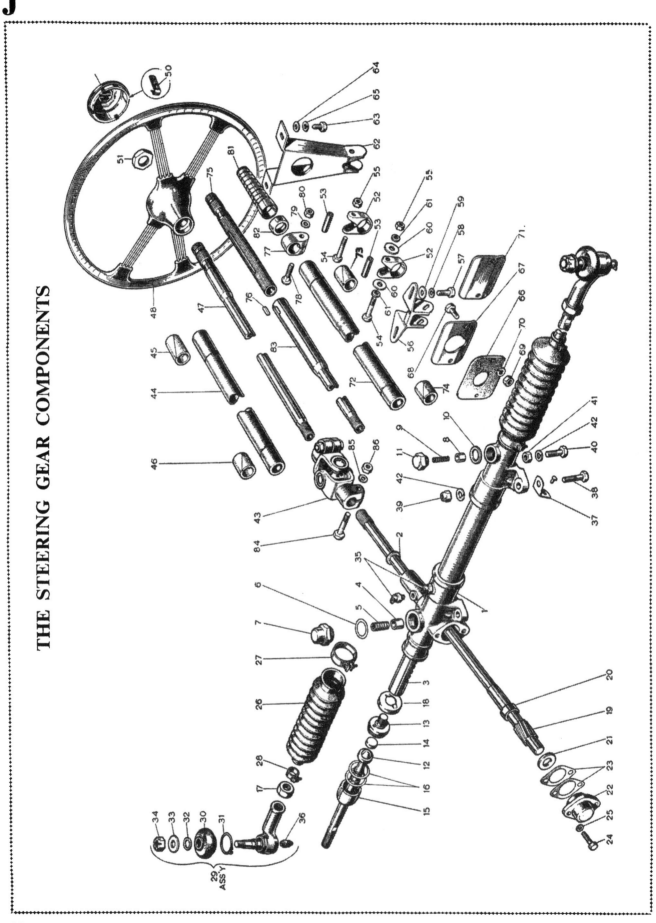

KEY TO THE STEERING GEAR COMPONENTS

No.	Description
1.	Housing assembly—rack. RHD.
2.	Seal—pinion shaft.
3.	Rack—steering.
4.	Pad—rack damper.
5.	Spring—rack damper.
6.	Shim—pad housing.
7.	Housing—rack damper.
8.	Pad—rack damper secondary.
9.	Spring—rack damper secondary.
10.	Washer—rack damper secondary.
11.	Housing—rack damper secondary.
12.	Rod—tie.
13.	Housing—male ball.
14.	Seat—ball.
15.	Housing—female ball.
16.	Shim—ball housing ·003 in.
17.	Locknut—tie-rod.
18.	Lock washer—tie-rod.
19.	Pinion—steering.
20.	Washer—thrust—upper pinion.
21.	Washer—thrust—lower pinion.
22.	Bearing—pinion tail.
23.	Shim ·005 in.—tail bearing.
24.	Screw—bearing to steering box.
25.	Washer—spring—bearing screw.
26.	Seal—rack.
27.	Clip assembly—large—seal.
28.	Clip assembly—small—seal.
29.	Socket assembly—ball.

No.	Description
30.	Boot—rubber.
31.	Clip—boot.
32.	Ring—boot clip.
33.	Washer—ball socket.
34.	Nut—ball socket.
35.	Greaser—pinion/rack.
36.	Greaser—ball socket.
37.	Shim—steering rack to brackets.
38.	Bolt—rack to bracket (front).
39.	Nut—rack to bracket (Nylon).
40.	Bolt—rack to bracket (rear).
41.	Nut—rack to bracket (rear).
42.	Washer—spring—rack to bracket.
43.	Universal joint—steering column.
44.	Tube—outer.
45.	Bush—felt—upper end.
46.	Bush—felt—lower end.
47.	Tube assembly—inner
48.	Wheel—steering.
49.	Cover—steering wheel.
50.	Spring clip—cover.
51.	Nut—steering wheel.
52.	Clamp—steering column.
53.	Distance-piece—clamp.
54.	Bolt—clamp.
55.	Nut—clamp bolt.
56.	Bracket (lower)—steering column.
57.	Screw—bracket to frame.
58.	Washer—spring—bracket to frame.

No.	Description
59.	Washer—plain—bracket to frame.
60.	Washer—plain—lower bracket to clamp.
61.	Washer—spring—lower bracket to clamp.
62.	Bracket—upper—steering column.
63.	Screw—bracket to body rail.
64.	Washer—plain.
65.	Washer—spring.
66.	Seal—rubber—column.
67.	Retainer—column seal.
68.	Screw—seal and retainer to dash.
69.	Nut—seal/retainer screw.
70.	Washer—spring—seal/retainer screw.
71.	Blanking plate.
72.	*Tube—column outer.
73.	*Bush—upper.
74.	*Bush—lower.
75.	*Top end—adjustable.
76.	*Key—top end.
77.	*Clamp—collar.
78.	*Bolt—clamp.
79.	*Washer—spring—clamp bolt.
80.	*Nut—clamp bolt.
81.	*Spring cover.
82.	*Cup—spring cover.
83.	*Tube assembly—inner.
84.	Bolt—universal joint.
85.	Washer—spring.
86.	Nut—universal joint bolt.

* Optional equipment.

GENERAL DESCRIPTION

The steering gear is of the direct-acting rack-and-pinion type, providing light and accurate control under all conditions.

It consists of a rack bar and toothed pinion, both working in the plain bearings of the housing.

No adjustment for bearing wear in the box is provided, except by the fitting of the necessary new parts.

When in new condition the backlash in the tooth engagement is hardly perceptible, i.e. ·001 to ·003 in. (·025 to ·075 mm.).

The steering mast is attached to the steering gearbox by a universal coupling.

MAINTENANCE

All working parts are immersed in oil. An oil gun nipple is provided in the centre of the box to replenish the oil, and a nipple on the pinion housing enables the upper end of the pinion shaft to be lubricated. (See page J.2.)

Felt bushes are fitted to the steering column. (See illustration on page J.2.) These are impregnated with oil and graphite, and no lubrication should be necessary, but if, after long periods, a dry squeak develops, this may be cured by a small application of oil.

Section J.1

REMOVING AND REPLACING THE STEERING WHEEL

Carefully prise the steering wheel cover from the hub of the wheel without chipping the material or the paint-work.

Unscrew the steering wheel nut and mark the wheel hub and column to ensure replacement in the original position. Pull off the wheel with a suitable tool.

When replacing the wheel, position it on the column splines in the original position to place the spokes equally about a horizontal datum line.

Tighten the nut to a torque wrench setting of 500 lb. in. (5·75 kg. m.).

The steering wheel on a car fitted with the optional adjustable steering column may be removed complete with the column extension if necessary.

Remove the clamping nut and bolt from the telescopic adjustment clamp and extend the column as far as possible. Contract the plated helical sleeve and clamp collar towards the steering wheel and extract the key which engages the splined shaft.

Withdraw the steering wheel and column extension.

Section J.2

REMOVING THE STEERING COLUMN

Withdraw the clamping bolt and nut securing the universal joint to the steering mast. Remove the nuts and clamp plate to release the draught excluding rubber from the toe-board at the lower end of the column.

Withdraw the two clamping bolts, nuts, spring and flat washers and distance tube which support the steering column, and withdraw the column complete with steering mast and steering wheel.

Section J.3

REMOVING STEERING COLUMN BUSHES

Remove the steering wheel and column assembly as detailed in Section J.2 and withdraw the mast from the outer column. Prise out the felt bushes.

New bushes should be soaked in graphite oil before reassembly.

Section J.4

REMOVING THE STEERING COLUMN UNIVERSAL JOINT

Bolts and nuts clamp the universal joint splines on the steering mast and steering pinion, and the bolts must be withdrawn completely to release the universal joint assembly.

Slacken the bolts supporting the steering column below the dash panel.

Withdraw the clamping bolts from the universal joint.

Move the steering column and steering mast assembly upwards to withdraw the steering mast from the universal joint.

Withdraw the universal joint from the steering pinion. When replacing the universal joint see Section J.9.

Section J.5

DISMANTLING THE UNIVERSAL JOINT

The Hardy Spicer joint has four needle roller bearings retained on a centre spider by circlips. The joints are packed with grease on assembly and there is no further provision for lubrication.

Remove any enamel and dirt from the snap rings and bearing races. Remove the snap rings by pinching the ears together and prising them out with a screwdriver.

If a ring does not slide readily from its groove, tap the end of the bearing race lightly to relieve the pressure against the bearing.

Hold the joint in one hand with the side of a yoke at the top and tap the radius of the yoke lightly with a copper hammer. The bearing should begin to emerge; turn the joint over and remove the bearing and needle rollers with the fingers. If necessary, tap the bearing race from the inside with a small-diameter bar, taking care not to damage the bearing face, or grip the needle bearing race in a vice and tap the yoke clear.

Repeat this operation for the opposite bearing.

One yoke can now be removed. Rest the two exposed trunnions on wood or lead blocks to protect their ground faces, and tap the top lug of the flange yoke to remove the bearing race.

Turn the yoke over and repeat the operation.

When reassembling, replace the cork gasket and gasket retainers on the spider journals, using a tubular drift. The spider journal shoulders should be shellacked prior to fitting the retainers to ensure a good oil seal.

Smear the walls of the races with grease and assemble the needle rollers to the bearing races and pack with grease.

Insert the spider in one yoke and, using a soft-nosed drift slightly smaller in diameter than the hole in the yoke, tap the bearings into position. It is essential that the bearing races are a light drive fit in the yoke trunnions.

Repeat this operation for the other bearings and replace the circlips, making sure that they are firmly located in their grooves. If the joint appears to bind, tap lightly with a wooden mallet to relieve any pressure by the bearings on the ends of the journals.

Section J.6

REMOVING AND REPLACING THE STEERING GEARBOX

The procedure detailed here will remove the steering rack from a completely assembled car. If the chassis frame front extension has been removed, the steering rack can be removed with the pinion in position.

Remove the steering rack damper and secondary damper assemblies.

Take out the two bolts and spring washers and withdraw the pinion tail bearing and shims and bottom thrust washer, placing a container to catch any oil that may drain from the steering rack. Support the front end of the car by placing jacks beneath the lower suspension arm spring pans, and remove the road wheels.

Remove the split pins and nuts and drive the tie-rod ball pins from the steering-arms. Turn the steering onto the left lock (R.H.D. cars) or right lock (L.H.D. cars). Withdraw the clamping nut and bolt from the universal joint on the pinion shaft and withdraw the pinion assembly. Remove the nuts and bolts securing the

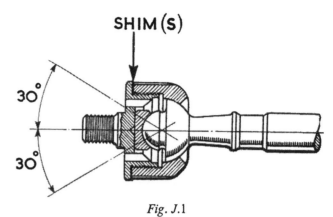

Fig. J.1
The assembly of a tie-rod ball joint

steering rack to the chassis frame, noting that the front bolts are fitted with self-locking nuts, and packing shims may be found between the rack and the frame brackets.

Move the steering assembly towards the centre of the car until the steering tie-rod is clear of the front extension plate, and withdraw the assembly downwards.

Replacing

The steering gearbox is assembled to the car by reversing the above procedure although special attention should be given to the instructions in Section J.9.

When re-engaging the pinion with the universal joint splines, ensure that the cut-away portion for the clamp bolt is aligned with the bolt hole.

Section J.7

DISMANTLING THE STEERING GEARBOX

If the steering rack assembly is removed complete with the steering pinion in position, remove the pinion as follows.

Remove the damper housing, spring, pad and shims from above the pinion housing.

Unscrew the secondary damper housing and remove complete with washer, spring and damper pad.

Withdraw the pinion tail bearing and shims and the pinion bottom thrust washer, placing a container to catch any oil that may drain from the steering rack. The top thrust washer will remain trapped behind the steering rack.

Unlock the tie-rod ball-end locknuts and remove the ball end assemblies. Release the rubber gaiter seal clips and remove the seals.

Secure the rack housing between suitable clamps in a vice and tap back the washers locking the tie-rod ball housings. Unscrew the ball housings with special tool 18G313 and remove the lock washers.

The steering rack may now be withdrawn from the housing.

Screw the ball seat housing from the ball joint caps, using the special 'C' spanner previously mentioned together with the special tie-rod pin spanner 18G312.

The shims and ball seats are now free to be removed; ensure that they are kept to their respective sides.

Section J.8

REASSEMBLING THE STEERING GEAR

Insert the ball end of the tie-rod in the female housing and assemble the ball seat, male seat housing and shims. Tighten the two housings together with special tools 18G312 and 18G313. The ball must be a reasonably tight sliding fit without play. Adjustment is carried out by varying the thickness of the shims between the ball housings. The shims are provided in thicknesses of ·003 and ·005 in. (·08 and ·13 mm.). When correctly adjusted fit a new lock washer to one end of the steering rack, then replace and tighten the ball housing with special tool 18G313. The ball housing must be locked in three places by the flange of the lock washer.

Insert the top thrust washer (the thick one) with the slotted side away from the pinion and insert the rack in its housing. Refit and adjust the other ball seat.

Refit the rubber gaiters and clips.

Replace the ball end locknuts and joint assemblies in their approximate original positions.

Fit a new pinion shaft felt seal.

If the chassis frame front extension is in position, the steering rack assembly should be positioned on its mounting brackets by reversing the procedure detailed in Section J.5 before refitting the pinion.

Replace the smaller thrust washer on the plain end of the pinion shaft.

Replace the shims and the pinion tail bearing, and secure them in position. Check the end-play of the pinion shaft, which should be between ·002 and ·005 in. (·05 and ·13 mm.). If necessary, the shims must be adjusted to give this degree of play.

To adjust the rack damper the plunger must be replaced in the cap and the cap screwed into position without the spring or shims until it is just possible to rotate the pinion shaft by drawing the rack through its housing. A feeler gauge is then used to measure the clearance between the hexagon of the plunger cap and its seating on the rack housing. To this figure must be added an additional clearance of ·002 to ·005 in. (·05 to ·13 mm.) to arrive at the correct thickness of shims which must be placed beneath the damper cap. The shims are ·003 in. (·08 mm.) thick.

Remove the damper cap and plunger and replace and tighten the assembly with the requisite number of ·003 in. (·08 mm.) shims as defined in the previous paragraph.

Replace the secondary damper without shims.

Pump approximately ½ pint (·28 litre) of Hypoid oil to Ref. B (page P.2) into the rack housing through the nipple provided, or release one of the outer rubber gaiter clips and pour the oil in through a funnel. Move the rack assembly backwards and forwards slowly to distribute the oil.

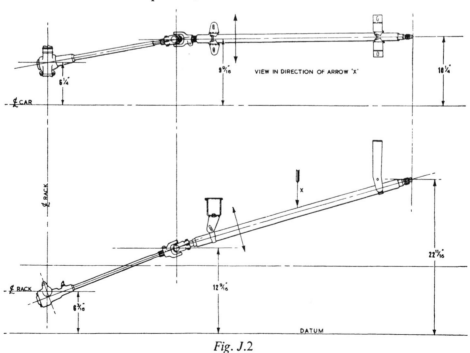

Fig. J.2
Steering-column alignment

Section J.9

STEERING COLUMN ALIGNMENT

When assembling the steering column or steering gearbox assembly to the car, care must be taken to ensure a free condition at the universal joint before the column or gearbox securing bolts are tightened. For the universal joint to be completely unloaded, the centre line of the steering column and the centre line of the steering rack pinion must pass through the centre of the universal joint spider when the assembly is viewed from above and from the side. Failure to ensure complete freedom at the universal joint will load the steering pinion upper bearing and cause extreme wear and steering stiffness.

To enable the assembly to be secured in the correct position, the attachment holes in the support bracket at the lower end of the steering column are slotted to permit up and down and sideways movement, and packing shims (see Fig. J.3) are fitted between the steering gearbox mounting bosses and the brackets on the front suspension member.

Tighten the universal joint clamp bolts.

With the steering column draught excluding rubber clamp plate and all column and rack securing bolts slack, position the universal joint and tighten the support bracket clamp bolts at the lower end of the column.

Should there be a gap between the gearbox bosses and mounting brackets, remove the bolts, pack with shims as required and replace and tighten the securing bolts.

To ensure complete alignment again slacken and retighten the steering column lower support bolt.

Tighten the upper support bracket bolt.

Section J.10

CHECKING AND ADJUSTING FRONT WHEEL ALIGNMENT

When checking the track width at the front and the rear of the front wheels, use a suitable trammel or any special proprietary alignment available.

The wheels should run parallel and have no toe-in.

See that the tyres are inflated to the correct pressures.

Set the wheels in the straight-ahead position.

Set the arms of a suitable trammel to the height of the hub centre on the outside of the wheels.

Place the trammel to the rear of the wheels and adjust the pointers to register with the wheel rims. Chalk the position of the pointers in each wheel rim and push the car forward one half-turn of the wheels. Take the front reading from the same marks on the rims. For the

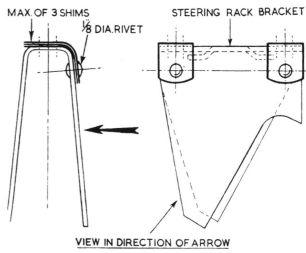

Fig. J.3

The location of the steering gearbox mounting bracket shims which are used to position the gearbox and assist in obtaining correct steering column alignment. When the necessary thickness of shims has been determined they are riveted to the chassis frame to prevent their loss

alignment to be correct the pointers should again register with the marks on the rims.

If adjustment is necessary, proceed as follows.

Slacken the locknuts at the ends of the short tie-rods and the clips securing the rubber gaiters to the tie-rods.

Use the spanner flats on the rods to rotate each of the tie-rods **equally** in the desired direction. These both have right-hand threads.

NOTE.—**To ensure that the steering gearbox is in the central position and that the steering geometry is correct, it is important that the tie-rods are adjusted to exactly equal lengths. This can be ascertained by measuring from the end of the flats to the locknuts.**

After adjustment retighten the ball joint locknuts and rubber gaiter clips and ensure that the machined undersides of the ball joints are in the same plane.

Section J.11

FITTING AN ADJUSTABLE STEERING COLUMN

Remove the steering wheel as detailed in Section J.1.

Remove the steering column assembly as detailed in Section J.2.

Fit the steering column to the car and tighten the clamp bolts.

Fit the steering wheel, locating it on the splines to bring the centre line of the spokes horizontal when the road wheels are in the straight-ahead position.

Section J.12

NYLON-SEATED BALL JOINTS

Nylon-seated ball joints, which are sealed in manufacture and therefore require no further lubrication, are being progressively introduced as alternatives to the conventional types which need lubricating at regular intervals.

It is essential that no dirt or abrasive matter should enter the nylon ball joint; in the event of a rubber boot being torn or damaged in service it is probable that the ball joint has been left exposed, and it is therefore important to renew both the ball joint and the boot.

If damage to the boot occurs whilst the steering side- or cross-rod is being removed in the workshop, only a new rubber boot need be fitted, provided the ball joint is clean. Smear the area adjacent to the joint with a little Dextragrease Super G.P. prior to assembling the boot.

SECTION K

THE FRONT SUSPENSION

General Description.

Maintenance.

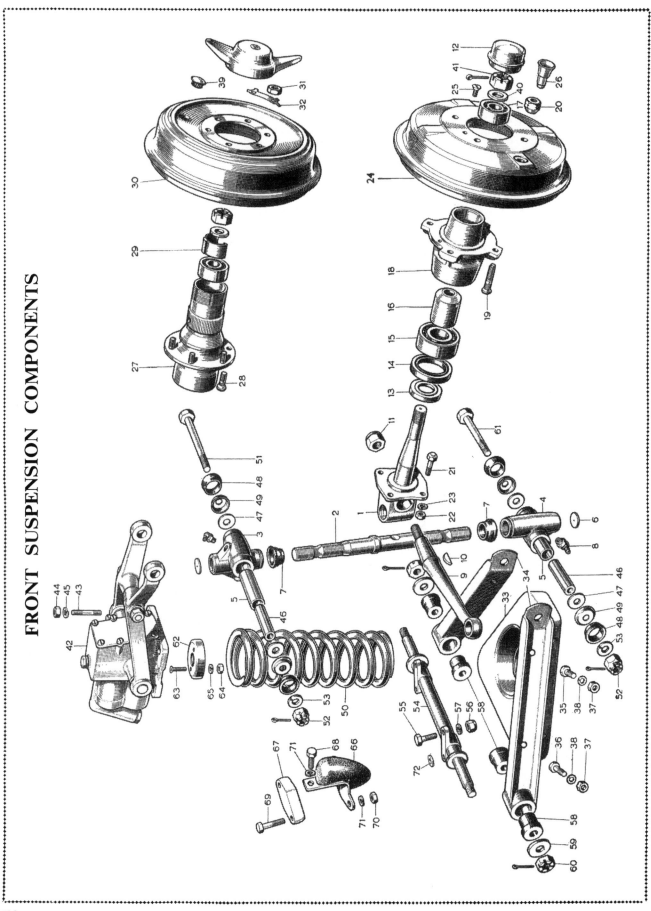

FRONT SUSPENSION COMPONENTS

KEY TO FRONT SUSPENSION COMPONENTS (Disc or Wire Wheel)

No.	Description	No.	Description	No.	Description
1.	Steering knuckle—L/H.	26.	Plug—large—brake-drum. DW.	50.	Spring—coil.
2.	Swivel pin—L/H.	27.	Hub assembly—front L/H. WW.	51.	Bolt—wishbone to link.
3.	Link—swivel pin—upper L/H.	28.	Stud. WW.	52.	Nut—castle—wishbone to link.
4.	Link—swivel pin—lower L/H.	29.	Grease retainer. WW.	53.	Washer—spring—wishbone to link.
5.	Bush.	30.	Drum—brake. WW.	54.	Wishbone pivot.
6.	Plate.	31.	Nut—drum to hub. WW.	55.	Bolt—pivot to member.
7.	Seal—swivel pin.	32.	Locking tab—drum to hub. WW.	56.	Nut—pivot to member bolt.
8.	Grease nipple—link.	33.	Spring pan assembly.	57.	Washer—spring—pivot to member bolt.
9.	Steering lever—L/H.	34.	Bottom wishbone assembly.	58.	Bush—bottom wishbone.
10.	Key—Woodruff No. 8—steering lever.	35.	Screw—spring pan to wishbone.	59.	Washer—wishbone pivot.
11.	Nut—steering lever.	36.	Screw—spring pan to wishbone.	60.	Nut—slotted—wishbone pivot.
12.	Grease-retaining cup. DW.	37.	Nut—spring pan to wishbone screw.	61.	Bolt—bottom wishbone to link.
13.	Distance washer—hub.	38.	Washer—spring pan to wishbone screw.	62.	Spigot—spring.
14.	Oil seal—hub.	39.	Plug—brake-drum—large. WW.	63.	Screw—spigot to member.
15.	Bearing—large—hub.	40.	Washer.	64.	Nut—spigot to member screw.
16.	Distance-piece—hub bearing.	41.	Nut—L/H thread.	65.	Washer—spigot to member screw.
17.	Bearing—small—hub.	42.	Hydraulic damper.	66.	Check rubber.
18.	Hub assembly—front.	43.	Stud—hydraulic damper to cross-member.	67.	Distance piece—check rubber.
19.	Stud—wheel. DW.	44.	Nut—hydraulic damper to cross-member stud.	68.	Screw—check rubber to member.
20.	Nut—wheel stud. DW.	45.	Washer—spring—hydraulic damper to cross-member.	69.	Bolt—check rubber to member.
21.	Bolt—brake backplate.	46.	Distance tube—link.	70.	Nut—check rubber to member—bolt.
22.	Nut—backplate bolt.	47.	Thrust washer—link.	71.	Washer—spring—check rubber to member.
23.	Washer—spring—backplate bolt.	48.	Seal—link.	72.	Washer—plain—under front outer head pivot to member bolt.
24.	Drum—brake. DW.	49.	Support—link seal.		
25.	Screw—countersunk—drum to hub. DW.				

GENERAL DESCRIPTION

The independent front suspension is the wishbone type with coil springing. The front wheels follow the road surface without influencing each other, and each wheel is permitted to rise and fall vertically. The suspension gives perfect stability with riding comfort and, by the combination of the direct-acting rack-and-pinion steering gear, it also provides light and accurate control under all conditions.

The inner mountings of the lower wishbones are fitted with flexing rubber bearings which require no lubrication and form a silent and resilient connection to the robust box-section chassis frame cross-member.

The steering swivel pins are of a special design, with the top and bottom bearings threaded to provide large areas and absorb both thrust and journal loads. The swivel pin threads are of opposite hand on each side of the car and are therefore not interchangeable. The steering connection from wheel to wheel is provided by the

steering gearbox rack bar and two short tie-rods, with ball joints at each end. The outer ball joints are fitted with grease gun nipples, but the inner ball sockets are enclosed in the telescopic rubber dust excluders and are automatically lubricated from the steering gearbox.

Section K.1

REMOVING THE FRONT SUSPENSION

Jack up the front of the car with a jack placed under the centre of the front cross-member until the front wheels are just clear of the ground.

Remove the front wheels.

Place the jack under each spring pan and lift until the hydraulic damper arms are just clear of the rebound rubbers.

Disconnect the hydraulic brake hose (Section M.13.)

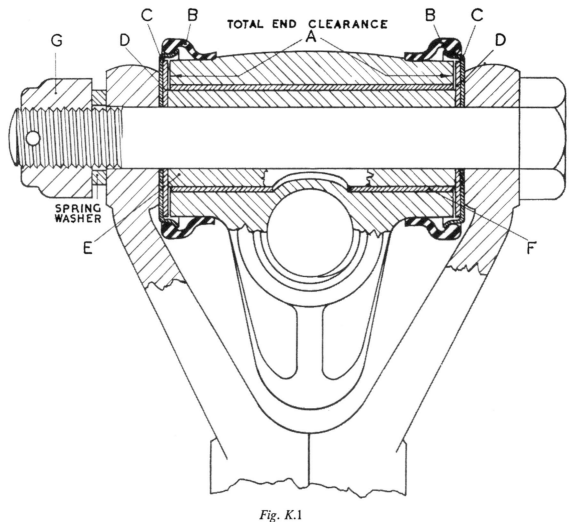

Fig. K.1

The assembly of the king pin swivel link

Slacken the steering tie-rod nuts and screw the tie-rods out of the steering ball joints, using a spanner on the flats on the rods.

Remove the cotters and nuts from the two outer fulcrum bolts. Draw out the bolts and take away the front hub and swivel pin units complete. (Take care of the thrust washers, rubber seals, retainers and fulcrum pins.)

Release the jacks from under the spring pans.

Press down the lower wishbone assemblies and remove the coil springs.

Remove the four bolts holding the spring pan to the levers.

Remove the cotters, nuts and washers from the ends of the inner lower fulcrum pin and slide off the levers and the rubber bushes.

Remove the bolts holding the lower fulcrum pins to the chassis cross-member.

Remove the nuts from studs securing the hydraulic dampers to the top of the suspension cross-member.

Inside the outer ends of the suspension cross-member will be found the coil spring locating plates. These are each attached by two bolts and nuts.

Section K.2

DISMANTLING THE SWIVEL PINS

Unscrew the upper and lower links from the ends of the swivel pins. The left-hand swivel pin has a left-hand thread at each end.

The stub axle is located by a collar on the swivel pin and the stem of the steering lever engaging a groove in the pin. To separate the two, the steering lever must be withdrawn from the stub axle, but this procedure is not advised unless absolutely necessary.

Section K.3

EXAMINING PARTS FOR WEAR

Examine the following parts before reassembling :—

Bushes for bottom wishbone

If these are split, perished, eccentric or oil-soaked, they should be renewed.

Bottom wishbone

Examine the end holes for elongation and the assembly for looseness. If there is any sign of slackness between the wishbone arms and the pan, separate the components and check the bolt holes for elongation. The bolt holes are $\frac{21}{64}$ in. (8·33 mm.) diameter.

Coil spring

Examine for cracks and check for tension, if necessary, to details in the General Data Section. Renew the springs if they are defective.

Swivel link assemblies

Check the swivel links. The dimension across the thrust faces should be 2·327 in. (59·11 mm.). If these are appreciably worn the assembly of link and bush should be renewed. If the bush only is worn, a new one should be pressed in and reamed and burnished to ·750 in. (19·05 mm.).

Note.—When pressing in this bush see that the hole in the bush faces the threaded bore. (See Fig. K.1.)

Check the threaded bores of the links on the swivel pins. When new, these are a free turning fit without slack. An appreciable amount of slack is permissible in these threaded bearings and they do not require renewal unless they are very slack.

Check the fulcrum pin distance tubes for scoring or wear. These should be 2·337 in. (59·36 mm.) long by ·7480 in. (19·00 mm.) diameter.

Examine the case-hardened thrust washers for ridges; the faces should be flat and parallel within ·0005 in. (·01 mm.).

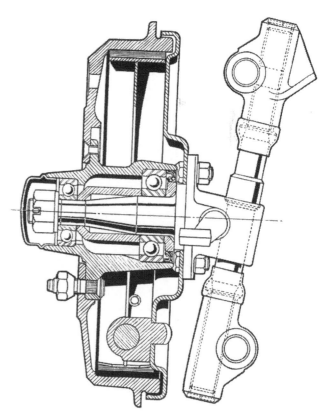

Fig. K.2.

A section through the front hub and brake drum (disc wheel type).

The thickness should be ·068 to ·065 in. (1·73 to 1·68 mm.), the bore ·510 to ·505 in. (12·95 to 12·83 mm.) and the outside diameter 1·25 in. (31·75 mm.).

When the swivel links, distance tubes and thrust washers are assembled, the total end clearance between the link and the thrust washers should be ·008 to ·013 in. (·2 to ·33 mm.). (See " A " Fig. K.1.)

Check that all grease nipples are clear.

Examine the rubber seals, and if these are perished or split, renew them.

Section K.4

HUB BALL BEARINGS

As far as possible, bearings which come under review during the overhaul of the car should be cleaned and inspected without being withdrawn from the housings to which they are fitted. Unnecessary withdrawal causes deterioration of the fitting surfaces, and may damage the bearing, whereas if bearings are examined in position and found still to be serviceable, they can be left undisturbed with advantage.

Rust on the exterior surfaces of the bearing is not detrimental unless the fit is affected, but if the tracks, balls or rollers are severely pitted, the running life of the bearing is at an end and a new one should be fitted.

Fig. K.3.
A right-hand front suspension (L.H.D. car).

Ball bearings should be cleaned thoroughly with paraffin. Bearings which have been washed in this manner should be rinsed thoroughly and dried, and should be immersed in mineral oil as soon as possible after they have been examined.

After cleaning, bearings should be examined for cracks in the races, chipped or broken balls, and worn or damaged tracks or cages.

If either race is cracked or chipped, or if the tracks have spalled or flaked, the bearing should be scrapped.

Section K.5

REPLACING THE FRONT SUSPENSION

Bolt up the coil spring top locating plates inside the front cross-member.

Replace the hydraulic dampers.

The dampers are interchangeable from side to side.

Bolt up the lower fulcrum pins. The two front outer bolts have their nuts uppermost and the six other bolts have their nuts below.

Fit the rubber bushes into the lower levers. These bushes will be found to be a loose fit in the lever, but when clamped up by the nut and washer will expand into their housing. These bushes do not rotate on their surfaces, the angular movement being taken up by the rubber itself flexing.

Special care should be taken when assembling these bushes to maintain a central location, so that the expansion of each half of the bush is equal.

To attain this, insert each bush so that it protrudes equally each side of the housing (see Fig. K.4), and then clamp up with the washer and nut and fit the cotter pins. When central, the outer flanges of the bushes should all be of equal proportions.

It is essential to clamp up the bushes when the lower suspension levers are set parallel with the ground to ensure even stresses on the bushes in service.

Fit the spring pans between the levers, but with the heads of the bolts inside the spring pan.

Do not tighten up the spring pan bolts solid, but leave them half a turn slack.

Press down the lower wishbone assemblies.

Smear each end of the coil springs with grease to prevent any slight squeaking in operation.

Push the coil springs up into the cross-member and over the locating plates.

Jack up the lower wishbone assemblies until they are approximately parallel to the ground.

Assemble the hub units and swivel pins as detailed in Sections K.6 and K.8.

K.6

NOTE.—The king pin bearing threads, the stub axles, and the stub axle nuts are right-hand-threaded on the right-hand side of the car and left-hand-threaded on the left-hand side.

Fit the front hub units to the suspension levers.

Ensure that the thrust washers, rubber seals and retainers are assembled in the right order. (See Fig. K.1.)

Lubricate these parts and the fulcrum pins during assembly and again afterwards with the grease gun, using the recommended lubricant as detailed on page P.2.

Do not tighten up the top or the bottom slotted nuts solid, but leave them half a turn slack.

Connect up the hydraulic brake hoses. See correct method as explained in Section M.13.

Screw the steering tie-rods into the outer steering ball joints. Screw the rods right in and then slack off five complete turns. This will give a rough wheel alignment and render subsequent accurate alignment easier.

Adjust and bleed the front brakes as detailed in Sections M.2 and M.3.

Fit the front wheels.

Bounce the front end of the car up and down a few times. This allows the suspension fulcrums to settle down.

Now tighten the spring pan bolts and then tighten and cotter up the outer fulcrum bolts.

Check and adjust the front wheel alignment.

Section K.6

REASSEMBLING THE SWIVEL PINS

The swivel pin assembly may be reassembled without difficulty by carrying out the removal instructions in the reverse order, provided the following points are given special attention:

(1) The swivel pin and links fitted to the left-hand side of the car have left-hand threads at each end and those fitted to the right-hand side have right-hand threads.

(2) The swivel pin links screw onto threads on each end of the swivel pin and the threads are waisted at their centre to avoid fouling the pivot bolts passing through the links. Before the pivot bolt is replaced the link must be correctly positioned on the thread.

First screw the link onto the swivel pin until the waisted portion of the pin lines up with the pivot bolt hole.

Place the pivot bolt in position in the link and screw the link to the extent of its maximum travel on the swivel pin thread; this is about three revolutions total. Screw the link back approximately one and a half times to obtain the maximum clearances for the pivot pin in each direction.

If the brake plate has been removed from the swivel pin assembly, the lower link must also be centralized in a similar manner before the brake plate is replaced and before the swivel pin is fitted to the suspension arm.

(3) Before the lower steering knuckle link is bolted in position ensure that both thrust washers and rubber seals are fitted correctly (see Fig. K.4) and make sure that the links have a total end clearance of ·008 to ·013 in. (·2 to ·33 mm.) between the end faces of the link and the thrust washers.

NOTE.—Be sure to locate the lower link assembly correctly because it cannot be set once the brake back-plate is fitted.

Section K.7

REMOVING THE BRAKE-DRUM AND HUB

Prise off the hub cover by inserting the flattened end of the wheel nut spanner in the depressions provided adjacent to the cover holding studs and giving a sideways twist—**not** a radial movement.

Slacken the wheel stud nuts.

Apply the hand brake and raise the car until the wheel to be operated on is clear of the ground.

Unscrew the stud nuts and remove the wheel.

Withdraw the two countersunk-headed screws and remove the brake-drum.

Remove the grease-retaining cap and withdraw the split pin from the stub axle nut and unscrew the nut, remembering that the axle on the left-hand side of the car has a left-hand thread.

Place the hub extractor (special tool 18G304 with adaptor bolts 18G304B in position and use the central extractor screw to withdraw the hub assembly.

IMPORTANT.—When the front hub has been removed the inner bearing, oil seal, and hub distance washer MUST be removed from the stub axle and replaced in the hub before it is refitted to the stub axle. If the hub is pressed on the shaft without first fitting the bearing and oil seal to it, the inner bearing will re-enter its housing but the oil seal will only be pushed farther from its correct position.

Section K.8

REPLACING THE FRONT HUB

If all grease has been cleaned from the hub and the bearings washed for examination, ensure that they are repacked with grease before the hub is reassembled.

Replace the bearing spacer with the chamfered side towards the small outer bearing and then press the large bearing into position. Replace the oil seal and distance washer. The metal face of the oil seal and the chamfered side of the distance washer are fitted away from the bearing.

Replace the hub on the stub axle shaft. Replace and tighten the hub nut, and replace grease-retaining cap.

Section K.9

REMOVING AND REPLACING THE FRONT COIL SPRING

Apply the hand brake and jack up the front end of the car until the wheels are clear of the ground, using a suitable jack placed under the centre of the front cross-member.

Remove the front wheel on the side affected.

Place an additional jack under the lower spring pan and jack up until the hydraulic damper levers are clear of the rebound rubber.

Remove the lower fulcrum bolt.

Swing up the hub unit and rest on a suitable block.

Release the jack from under the spring pan, press

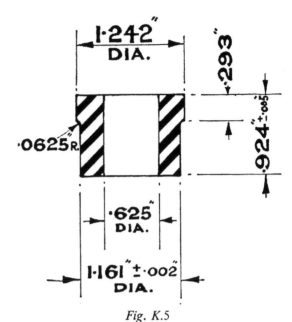

Fig. K.5

The dimensions of the lower wishbone bushes when in new condition

down the lower wishbone assembly and remove the coil spring.

Replacement is carried out in the reverse manner to that detailed for removal.

NOTE.—Take care that the thrust washers, rubber seals, and retainers are assembled in the right order. (See Fig. K.1.)

Lubricate these parts and the fulcrum pins during and after assembly.

Smear each end of the coil spring with grease and ensure that the upper end of the spring is correctly located.

Section K.10

FITTING NEW RUBBER BUSHES

Remove the coil springs as detailed in Section K.9.

Remove the four bolts holding the spring pan to the levers.

Remove the cotters, nuts and washers from the ends of the inner lower fulcrum pin and slide off the levers and the rubber bushes.

Fit the new rubber bushes into the levers. These will be found to be quite a loose fit in the lever, but when clamped up by the nut and washer will expand into their housing. These bushes do not rotate on their surfaces, the angular movement being taken by the rubber deflecting torsionally in itself. Special care should be taken when assembling these bushes to maintain a central location, so that the expansion of each half of the bush is equal.

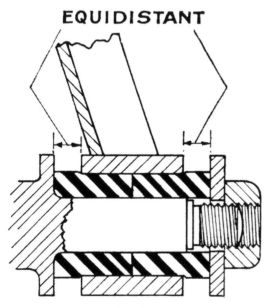

EQUIDISTANT

Fig. K.4

The correct method of clamping the rubber bushes of the lower suspension arm

To attain this, insert each bush so that it protrudes equally each side of the housing (see Fig. K.4), and then clamp up with the washer and nut. When central, the outer flanges of the bushes should be of equal proportions.

It is essential to clamp up the bushes when the suspension levers are set parallel with the ground to ensure even stresses on the bushes.

Now fit the spring pan between the levers, but with the heads of the bolts inside the spring pan.

Do not tighten up the spring pan bolts solid, but leave them half a turn slack.

Press down the lower wishbone assembly.

Smear each end of the coil spring with grease, and push the spring up into the front cross-member and over its top locating plate.

Jack up the lower wishbone assembly until it is approximately parallel to the ground.

Swing down the hub unit and fit the lower fulcrum bolt.

Note.—Take care that the thrust washers, rubber seals and retainers are assembled in the right order. (See Fig. K.4.)

Lubricate these and the fulcrum pin during and after assembly with the grease gun.

Remove the jack from under the wishbone assembly.

Finally tighten up the spring pan bolts, and insert the split cotter pins in all castellated nuts.

Section K.11

MODIFIED FRONT COIL SPRINGS

Modified front coil springs are introduced at Car No. 15152. The new springs (Part No. AHH5546) are interchangeable on earlier cars in pairs only.

Notes

SECTION KK

THE FRONT SUSPENSION

(MGA 1600)

Section KK.1

REMOVING AND DISMANTLING A FRONT HUB

Apply the hand brake and raise the front of the car until the wheel to be operated on is clear of the ground. Remove the road wheel.

Remove the brake unit as detailed in Section MM.2.

Withdraw the grease retainer and extract the split pin from the stub axle nut and unscrew the nut, remembering that the stub axle on the left-hand side of the car has a left-hand thread.

Draw off the hub and brake disc assembly, using special tool 18G363 (wire wheels) or 18G304 with adaptors 18G304B and 18G304J (pressed-steel wheels).

Fig. KK.1

Withdrawing a front hub (wire wheel type), using special tool 18G363. Special tool 18G304 should be used on vehicles fitted with pressed-steel wheels

The brake disc can now be removed from the hub, if necessary, by removing the four nuts and spring washers.

Tap out the small bearing with a drift; remove the spacer tube and tap out the large bearing and oil seal.

IMPORTANT.—If the inner bearing, oil seal, and distance washer remain on the stub axle as the hub is removed they must be replaced in the hub before it is fitted to the stub axle to ensure that the oil seal is in its correct position.

*KK.*2

Section KK.2

REASSEMBLING AND REPLACING A FRONT HUB

If all the grease has been cleaned from the hub and the bearings washed for examination ensure that they are repacked with grease before the hub is reassembled.

Replace the outer bearing and the bearing distance piece with the chamfered side towards the outer bearing and then press the large (inner) bearing into position. Replace the oil seal and distance washer. The metal face

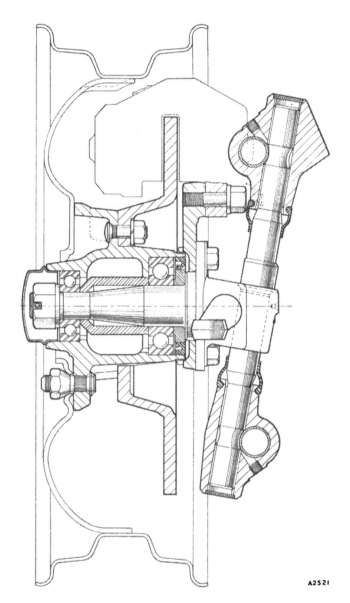

Fig. KK.2

A section of the front hub (pressed-steel wheel type) and brake disc. The brake unit is shown out of position for clarity

of the oil seal and the recessed side of the distance washer are fitted away from the bearing.

Reassembly is then a reversal of the instructions given in Section KK.1.

NOTE.—The brake unit securing bolts must be tightened to the correct torque figure and reference should be made to Section MM.3. If the brake disc has been disturbed the run-out must be checked (see Section MM.9).

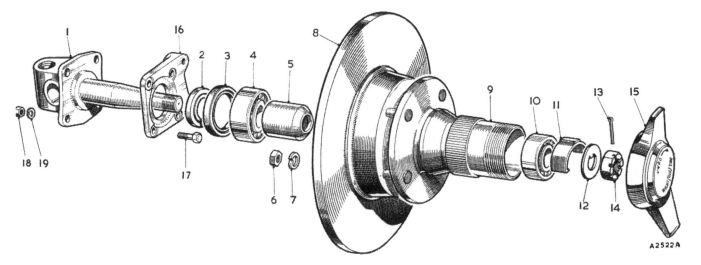

*Fig. KK.*3

The front hub components (wire wheel type)

1. Steering knuckle.
2. Distance washer.
3. Oil seal.
4. Large (inner) bearing.
5. Distance piece.
6. Disc securing nut.
7. Spring washer.

8. Brake disc.
9. Hub assembly.
10. Small (outer) bearing.
11. Grease retainer
12. Washer.
13. Split pin.
14. Hub securing nut.

15. Hub cap.
16. Adaptor plate.
17. Adaptor plate bolt.
18. Nut for bolt.
19. Spring washer.

Notes

SECTION KKK

THE FRONT SUSPENSION
(MGA 1600 fitted with Dunlop disc brakes)

Section KKK.1

REMOVING AND DISMANTLING A FRONT HUB

Apply the hand brake and raise the front of the car until the wheel to be operated on is clear of the ground.

Remove the wheel.

Remove the wheel brake unit as detailed in Section MMM.7.

Withdraw the grease retainer cap using service tool 18G568.

Extract the split pin from the stub axle nut and remove the nut, remembering that the stub axle on the left-hand side of the car has a left-hand thread.

Draw off the hub and brake disc assembly, using service tool 18G304. The brake disc can now be removed from the hub by removing the four securing screws and spring washers.

Remove the distance washer, which will have remained on the stub axle.

The centre of the outer hub bearing may now be withdrawn together with the shims which are fitted between the bearing and the distance piece.

Remove the oil seal and draw out the centre of the inner bearing and the bearing distance piece.

Place the hub on a press with the outer end downwards and press out the outer bearing ring. Press out the inner bearing ring in the same manner with the inner end of the hub downwards.

Fig. KKK.1

Withdrawing the front hub, using tool 18G304

Section KKK.2

REASSEMBLING AND REPLACING A FRONT HUB

If all grease has been cleaned from the hub and the bearings washed for examination, ensure that they are repacked with grease before the hub is reassembled.

Press the two bearing outer rings into the hub. Insert the bearing distance piece. Fit the inner bearing centre, the oil seal, and the distance washer, with the metal face of the oil seal and the chamfered side of the distance washer away from the bearing.

Mount the assembly on the stub axle shaft and fit the adjusting shims and outer bearing centre. Adjust the bearing end-float if necessary, and, finally, lock up as detailed in Section KKK.3.

Pack the assembly with grease and replace the grease-retaining cap. Replace the wheel brake unit as detailed in Section MMM.10.

Section KKK.3

ADJUSTING THE FRONT HUB BEARINGS

The end-float in the hub bearings must be checked and adjusted whenever the hub has been dismantled for attention or when the play in the hub bearings becomes excessive. The end-float is adjustable by means of shims situated between the outer bearing and the bearing distance piece.

Proceed as follows to obtain the correct setting:

(1) Assemble the hub, **using no shims,** and mount the assembly on the stub axle. Fit the stub axle nut and washer and tighten the nut until the hub bearings bind. This will pull the outer rings of the bearings fully against their locating flanges inside the hub.

(2) Remove the stub axle nut and washer and pull out the centre of the outer bearing. Insert a sufficient thickness of shims **to produce an excessive amount of end-float** and note the total thickness of the shims used. Fit the bearing centre, stub axle nut, and washer and tighten the nut.

(3) Measure accurately the total amount of end-float in the bearings. Remove the stub axle nut, washer, and outer bearing centre. Reduce the number of shims to a thickness which will give an end-float of between ·002 and ·004 in. (·051 and ·102 mm.).

(4) Replace the stub axle nut and washer and tighten the nut to a torque wrench reading of 40 to 70 lb. ft. (5·33 to 9·68 kg. m.). Latitude for the torque wrench reading is given so that the nut can be tightened sufficiently to align a castellation with the stub axle split pin hole. Insert a new split pin.

Section KKK.4

FITTING THE ANTI-ROLL BAR

NOTE.—Andrex Dampers and anti-roll bar equipment MUST NOT be used simultaneously.

Place a jack under the centre of the front cross-member and lift the front of the car; support the chassis side-members on stands. Remove the bumper bar and the front apron and the four body holding bolts on the front extension. Unscrew and remove the eight nuts and bolts securing the front extension to the chassis and remove the extension.

Locate the anti-roll bar, the split bushes, and the bush housings on the front extension cradles and ensure that the washers on the bar are interposed between the locating plates and the flange on the bushes. Secure the bush housings to the extension with the four $\frac{7}{16}$ in. bolts, spring washers, and nuts.

Refit the front extension to the chassis and secure the body to the body plates on the extension.

Fit the fork end of the left-hand link and the right-hand link to the appropriate ends of the anti-roll bar, insert a $\frac{7}{16}$ in. washer between each side of the Metalastik bushes and the fork ends, and secure the links with the $\frac{7}{16}$ in. clamp bolts and Aerotight nuts. Locate the ball end of each link in the appropriate wishbone and spring pan assembly and secure them with the $\frac{1}{2}$ in. spring washers and nuts.

Replace the front apron and the bumper bar.

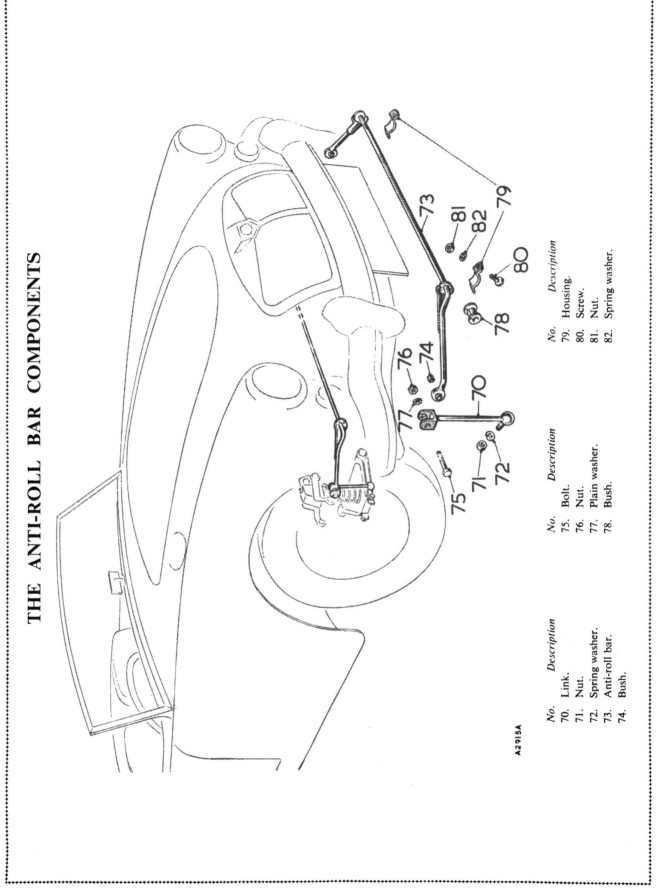

THE ANTI-ROLL BAR COMPONENTS

A2915A

No.	Description
70.	Link.
71.	Nut.
72.	Spring washer.
73.	Anti-roll bar.
74.	Bush.

No.	Description
75.	Bolt.
76.	Nut.
77.	Plain washer.
78.	Bush.

No.	Description
79.	Housing.
80.	Screw.
81.	Nut.
82.	Spring washer.

SECTION L

THE HYDRAULIC DAMPERS

Maintenance.

MAINTENANCE

The maintenance of the hydraulic dampers, when in position on the vehicle, is confined to examination for leakage and examination of the anchorage to the chassis to ensure that the fixing bolts are tight. No adjustment of the hydraulic dampers is required or provided. They are accurately set before leaving the manufacturer to give the amount of damping most suitable for the car. **Any attempt to dismantle the assembly will seriously affect the operation and performance.**

Section L.1

TOPPING UP

The fluid level of the front hydraulic dampers should be topped up by removing the filler plug and filling up to the bottom of the filler plug hole. **Use Armstrong Super (Thin) Shock Absorber Fluid No. 624.** (If this fluid is not available, any good-quality mineral oil to Specification S.A.E. 20/20W should be used, but this alternative is not suitable for low-temperature operation.)

Before removing the filler cap, which is located on the top of the damper, carefully wipe the exterior, as it is of utmost importance that no dirt whatever enters through the filler hole.

On no account neglect the operation of topping up the damper fluid because if the low-pressure chamber of the unit is allowed to become empty, air will enter the pressure cylinders and the action of the damper will be impaired.

The rear dampers must be removed from the chassis frame (see Section L.2) for topping up of the fluid.

Fig. L.1

A front damper, showing the filler plug

Section L.2

REMOVING AND REPLACING REAR DAMPERS

Jack up the rear of the car below the axle or rear springs and remove the rear wheel.

Remove the nut and spring washers securing the damper arm to the bracket on the rear spring.

Remove the nuts and spring and flat washers from the two bolts securing the damper to the chassis side-member and withdraw the damper.

When replacing the damper, it is advisable to work the lever arm up and down a few times through its full stroke to expel trapped air from the pressure chambers.

NOTE.—**When handling hydraulic dampers that have been removed from the chassis for any purpose, it is important to keep the assemblies upright as far as possible, otherwise air may enter the operating chamber, resulting in free movement.**

Section L.3

REMOVING A FRONT DAMPER

Jack up the car under the lower wishbone spring pan until the wheel is clear of the ground.

Remove the wheel and take out the swivel pin top pivot bolt. Swing out the hub unit clear of the upper wishbone and support it on a suitable stand to prevent straining the brake hose. Unscrew the four nuts holding the damper to the chassis frame.

Section L.4

TESTING THE DAMPERS

If the hydraulic dampers do not appear to function satisfactorily, the resistance may be roughly checked by bouncing each corner of the car up and down. A uniform movement indicates that no attention is required, but if the resistance is erratic or free movement of the car is felt, the damper should be removed for checking and topping up.

Indication of their resistance can be obtained by carrying out the following check.

Bolt the damper, in an upright position, to a plate held in a vice.

Move the lever arm up and down through its complete stroke. A moderate resistance throughout the full stroke should be felt. If the resistance is erratic, and free movement in the lever arm is noted, it may indicate lack of fluid.

While adding fluid the lever arm must be worked throughout its full stroke to expel any air that may be present in the operating chamber.

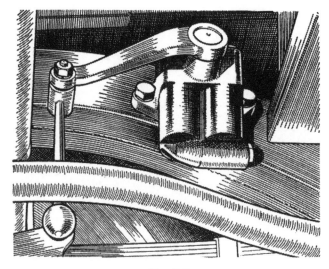

Fig. L.2

Rear dampers must be removed by unscrewing the two securing bolts and disconnecting the lower end of the link from the rear spring bracket

If the addition of fluid gives no improvement a new damper should be fitted.

Too much resistance, i.e. when it is not possible to move the lever arm by hand, indicates a broken internal part or a seized piston.

As it is essential for the dampers to apply the correct restraining action on the suspension, they should be checked whenever there is any doubt regarding their functioning.

The arms should not be removed from the dampers at any time as it is essential that they should be assembled to the damper shaft in the right relation to the damper cam lever so that there is the full range of movement on either side of the centre-line.

It must be clearly understood that there is no provision for adjusting the setting of the dampers, and if they are in any way defective they must be returned to the manufacturers for attention.

Notes

SECTION M

THE BRAKING SYSTEM

General Description.

Maintenance.

The Master Cylinder and Supply Tank Assembly.

M

THE HAND BRAKE

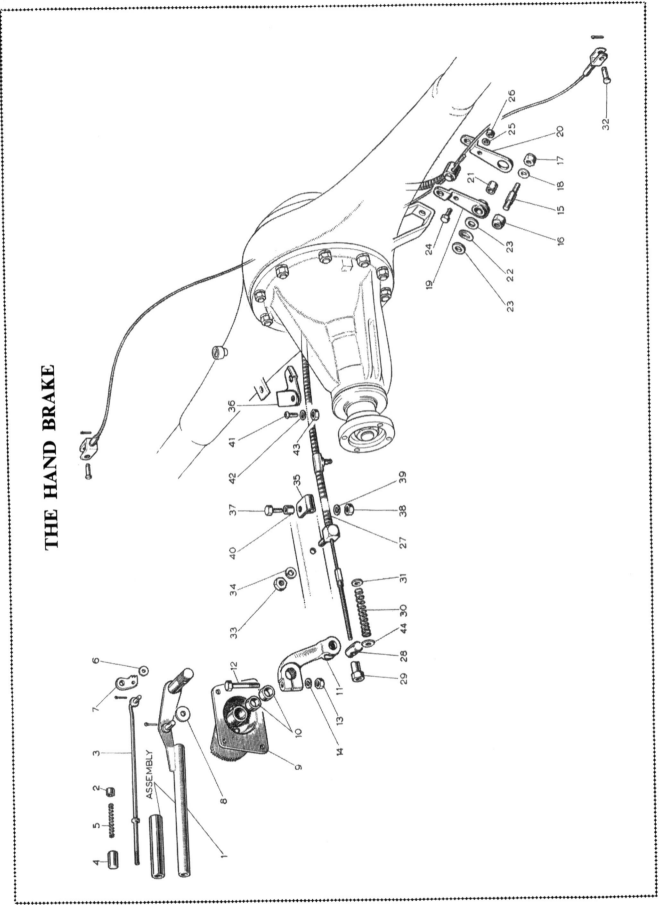

KEY TO THE HAND BRAKE

No.	Description
1.	Handle plate and shaft assembly.
2.	Bush—handle plate and shaft.
3.	Pawl rod assembly.
4.	Knob—pawl rod.
5.	Spring—pawl rod.
6.	Washer—plain.
7.	Pawl.
8.	Washer.
9.	Fulcrum and ratchet assembly.
10.	Bush.
11.	Operating lever—hand brake.
12.	Bolt—operating lever.
13.	Nut—operating lever.
14.	Washer—spring.
15.	Fulcrum—hand brake compensator.

No.	Description
16.	Nut—fulcrum to axle.
17.	Nut—compensator to fulcrum.
18.	Washer—compensator to fulcrum.
19.	Lever—inner compensating.
20.	Lever—outer compensating.
21.	Bush—compensating lever.
22.	Washer—anti-rattle—compensating.
23.	Washer—plain—compensating lever.
24.	Screw—compensating lever.
25.	Washer—spring—compensating lever.
26.	Nut—compensating lever.
27.	Cable—hand brake.
28.	Trunnion—cable.
29.	Adjuster nut.
30.	Spring—cable.

No.	Description
31.	Washer—plain—cable spring.
32.	Clevis pin.
33.	Abutment nut.
34.	Washer—spring.
35.	Clip—cable to battery carrier.
36.	Clip—cable to 3-way piece—axle brackets.
37.	Screw—clip.
38.	Nut—clip screw.
39.	Washer—spring—battery carrier clip.
40.	Distance tube—battery carrier clip.
41.	Screw—clip—3-way piece.
42.	Washer—spring—3-way piece clip screw.
43.	Nut—3-way piece clip screw.
44.	Washer—plain—cable spring front.

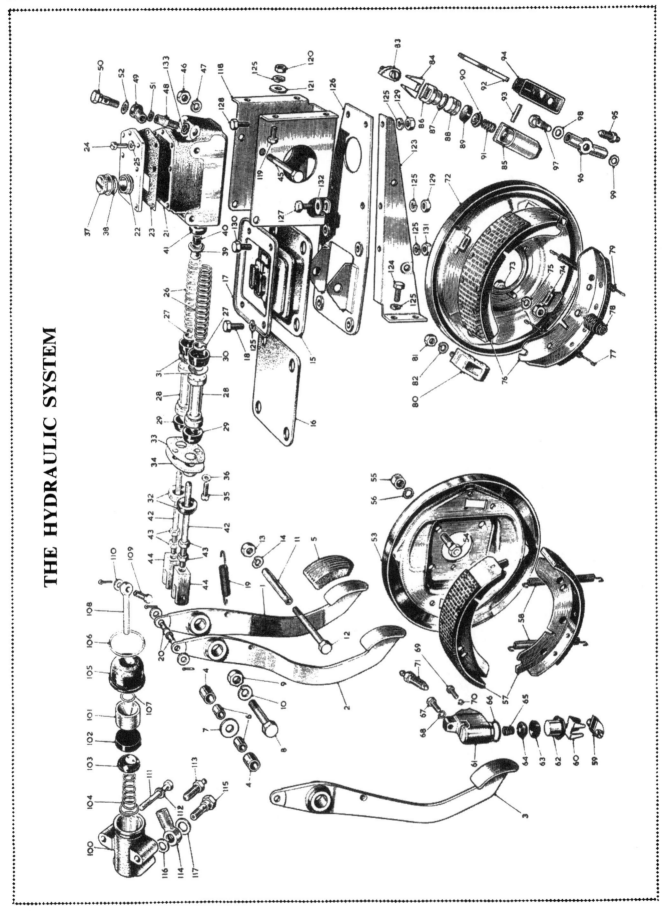

THE HYDRAULIC SYSTEM

M

M.4

KEY TO THE HYDRAULIC SYSTEM

No.	Description	No.	Description	No.	Description
1.	Brake pedal.	45.	Bolt—master cylinder to box.	89.	Cup—piston.
2.	Clutch pedal—right-hand drive.	46.	Nut—master cylinder to box bolt.	90.	Filler cup.
3.	Clutch pedal—left-hand drive.	47.	Washer—spring.	91.	Spring—filler.
4.	Bush.	48.	Adaptor—master cylinder.	92.	Lever—hand brake.
5.	Rubber pad—pedal.	49.	Banjo—master cylinder.	93.	Pin—lever.
6.	Distance-piece—pedal.	50.	Bolt—banjo.	94.	Boot—hydraulic cylinder.
7.	Distance washer—pedal.	51.	Gasket—banjo connection.	95.	Screw—bleeder.
8.	Bolt—pedal bracket.	52.	Gasket—banjo connection.	96.	Banjo connection—wheel cylinder.
9.	Nut—pedal bracket bolt.	53.	Plate—L/H front brake.	97.	Bolt—banjo connection.
10.	Washer—spring.	54.	Bolt—brake back-plate.	98.	Gasket—banjo connection—large.
11.	Distance tube—pedal stop.	55.	Nut—brake back-plate.	99.	Gasket—banjo connection—small.
12.	Distance tube—bolt.	56.	Washer—spring—brake back-plate.	100.	Body.
13.	Nut—distance tube bolt.	57.	Shoe—lined—brake.	101.	Piston.
14.	Spring washer.	58.	Spring—shoe pull-off.	102.	Cup—piston.
15.	Fume excluder—brake and clutch pedal.	59.	Adjuster.	103.	Filler—piston cup.
16.	Rubber—blanking piece.	60.	Mask—adjuster.	104.	Spring—cup filler.
17.	Cover—blanking piece.	61.	Body—L/H.	105.	Boot.
18.	Cover screw.	62.	Piston and dust cover.	106.	Clip—large—boot.
19.	Spring—pedal pull-off.	63.	Cup—piston.	107.	Clip—small—boot.
20.	Clevis pin.	64.	Filler—piston cap.	108.	Push-rod.
21.	Body.	65.	Spring—filler.	109.	Clevis pin—slave cylinder to clutch fork.
22.	Cover—body.	66.	Sealing ring.	110.	Washer—plain.
23.	Gasket—cover.	67.	Bolt—cylinder to brake plate ($\frac{5}{16}''$).	111.	Bolt—slave cylinder to gearbox.
24.	Screw—cover to body.	68.	Spring washer—cylinder bolt.	112.	Washer—spring.
25.	Washer—shakeproof.	69.	Bolt—cylinder to brake plate ($\frac{1}{4}''$).	113.	Bleeder screw.
26.	Spring—piston return.	70.	Spring washer—cylinder bolt.	114.	Banjo—slave cylinder.
27.	Retainer—spring.	71.	Screw—bleeder.	115.	Bolt—banjo—slave cylinder.
28.	Piston.	72.	Plate—L/H rear brake.	116.	Gasket—banjo.
29.	Cup—primary—piston.	73.	Bolt—brake back-plate.	117.	Gasket—banjo.
30.	Cup—secondary—piston.	74.	Nut—brake back-plate.	118.	Master cylinder box.
31.	Washer—primary clip to piston.	75.	Washer—spring—brake back-plate.	119.	Screw box—master cylinder.
32.	Boot—push-rod.	76.	Shoe—lined—brake.	120.	Nut.
33.	Gasket—boot fixing plate.	77.	Spring—shoe pull-off.	121.	Washer—plain.
34.	Plate—boot fixing.	78.	Spring—shoe steady.	123.	Support bracket—L/H master cylinder box.
35.	Screw—plate.	79.	Spring—shoe tension.	124.	Screw—bracket to topping plate.
36.	Washer—shakeproof.	80.	Abutment strip—brake-shoe.	125.	Washer—spring.
37.	Cap—filler.	81.	Nut—abutment strip.	126.	Base plate assembly.
38.	Seal.	82.	Washer—spring.	127.	Screw.
39.	Body—valve.	83.	Adjuster.	128.	Screw—box to bracket rear.
40.	Cup.	84.	Mask—adjuster.	129.	Nut.
41.	Washer.	85.	Body—with abutment strip.	130.	Screw—bracket to base plate.
42.	Push-rod.	86.	Piston—with dust cover.	131.	Nut—bracket to base screw.
43.	Nut—locking.	87.	Seal.	132.	Plain washer.
44.	Yoke—push-rod—to pedal.	88.	Piston—hydraulic.	133.	Gasket—adaptor.

GENERAL DESCRIPTION

The Lockheed hydraulic equipment includes a master cylinder and supply tank assembly in which two separate cylinders are supplied with fluid from a common supply tank. Hydraulic pressure generated in one of the cylinders operates the brakes through the wheel cylinders, while pressure in the other operates the clutch withdrawal mechanism through a slave cylinder, piston, and push-rod connected to the clutch lever.

The master cylinder and supply tank assembly is mounted on the engine side of the bulkhead just above the level of the pedal pads.

Steel pipe lines, unions and flexible hoses convey the hydraulic pressure from one of the master cylinders to each wheel cylinder and from the other to the clutch slave cylinder.

Each brake-shoe in the front drums has a separate wheel cylinder, thus providing two leading shoes. In the rear drums a single wheel cylinder, operated both hydraulically and mechanically, floats on the brake plate and operates the two shoes, giving one leading and one trailing shoe in either direction of rotation to provide adequate braking in reverse.

MAINTENANCE

Periodically examine the quantity of brake fluid in the master cylinder. It should never be less than half-full or closer than $\frac{1}{2}$ in. (13 mm.) to the bottom of the filler neck. The necessity of frequent topping up is an indication of over-filling or a leak in the system, which should at once be traced and rectified.

Adjust the brake-shoes to compensate for wear of the linings. The need for this is shown by the pedal going down almost to the floorboards before solid resistance is felt. For brake-shoe adjustments see Section M.2.

Adjustment of the brake-shoes in the manner indicated also adjusts the hand brake automatically, and no separate adjustment is required or permitted.

THE MASTER CYLINDER AND SUPPLY TANK ASSEMBLY

The brake master cylinder

Within the cylinder is a piston, backed by a rubber cup, normally held in the " off " position by a piston return spring. Immediately in front of the cup, when it is in the " off " position, is a compensating orifice connecting the cylinder with the fluid supply. This port allows free compensation for any expansion or contraction of the fluid, thus ensuring that the system is constantly filled; it also serves as a release for additional fluid drawn into the cylinder during brake applications. Pressure is applied to the piston by means of the push-rod attached to the brake pedal. The push-rod length is adjustable and should give a slight clearance when the

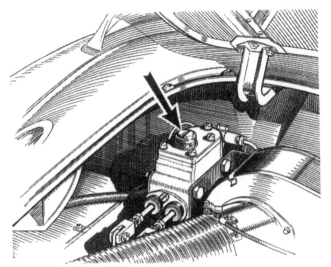

Fig. M.1.
The master cylinder filler cap.

system is at rest to allow the piston to return fully against its stop. Without this clearance the main cup will cover the by-pass port, causing pressure to build up within the system, and produce binding of the brakes on all wheels. The reduced skirt of the piston forms an annular space which is filled with fluid from the supply tank via the feed hole. Leakage of fluid from the open end of the cylinder is prevented by the secondary cup fitted to the flange end of the piston. On releasing the brake pedal, after application, the piston is returned quickly to its stop by the return spring, thus creating a vacuum in the cylinder; this vacuum causes the main cup to collapse and pass fluid through the small holes in the piston head from the annular space formed by the piston skirt. This additional fluid finds its way back to the reserve supply under the action of the brake return springs, when the system finally comes to rest, through the outlet valve and compensating orifice. If the compensating orifice is covered by the piston cup when the system is at rest, pressure will build up as a result of the brake application. The combination inlet and outlet check valve in the head of the cylinder is provided to allow the passage of fluid under pressure from the master piston into the pipe lines and control its return into the cylinder, so that a small pressure of approximately 8 lb. per square inch (·56 kg./cm.²) is maintained in the pipe lines to ensure that the cups of the wheel cylinders are kept expanded; it also prevents fluid pumped out from the cylinder when " bleeding " from returning to the cylinder, thus ensuring a fresh charge being delivered at each stroke of the pedal.

The clutch master cylinder

The components and operation of the clutch master cylinder are, in the main, similar to those of the brake master cylinder, but with one important difference: the combination inlet and outlet check valve used in the brake cylinder is not incorporated in the clutch cylinder, and therefore no pressure is maintained in the clutch line when the clutch pedal is released.

Section M.1

ADJUSTING THE BRAKE PEDAL

The correct amount of free movement between the master cylinder push-rod and piston is set during erection of the vehicle, and should never need alteration.

In the event of the adjustment having been disturbed, reset the effective length of the rod connecting the cylinder to the pedal until the pedal pad can be depressed approximately ½ in. (13 mm.) before the piston begins to move. The clearance can be felt if the pedal is depressed by hand.

Adjusting the clutch pedal
See Section E.

Section M.2

BRAKE-SHOE ADJUSTMENTS

As the linings wear, the pedal will travel farther before the brakes come into action. When the travel becomes excessive the brake-shoes should be adjusted.

A hole sealed with a rubber plug is provided in the wheel and in the drum to allow adjustment without removal of the wheel. The hole in the drum is sealed by a neoprene tubular seal between the wheel and drum. When the wheel is replaced after removal for any purpose, take care to refit it with the holes in the wheel and drum in line and with both seals in position.

Front shoe adjustment
Jack up the front of the car and remove the wheel disc and rubber plug from the hole in the wheel.

Turn the wheel until one of the two adjustment screws is visible through the hole in the wheel and drum.

Insert a screwdriver and turn the adjustment screw in a clockwise direction until the drum is locked and then turn it anti-clockwise one notch. Rotate the drum until the other screw is visible and repeat the adjustment.

The drum should then be free to rotate without the shoes rubbing and the adjustment on that wheel is complete.

Rear shoe adjustment
The procedure is similar to that detailed for the front wheels except that there is only one adjuster controlling both shoes and hand brake.

Section M.3

BLEEDING THE SYSTEM
(Expelling Air)

Bleeding the system is not a routine maintenance job, and should only be necessary when some portion of the hydraulic equipment has been disconnected or the fluid drained off.

Fill the master cylinder with Lockheed Genuine Brake Fluid (if this fluid is not available an alternative fluid conforming to S.A.E. Specification No. 70.R1 should be used) and keep it at least half-full throughout the operation, otherwise air will be drawn into the system, necessitating a fresh start.

Attach the bleeder tube to the wheel cylinder bleeder screw and allow the free end of the tube to be submerged in a small quantity of fluid in a clean glass jar.

Open the bleeder screw one full turn.

Depress the brake pedal quickly, and allow it to return

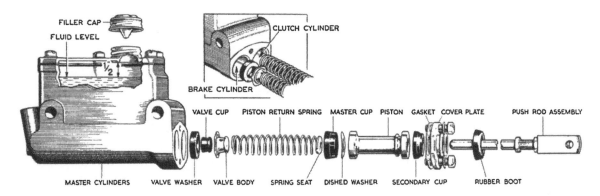

Fig. M.2

Illustrating the master cylinders for brake and clutch operation, and their components. Note that no valve is used in the clutch master cylinder

without assistance. Repeat this pumping action with a slight pause before each depression of the pedal.

Watch the flow of fluid into the glass jar, and when air bubbles cease to appear, hold the pedal firmly against the floorboards while the bleeder screw is securely tightened.

Repeat the operation on each wheel.

NOTE.—Clean fluid bled from the system must be allowed to stand until it is clear of air bubbles before it is used again. Dirty fluid should be discarded.

Section M.4

REMOVING THE MASTER CYLINDER ASSEMBLY

Extract the push-rod split pins and clevis pins.

Unscrew the union nut and disconnect the clutch pipe from the rear end of the master cylinder.

Unscrew the brake pipe at the three-way union on the chassis side-member.

Remove the bolt securing the brake pipe clip to the front of the mounting plate.

Remove the two bolts passing through the bracket and master cylinder and lift out the assembly complete with the brake pipe and clip.

Section M.5

DISMANTLING THE MASTER CYLINDER AND SUPPLY TANK ASSEMBLY

Remove the filler cap and drain the Lockheed fluid from the unit.

Unscrew the two retaining screws and remove the end cover.

Withdraw the two pistons, piston cups, return springs and one valve cup assembly. Note that no valve assembly is fitted to the clutch cylinder.

To remove the secondary cups from the pistons, carefully stretch each one over the end flange of the piston, using only the fingers.

Section M.6

ASSEMBLING THE BRAKE MASTER CYLINDER

Clean all parts thoroughly, using Lockheed Genuine Brake Fluid for all rubber components. All traces of petrol (gasoline), paraffin (kerosene) or trichlor-ethylene used for cleaning the metal parts must be removed before assembly.

Examine all the rubber parts for damage or distortion. It is usually advisable to renew the rubbers when re-building the cylinder. Dip all the internal parts in brake fluid and assemble them wet.

Stretch the secondary cup over the end flange of the piston with the lip of the cup facing towards the opposite end of the piston. When the cup is in its groove, work it round gently with the fingers to make sure it is correctly seated.

Fit the valve washer, valve cup and body onto the return spring, and insert the spring, valve first, into the cylinder. See that the spring retainer is in position.

Insert the master cup, lip first, taking care not to damage or turn back the lip, and press it down onto the spring retainer.

Insert the piston, taking care not to damage or turn back the lip of the secondary cup.

Push the piston down the bore and replace the end plate, rubber boots and push-rods.

Test the master cylinder by filling the tank and by pushing the piston down the bore and allowing it to return; after one or two applications fluid should flow from the outlet.

Assembling the clutch master cylinder

The procedure is similar to that for the brake cylinder, omitting the valve assembly.

Section M.7

REPLACING THE MASTER CYLINDER AND SUPPLY TANK ASSEMBLY

The replacement procedure is the reverse of the removal instructions given in Section M.4 with the following additions:

Replace the union in its correct position and fully tighten before replacing the master cylinder in its housing.

Check the shoe adjustment as detailed in Section M.2.

Connect the fluid pipes and bleed the system (including the slave cylinder) as in Section M.3.

Check the system for leaks with the brakes fully applied.

Section M.8

BRAKE ASSEMBLY

Two leading shoes are incorporated in the front wheel braking system and take the greater percentage of the braking load. The rear brakes are of the leading and trailing shoe type, giving the advantage of equal braking action when the brakes are used in reverse.

All the shoes have a floating anchorage, each front shoe utilizing the closed end of the other shoe actuating cylinder as its abutment. The two rear shoes share one common abutment stop.

The hand brake lever operates the rear brakes mechanically through a linkage operating on the piston of the rear wheel cylinder, which is made in two halves. The outer half of the piston applies the leading shoes when actuated by a lever pivoted in the cylinder body. The trailing shoe is applied by the movement of the cylinder body, which slides on the brake plate as a result of the reaction of the mechanically operated lever on the pivot.

When operated hydraulically the inner half of the piston is forced outwards, carrying with it the outer half, thus applying the leading shoe, and the trailing shoe is applied by the floating cylinder body as a result of the reaction of the fluid pressure on the body.

Section M.9

REMOVING A WHEEL CYLINDER

Front cylinders

Raise the front of the car and remove the hub cap and road wheel.

Remove the brake-drum as detailed in Section K.7.

Draw the brake-shoes apart until the assembly can be removed from the back-plate.

Release the flexible hose as detailed in Section M.13.

Unscrew the hose from the wheel cylinder.

Unscrew the unions and remove the link pipe from both cylinders.

Unscrew the set bolts securing the cylinders to the back-plates and remove the cylinders.

Rear cylinder

Raise the rear of the car and remove the wheel.

Remove the brake-drum as detailed in Section H.1.

Turn and withdraw the brake-shoe steady springs.

Draw the brake-shoes apart until they can be removed from the brake-plate.

Unscrew the pipe union from the cylinder, noting the positions of the copper washers.

Remove the clevis pin from the hand brake cable yoke to disconnect the cable from the lever on the cylinder.

Remove the rubber boot.

Slide the cylinder upwards, push the lower end through the back-plate and slide the cylinder downwards and away from the back-plate.

Section M.10

DISMANTLING A WHEEL CYLINDER

Front cylinders

Withdraw the piston, the rubber cup, the cup filler and the spring.

Rear cylinders

Tap out the hand brake lever pivot pin and withdraw the lever. Withdraw the upper half of the piston, the rubber cup, the cup filler and the spring.

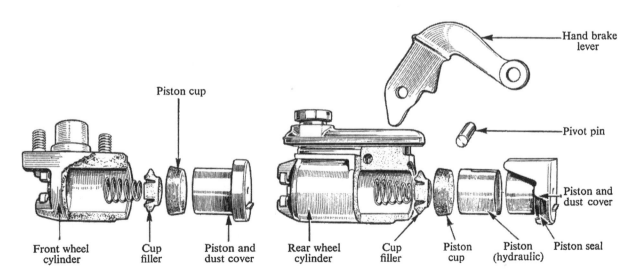

Fig. M.3.

Front and rear wheel cylinder components.

M.9

Fig. M.4.
A front brake assembly.

Section M.11

ASSEMBLING A WHEEL CYLINDER

Clean all parts thoroughly, using only Lockheed hydraulic brake fluid for the rubber components. All traces of petrol, paraffin or trichlorethylene used for cleaning the metal parts must be completely removed before assembly.

Examine the rubber cups for damage, wear, or distortion. Dip all parts in brake fluid and assemble wet.

Insert the cup filler and spring, and the rubber cup concave side first.

Replace the piston, and in the case of the rear cylinders insert the hand brake lever and its pivot pin.

Section M.12

REPLACING A WHEEL CYLINDER

The procedure for replacing the wheel cylinder is a reversal of the sequence of operations given in Section M.9, but attention must be given to the following important points:—

Front

The front brake wheel cylinders are interchangeable but the link pipe banjo unions must be fitted to them so that the flexible hose is connected to the rear cylinder and the bleeder screw to the forward cylinder. The link pipe must pass below the centre line of the stub axle.

The brake-shoes are interchangeable but the recessed ends must engage the " Micram " shoe adjusters.

Rear

The wheel cylinder must be fitted on the rear side of the axle casing with the bleeder screw pointing downwards.

The brake-shoes are interchangeable but the recessed end of the upper or leading shoe must engage the " Micram " shoe adjuster. The other shoe should also be fitted with its recessed end against the wheel cylinder.

Section M.13

REMOVING THE FLEXIBLE HOSE

Do not attempt to release a flexible hose by turning either end with a spanner. It should be removed as follows:—

Unscrew the metal pipe line union nut from its connection to the hose.

Remove the locknut securing the flexible hose union to the bracket and unscrew the hose from the wheel cylinder. Note that a distance-piece is fitted at the rear of the bracket securing the front hoses.

Section M.14

REMOVING AND REPLACING BRAKE-SHOE ASSEMBLIES

Jack up the car and remove the wheel.

Remove the brake-drum as detailed in Section K.7 (*Front*) and Section H.1 (*Rear*).

Turn and withdraw the steady springs (rear only).

Draw the shoes apart until they can be removed from the back-plate.

Replacement is a reversal of the above procedure, but note the correct fitting of the shoes and springs.

Front springs

Both springs are fitted between the shoes and the back-plate.

The shoes are fitted with the recessed ends on the adjusters.

The rear spring is fitted to the rear hole in the upper shoe and the inner of the two holes in the lower shoe.

Rear springs

Both springs are fitted between the shoes and the back-plate.

The lighter spring is fitted at the abutment end of the shoes.

Both springs are fitted to the end holes in the shoes.

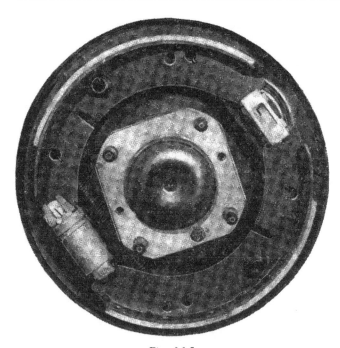

Fig. M.5
A rear brake assembly.

Section M.15

RELINING THE BRAKE-SHOES

Owing to the need for the brake linings to be finished so that they are perfectly concentric with the brake-drums, special precautions must be taken when relining the shoes.

It is imperative that all brake linings should be of the same make, grade, and condition to ensure even braking.

When brake linings are in need of renewal they must always be replaced in axle sets, and the relining of the shoes in one brake-drum must be avoided.

Any variations from this will give an unequal and unsatisfactory braking performance.

After riveting the new brake linings to the brake-shoes it is essential that any high-spots should be removed before replacement of the backplate assembly.

When new shoes and linings are fitted it must be appreciated that considerable adjustment has to be made on the foot brake mechanism, and it is necessary to return the Micram adjusters to their fully anti-clockwise position before attempting to refit the brake-drums over the new linings. The hand brake must also be in the fully released position.

Do not allow grease, paint, oil or brake fluid to come into contact with the brake linings.

Do not clean the rubber parts with anything other than Lockheed hydraulic brake fluid. All traces of petrol, paraffin, etc., used for cleaning metal parts must be removed before reassembly.

Do not allow the fluid in the master cylinder and supply tank assembly to fall below the half-full mark. When full the fluid should be $\frac{1}{2}$ in. (13 mm.) below the bottom of the filler neck, with the brakes in the 'off' position.

IMPORTANT.—**Do not use any substitute for Lockheed Super Heavy Duty Brake Fluid unless this is completely unobtainable**: in such conditions use a fluid to Specification S.A.E. 70.R3.

Section M.16

HAND BRAKE CABLE

The amount of slack in the cable and therefore the amount of movement of the lever is adjustable by means of the nut on the screwed end of the cable located beneath the car in the centre.

To remove

Unscrew and remove the adjustment nut; withdraw the end of the cable from the lower end of the lever and remove the spring.

Disconnect the clips securing the cable assembly to the body.

Remove the clevis pins from the levers on the wheel cylinders.

Unscrew the two nuts on the axle balance lever; separate the two halves of the lever and remove the cable and trunnion.

Section M.17

BRAKING IRREGULARITIES AND THEIR CAUSES

Pedal travel excessive (requires pumping)
(1) Brake-shoes require adjusting.
(2) Leak at one or more joints.
(3) Master cylinder cup worn.

Pedal feels springy
(1) System requires bleeding.
(2) Linings not bedded in.
(3) Master cylinder fixing loose.
(4) Master cylinder cup worn.

Brakes inefficient
(1) Shoes not correctly adjusted.
(2) Linings not bedded in.
(3) Linings greasy.
(4) Linings wrong quality.
(5) Drums badly scored.
(6) Linings badly worn.
(7) Wrongly fitted cup fillers.

Brakes grab

(1) Shoes require adjusting.
(2) Drums distorted.
(3) Greasy linings.
(4) Broken or loose road spring
(5) Scored drums.
(6) Worn suspension linkage.

Brakes drag

(1) Shoes incorrectly adjusted.
(2) Shoe springs weak or broken.
(3) Pedal spring weak or broken.
(4) Hand brake mechanism seized.
(5) Wheel cylinder piston seized.
(6) Locked pipe line.
(7) Filler cap vent hole choked.

Brakes remain on

(1) Shoes over-adjusted.
(2) Hand brake over-adjusted.
(3) No free movement on pedal.
(4) Compensator port in master cylinder covered by swollen rubber cup, or incorrect adjustment of push-rod.
(5) Swollen wheel cylinder cups.
(6) Choked flexible hose.

Unbalanced braking

(1) Greasy linings.
(2) Distorted drums.
(3) Tyres unevenly inflated.
(4) Brake-plate loose on the axle.
(5) Worn steering connections.
(6) Worn suspension linkage.
(7) Different types or grades of lining fitted.

SECTION MM

THE BRAKING SYSTEM

(MGA 1600 and MGA 1600 [Mk. II])

General description.

Maintenance.

GENERAL DESCRIPTION

The braking system consists of two Lockheed calliper-type disc brakes on the front wheels with conventional Lockheed drum brakes at the rear. The rear brakes are similar to those described in Section M and are serviced in the same way.

Front brake units

Each front wheel brake unit comprises a hub-mounted disc rotating with the wheel and a braking unit rigidly attached to the swivel pin. The brake unit consists of a calliper manufactured in two halves—the mounting half and the rim half—which are held together by three bolts. A cylinder in each calliper half houses a self-adjusting hydraulic piston, a fluid seal, a dust seal, and a seal retainer. Each piston is located on a guide post securely held in the back of each cylinder. A friction stop and a sleeve are permanently located in each piston. The pistons are interchangeable side for side.

The friction pad assemblies are fitted adjacent to the pistons and are retained in position by a retainer spring and pin.

Fluid pressure generated in the master cylinder enters the mounting half of each calliper and passes through the internal fluid ports into the rim half. An even pressure is therefore exerted on both hydraulic pistons, moving them along the cylinder bores until the friction pad assemblies contact the disc. In order to compensate for wear of the pads the pistons move progressively along each corresponding guide post, and the friction stops, which grip the posts, provide a positive datum to which the pistons return. The movement of the piston deflects the fluid seal in the cylinder bore, and on releasing the pressure the piston moves back into its original position (see Fig. MM.2), thus providing the required clearance for the friction pads.

MAINTENANCE

Periodically examine the quantity of brake fluid in the master cylinder. The correct level is $\frac{1}{2}$ in. (13 mm.) below the bottom of the filler neck. The necessity for very frequent topping up is an indication of overfilling or of a leak in the system which should be traced and rectified at once.

Adjust the rear brake-shoes to compensate for wear of the linings. The need for this will be indicated by excessive pedal travel. For rear brake-shoe adjustment see Section M.2. The front disc brakes automatically compensate for wear of the brake pads. The pads should be checked for wear every 6,000 miles (10000 km.) by visual observation and measurement. When wear has reduced the pads to the minimum permissible thickness of $\frac{1}{16}$ in. (1·59 mm.) the pads must be renewed.

MM.2

Section MM.1

PRIMING AND BLEEDING THE BRAKE SYSTEM (EXPELLING AIR)

The following procedure should be adopted either for initial priming of the system or to bleed in service if air has been permitted to enter the system. Air may enter the system if pipe connections become loose or if the level of fluid in the reservoir is allowed to fall below the recommended level. During the bleeding operation it is important that the reservoir is kept at least half-full to avoid drawing air into the system.

(1) Check that all connections are tightened and all bleed screws closed.
(2) Fill the reservoir with brake fluid. The use of Lockheed Super Heavy Duty Brake Fluid is recommended, but if this is not available an alternative fluid conforming to Specification S.A.E. 70.R3 should be used.
(3) Attach the bleeder tube to the bleed screw on the near-side rear brake and immerse the open end of the tube in a small quantity of brake fluid contained in a clean glass jar. Slacken the bleed screw and operate the brake pedal slowly backwards and

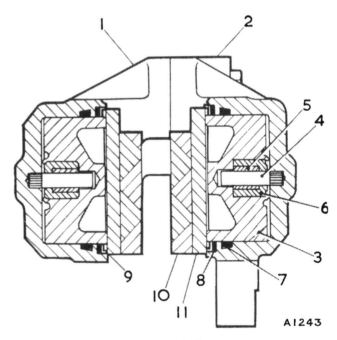

Fig. MM.1

A disc brake in section

1. Calliper—rim half.	7. Fluid seal.
2. Calliper—mounting half.	8. Dust seal.
3. Hydraulic piston.	9. Dust seal retainer.
4. Guide post.	10. Friction pad.
5. Friction stop.	11. Pad backing plate.
6. Sleeve.	

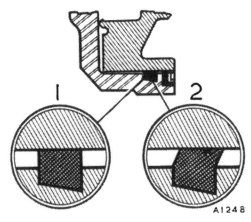

Fig. MM.2

As the piston moves outwards from its static position (1) the fluid seal is deflected (2). When the hydraulic pressure is released the fluid seal regains its static position, moving the piston back and providing clearance at the friction pads

forwards through its full stroke until fluid pumped into the jar is completely free from air bubbles. Close the bleed screw on a down stroke of the pedal. Release the pedal.

(4) Repeat on the off-side rear brake.

(5) Two bleeder screws are fitted to each front disc brake unit. Attach the bleeder tube to the inner bleed screw on the near-side brake unit. Slacken the bleed screw and bleed as described above. Repeat the operation on the outer bleed screw.

(6) Bleed the off-side front brake in the same way.

(7) Top up the fluid level in the reservoir.

(8) If the disc brake callipers have been disturbed it will be necessary to pump the brake pedal several times to restore the automatic adjustment of the friction pad clearance.

(9) Apply a normal working load on the brake pedal for a period of two or three minutes and examine the entire system for leaks.

NOTE.—Clean fluid bled from the system must be allowed to stand until it is clear of air bubbles before it is used again. Dirty fluid should be discarded.

Section MM.2

REMOVING A BRAKE UNIT

Apply the hand brake, place chocks under the rear wheels, and jack up the front of the vehicle. Remove the road wheel. Withdraw the brake friction pads as detailed in Section MM.6.

Remove the two nuts and spring washers securing the brake hose support bracket and remove the bracket.

Unscrew the two calliper securing bolts and swing the calliper clear of the disc. **Do not disconnect the fluid hose if the calliper is to be subsequently dismantled.** Support the brake unit carefully to avoid straining the hose.

Section MM.3

REPLACING A BRAKE UNIT

The replacement procedure is a reversal of the instructions given in Section MM.2. It is essential that the two calliper securing bolts are tightened to the correct torque figure of 45 to 50 lb. ft. (6·22 to 6·91 kg. m.). If the unit has been dismantled or the brake hose uncoupled the system must be bled as described in Section MM.1.

Section MM.4

DISMANTLING A BRAKE UNIT

Remove the brake unit from the swivel pin as detailed in Section MM.2, leaving the brake hose in position to enable the pistons to be removed from the cylinders, using hydraulic pressure.

Only one piston can be removed at a time and this must be replaced before the other piston can be withdrawn. Using a suitable clamp, retain one piston in the calliper and remove the other piston by gently applying the foot brake until the piston has been pushed out far enough to allow removal by hand. It is advisable to carry out this operation over a receptacle to catch the fluid as the piston is withdrawn.

With the aid of a blunt-nosed tool remove the fluid seal from the cylinder bore, taking great care not to damage the base of the seal groove or cylinder bore.

Should the dust seal require renewal, the seal retainer can be removed by placing the blade of a screwdriver between the retainer and seal and carefully prising the retainer from the mouth of the bore.

After assembling the piston into its cylinder as described in Section MM.5 the other piston can be removed by the method described above.

IMPORTANT.—The two halves of the calliper should not be separated unless it is absolutely essential. If separation cannot be avoided then the following points must be borne in mind during assembly.

(1) **New bolts, lock plates, and fluid channel seal must be used. The bolts are manufactured from special high tensile steel and only the correct replacements should be employed. Failure to use the correct bolts could have serious results.**

(2) **Ensure that the calliper faces are clean and that the threaded bolt holes are thoroughly dry. Make**

certain that the new fluid channel seal is correctly located in the recessed face before assembling the two calliper halves.

(3) **The bolt threads must be perfectly dry on assembly and they must be tightened to the correct torque readings detailed below.**

Large bolts	.. ..	65 lb. ft. (8·99 kg. m.)
Small bolt	.. ..	10 lb. ft. (1·38 kg. m.)

Section MM.5

ASSEMBLING A BRAKE UNIT

Thoroughly lubricate a new fluid seal with Lockheed Disc Brake Lubricant (the seal must be dry before applying this lubricant) and locate it in its groove in the cylinder.

Gently work round the seal with the fingers to ensure correct seating.

Ensure that the piston and bore are quite clean; if it is necessary to clean either, use only methylated spirit or Lockheed Brake Fluid.

Coat the piston with Lockheed Disc Brake Lubricant. Open the appropriate bleed screw and offer up the piston to the calliper body. Turn the piston round until the portion which has been machined away from the outer face of the piston is adjacent to the lower end of the calliper (i.e. the end opposite the bleeder screws—see Fig. MM.4) and locate the piston squarely in the mouth of the bore. With the aid of a clamp press the piston fully home. **Great care must be taken to ensure that the piston is not allowed to tilt at any time during this operation.**

Lubricate a new dust seal and dust seal retainer with

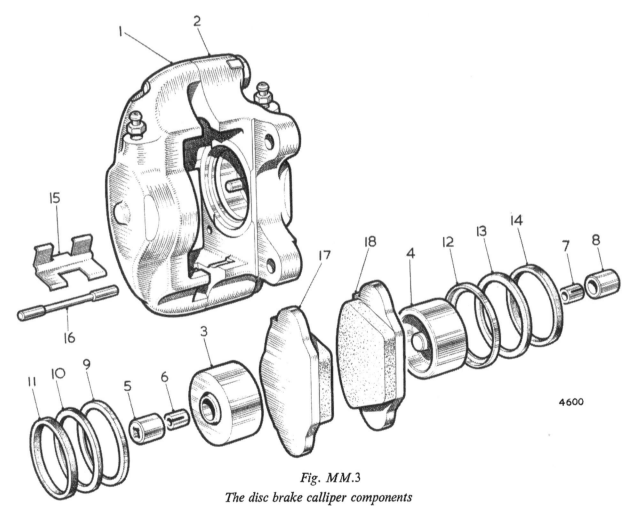

Fig. MM.3

The disc brake calliper components

1. Calliper—rim half.	7. Friction stop.	13. Dust seal.
2. Calliper—mounting half.	8. Sleeve.	14. Fluid seal.
3. Hydraulic piston.	9. Dust seal retainer.	15. Retainer spring.
4. Hydraulic piston.	10. Dust seal.	16. Retainer pin.
5. Sleeve.	11. Fluid seal.	17. Friction pad.
6. Friction stop.	12. Dust seal retainer.	18. Friction pad.

NOTE.—Items 5, 6, 7, and 8 are locked in position during manufacture and cannot be removed for service purposes.

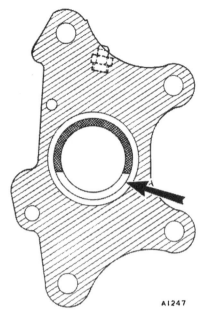

A1247

Fig. MM.4

The cut-away portion of the piston (arrowed) must be located at the lower end of the calliper (i.e. the end opposite the bleeder screws)

Lockheed Disc Brake Lubricant (the seal must be dry before applying the lubricant). Fit the seal in the mouth of the bore, followed by the retainer (with the recessed side outermost). Press the retainer fully home, using a suitable clamp and disc. Earlier types of calliper are fitted with a dust seal which has a 'T'-shaped cross-section; this seal fits inside the retainer. The later seal is interchangeable with the earlier type. Place a suitable clamp on the piston and restore the fluid in the system by gently depressing the foot brake until the fluid flows from the open bleeder screw. Lock the bleeder screw.

Repeat the dismantling and assembling operations on the other piston if necessary. Replace the calliper, fit the friction pads (Section MM.7), and bleed the system (Section MM.1).

NOTE.—Throughout the whole assembly operation it is essential that the clamp fitted to the opposite piston on dismantling is not removed.

Section MM.6

REMOVING THE FRICTION PADS

Jack up the front of the vehicle and remove the appropriate road wheel.

Push in the spring clip retaining the pads with a screwdriver and remove the locating pin (see Fig. MM.5).

The pads are now free to be withdrawn from the brake unit.

Thoroughly clean the surrounding area of the calliper.

Section MM.7

REPLACING THE FRICTION PADS

Where the original friction pads are to be refitted it is only necessary to reverse the instructions given in Section MM.6.

If wear has reduced the pads to the minimum permissible thickness of $\frac{1}{16}$ in. (1·59 mm.) the pads must be renewed. Press the piston assemblies with a suitable clamp to the base of the cylinder bores against the resistance offered by the friction stop and sleeve. The fluid level in the master cylinder must be observed at regular intervals throughout this operation as the displaced fluid returning to the reservoir may overflow.

Check that the portion which has been machined away from the outer face of each piston is correctly positioned (see Fig. MM.4). Insert the new friction pad assemblies (which are interchangeable side for side), replace the spring clip, and fit the locating pin. **Make certain that the clip is centrally located in the recessed portion of the pin.** Pump the brake pedal several times to readjust the brake.

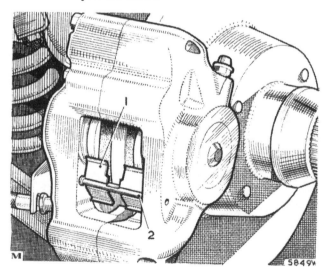

Fig. MM.5

Depress the retainer spring (1) and withdraw the pin (2) to gain access to the friction pads

An improved type of brake pad was introduced on the 'MGA 1600' at Car No. 78106 (wire wheels) and Car No. 78144 (disc wheels). These pads are available in sets only. If the improved pads are fitted to earlier cars it is essential that both right- and left-hand brakes are equipped. The later-type pads are colour-coded red.

Section MM.8

REMOVING A BRAKE DISC

Remove the brake unit as detailed in Section MM.2 and withdraw the hub by the method described in Section KK.1.

The disc is separated from the hub by removing the four securing nuts and washers.

Section MM.9

REPLACING A BRAKE DISC

Assemble the brake disc to the hub by a reversal of the instructions given in Section MM.8 and fit the assembly to the vehicle.

Check the disc for true rotation by clamping a dial indicator to a suitable fixed point on the vehicle with the needle pad bearing on the face of the disc. Run-out must not exceed ·003 in. (·076 mm.), and in the event of the value being exceeded the components should be examined for damage and, if necessary, renewed.

Replace the brake unit as detailed in Section MM.3.

A certain amount of concentric and even scoring of the disc faces is not detrimental to the satisfactory operation of the brakes.

If it is found necessary to regrind the disc faces they can be ground up to a maximum of only ·040 in. (1·016 mm.) off the original thickness of ·380 to ·370 in. (9·65 to 9·4 mm.). This may be ground off equally each side, or more on one side than the other, provided that the total reduction does not exceed the maximum limit of ·040 in. (1·016 mm.). The reground surface must not exceed 63 micro-in.

After grinding, the faces must run true to within a total clock reading of ·003 in. (·076 mm.) and the thickness must be parallel to within ·001 in. (·0254 mm.) clock reading.

Section MM.10

MODIFIED CALLIPER UNITS

A modified disc brake calliper dust seal and retainer (Fig. MM.3, items 9, 10, 12, 13) was introduced at Chassis No. 103261 (disc wheels) and Chassis No. 103834 (wire wheels).

The new seal is 'U'-shaped in cross-section and is retained in the counterbore of the calliper by a retainer having an 'L'-shaped cross-section. The seal is carried within the retainer flange and abuts the face of the calliper counterbore.

Section MM.11

DISC BRAKE DUST COVERS

Disc brake covers which reduce inner pad wear have been fitted to cars from Chassis No. 102589 (disc wheels) and Chassis No. 102929 (wire wheels). The covers may be fitted in sets to earlier cars.

To fit the covers proceed as follows:

(1) Remove the front hub assembly as detailed in Section KK.1.

(2) Remove the adaptor plate from the steering knuckle and discard the adaptor plate bolts, washers, and nuts.

(3) Refit the adaptor plate to the steering knuckle, using bolts (Part No. HBZ0611). Assemble the distance washers (Part No. BTB386) and cover (Part No. BTB384 [R.H.] or BTB385 [L.H.]) to the bolts and secure the bolts with spring washers (Part No. LWZ206) and nuts (Part No. FNZ506). Tighten the nuts to between 35 and 40 lb. ft. (4·83 and 5·52 kg. m.).

(4) Reassemble the front hub as detailed in Section KK.2 and ensure that the calliper securing bolts are tightened to between 45 and 50 lb. ft. (6·2 and 6·9 kg. m.).

SECTION MMM

THE BRAKING SYSTEM
(MGA 1600 fitted with Dunlop disc brakes)

THE BRAKING SYSTEM

MMM

GENERAL DESCRIPTION

The braking system consists of four Dunlop calliper-type disc brakes hydraulically controlled by means of a foot-operated master cylinder.

Steel pipe lines, unions, and flexible hoses convey the hydraulic pressure from the master cylinder to each wheel cylinder.

The cable-actuated hand brake mechanism is entirely separate in operation from the hydraulic system and operates on the rear wheels only. Each brake consists of two carriers to which friction pads are riveted. The carriers are mounted on the top of the rear callipers, one each side of the disc, by means of hinge bolts.

BRAKE UNITS

Each wheel brake unit comprises a hub-mounted disc rotating with the wheel and a braking unit rigidly attached to the axle at the rear and to the swivel pin at the front. The brake unit consists of a calliper which straddles the disc and houses a pair of rectangular friction pad assemblies. Cylinder blocks bolted to the outer faces of each calliper accommodate piston assemblies keyed to the friction pad and securing plate assemblies. A spigot formed on the outer face of each piston locates in the bore of a backing plate with an integral boss grooved to accommodate the collar of a flexible rubber dust seal. When the piston is assembled to the cylinder block the seal engages a lip on the block face and so protects the assembly from intrusion of moisture and foreign matter. The central blind bore of the piston inner face accommodates the end of a retractor pin and its friction bush. A piston seal is located between the piston inner face and a plate secured to the piston by peen-locked screws. The piston assembly when pressed into the cylinder bore locates on the retractor pin assembly, which is peened into the base of the cylinder bore. This assembly comprises a retractor stop bush, two spring washers, a dished cap, and the retractor pin; it functions as a return spring and maintains the 'brake-off' working clearance of approximately ·008 to ·010 in. (·203 to ·254 mm.) between the pads and the disc throughout the life of the pads.

MASTER CYLINDER

The components of the master cylinder are housed within the bore of a cylinder body with an integral reservoir. The reservoir is fitted with a detachable cover which incorporates the filler orifice and is secured by means of six bolts and spring washers. A fluidtight joint is maintained by a cork gasket between the cover and reservoir faces. The enclosed end of the cylinder is bored to provide communication between the reservoir and the cylinder; a housing for an outlet connection is provided by an internally threaded boss integral with the cylinder. Formed around the opposite end of the cylinder is a flange with two holes for the master cylinder attachment bolts. In the unloaded condition a spring-loaded piston carrying a rubber 'O' ring in a groove is held against the under side of a dished washer retained by a circlip at the head of the cylinder. A hemispherically ended push-rod seats in a similarly formed recess at the head of the piston. The head of the master cylinder is shrouded by a rubber dust excluder, the lip of which seats in a groove in the cylinder body.

A cylindrical spring support is fitted around the inner end of the piston and a small drilling in the end of the support is engaged by the stem of a valve. The larger-diameter head of the valve is located in a central blind bore in the piston. The valve passes through the bore of a vented spring support and protrudes into the fluid passage which communicates with the reservoir. Interposed between the spring support and an integral flange formed on the valve is a small coiled spring. A rubber seal is fitted between the end of the cylinder body and the under side of the valve flange. This assembly forms a recuperation valve which controls fluid flow to and from the reservoir.

When the foot pedal is in the 'off' position the master cylinder is fully extended and the valve is held clear of the base of the cylinder by the action of the main spring. In this condition the master cylinder is in fluid communication with the reservoir, thus permitting recuperation of any fluid loss sustained, particularly during the priming and bleeding operation of the brake system.

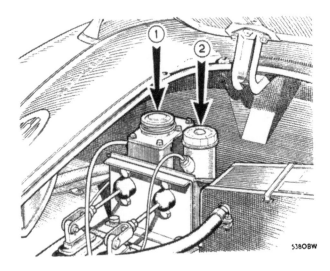

Fig. MMM.1

Periodically examine the quantity of fluid in the brake master cylinder reservoir (arrow 1). The clutch master cylinder reservoir is indicated by arrow 2

M.G. 'MGA'. Issue 1. 29723

MMM.2

When a load is applied to the foot pedal the piston moves down the cylinder against the compression of the main spring. Immediately this movement is in excess of the valve clearance the valve closes under the influence of its spring and isolates the reservoir. Further loading of the pedal results in the discharge of fluid under pressure from the outlet connection via the pipe lines to the brake system.

Removal of the load from the pedal reverses the sequence; the action of the main spring returns the master cylinder to the extended position and restores the open condition between the cylinder and reservoir previously described.

MAINTENANCE

Periodically examine the quantity of brake fluid in the master cylinder. It should never be less than half-full nor closer than ½ in. (13 mm.) to the bottom of the filler neck. The necessity for frequent topping up is an indication of overfilling or of a leak in the system which should be traced and rectified at once.

If the travel of the hand brake lever becomes excessive the mechanism should be adjusted as detailed in Section MMM.14.

The friction pads should be checked for wear every 6,000 miles (10000 km.) by visual observation and measurement. When wear has reduced the pads to the minimum permissible thickness of ·25 in. (6·35 mm.) the pads must be renewed.

Every 3,000 miles (5000 km.) apply three or four strokes of the grease gun filled with grease to Ref. C (page P.2) to the nipple provided on the brake cable.

Section MMM.1

ADJUSTING THE BRAKE PEDAL

The correct amount of free movement between the master cylinder push-rod and piston is set during the erection of the vehicle and should not require adjustment during normal service.

In the event of the adjustment having been disturbed a check should be made to ensure that there is no preloading of the master cylinder piston when the brake pedal is in the fully 'off' position. In this position the piston should be held against the dished washer at the head of the master cylinder unit by the pressure of the piston return spring, thus forming a return stop, and a free axial movement of approximately ·015 to ·020 in. (·381 to ·508 mm.) should be felt at the master cylinder push-rod. If necessary, the effective length of the push-rod should be reset to this figure.

Section MMM.2

PRIMING AND BLEEDING THE BRAKE SYSTEM
(Expelling Air)

The following procedure should be adopted either for initial priming of the system or to bleed in service if air has been permitted to enter the system. Air may enter the system if pipe connections become loose or if the level of fluid in the reservoir is allowed to fall below the recommended level. During the bleeding operation it is important that the reservoir be kept at least half-full to avoid drawing air into the system.

(1) Check that all connections are tightened and all bleed screws closed.
(2) Fill the reservoir with brake fluid. The use of Wakefield Crimson Brake Fluid is recommended, but if this is not available an alternative fluid conforming to Specification S.A.E. 70.R1 should be used.
(3) Attach the bleeder tube to the bleed screw on the near-side rear brake and immerse the open end of the tube in a small quantity of brake fluid contained in a clean glass jar. Slacken the bleed screw and operate the brake pedal slowly backwards and forwards through its full stroke until fluid pumped into the jar is reasonably free from air bubbles. Keep the pedal depressed and close the bleed screw. Release the pedal.
(4) Repeat for each brake in turn.
(5) Repeat the complete bleeding sequence until the brake fluid pumped into the jar is completely free from air bubbles.
(6) Lock all bleed screws and top up the fluid level in the reservoir.
(7) Apply a normal working load on the brake pedal for a period of two or three minutes and examine the entire system for leaks.

NOTE.—Clean fluid bled from the system must be allowed to stand until it is clear of air bubbles before it is used again. Dirty fluid should be discarded.

Section MMM.3

REMOVING THE MASTER CYLINDER

Remove the split pin and washer and withdraw the clevis pin from the push-rod yoke. Remove the push-rod.

Remove the two bolts, nuts, and washers securing the front end of the master cylinder to the mounting plate and disconnect the brake pipe at the rear of the cylinder. On right-hand-drive vehicles this operation will be eased if the brake pipe securing clip on the bulkhead is released first.

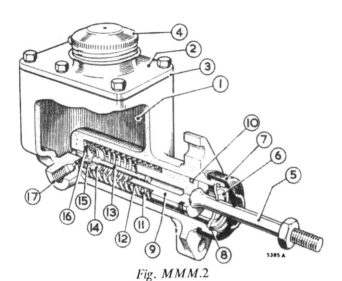

Fig. MMM.2

The brake master cylinder components

1.	Reservoir.	10. 'O' ring.
2.	Cover.	11. Return spring.
3.	Cork gasket.	12. Spring support.
4.	Filler cap.	13. Valve.
5.	Push-rod.	14. Spring support.
6.	Dished washer.	15. Valve spring.
7.	Dust excluder.	16. Seal.
8.	Circlip.	17. Outlet connection.
9.	Piston.	

Section MMM.4

DISMANTLING THE MASTER CYLINDER

Remove the master cylinder filler cap and drain the brake fluid from the unit.

Ease the dust excluder clear of the head of the master cylinder. Remove the retaining circlip with a suitable pair of pliers and withdraw the push-rod complete with dished washer. Draw out the piston and remove the rubber 'O' ring. The valve assembly complete with springs and supports can then be extracted and the valve sealing ring removed from the seal bush.

Section MMM.5

ASSEMBLING THE MASTER CYLINDER

Clean all parts thoroughly, using only the recommended brake fluid for all rubber components. All traces of petrol (gasoline), paraffin (kerosene), or trichlorethylene used for cleaning the metal parts must be removed before assembly.

Examine all the rubber parts for damage or distortion. It is usually advisable to renew the rubbers when rebuilding the cylinder. Dip all the internal parts in brake fluid and assemble them wet. Fit the valve seal around the seal bush and the 'O' ring in the groove on the piston.

Place the seal bush in position on the valve stem and insert the piston into the spring support, ensuring that the head of the valve engages the piston bore. Slide the complete assembly into the cylinder body, taking particular care not to damage or twist the 'O' ring.

Position the push-rod and depress the piston sufficiently to allow the dished washer to seat on the shoulder at the head of the cylinder. Fit the circlip and check that it fully engages in the groove.

Fill the dust excluder with Wakefield No. 3 Rubber Grease and reseat the excluder around the head of the master cylinder.

Section MMM.6

REPLACING THE MASTER CYLINDER

The replacement procedure is the reverse of the removal instructions given in Section MMM.3.

After replacement, bleed the brake system as detailed in Section MMM.2. Finally, check for leaks with the brakes fully applied.

Section MMM.7

REMOVING A BRAKE UNIT

Front

Unscrew the brake pipe union nut below its support bracket and disconnect and blank off the pipe. Remove the two nuts securing the brake hose support bracket and remove the bracket.

Fig. MMM.3
A front disc brake calliper

M.G. 'MGA'. Issue 2. 34279

Unscrew the two calliper retaining bolts and remove the calliper assembly complete with cylinders.

Take care not to misplace the shims which are fitted behind the mounting lugs on the calliper body. **The shims must be retained for reassembly and replaced in their original positions.**

Rear

Unscrew the fluid supply pipe union (below the inner cylinder block) and disconnect and blank off the pipe. Remove the split pin and clevis pin from the hand brake cable yoke to disconnect the cable from the calliper lever.

Tap back the tab washers and unscrew the two set screws securing the calliper to the mounting flange on the axle. The calliper complete with parking mechanism may now be removed from the vehicle.

The shims taken from behind the calliper body mounting lugs must be retained and **replaced in their original positions on reassembly.**

Section MMM.8

DISMANTLING A BRAKE UNIT

The brake must be thoroughly cleaned before proceeding with the dismantling. It is recommended that a new dust seal should be fitted whenever the unit is dismantled.

Withdraw the brake pads as described in Section MMM.11. Disconnect and blank off the supply pipe (if the unit is being dismantled on the vehicle) and remove the bridge pipe.

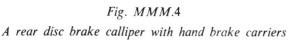

Fig. MMM.4

A rear disc brake calliper with hand brake carriers

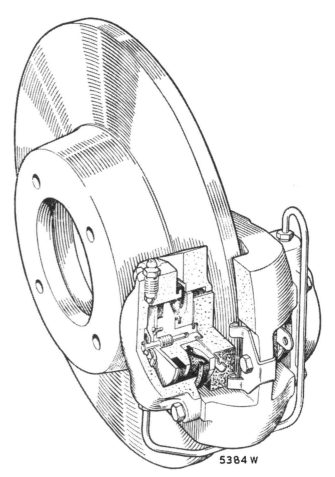

Fig. MMM.5

Sectioned view of a brake calliper

Remove the bolts securing the cylinder blocks to the calliper and withdraw the cylinder blocks.

Disengage the dust seal from the lip on the cylinder block face, connect the cylinder to a source of fluid supply, and apply pressure to eject the piston assembly. Remove the screws securing the plate to the piston, lift off the plate and piston seal, and withdraw the retractor bush from within the piston bore. Carefully cut away and discard the dust seal.

Support the backing plate on a bush of sufficient bore just to accommodate the piston; with a suitable tubular distance piece placed against the end of the piston spigot around the shouldered head press out the piston from the backing plate. Care must be taken during the operation to avoid damaging the piston.

Section MMM.9

ASSEMBLING A BRAKE UNIT

Clean all components thoroughly, using only the recommended brake fluid for all rubber parts.

Engage the collar of a new dust seal with the lip on

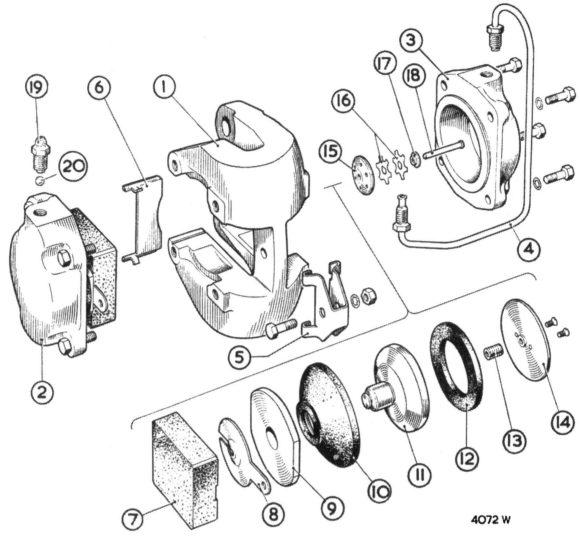

Fig. MMM.6

The disc brake calliper components

1. Calliper.	6. Support plate.	11. Piston.	16. Spring washer.
2. Cylinder block.	7. Friction pad.	12. Piston seal.	17. Retractor stop bush.
3. Cylinder block.	8. Securing plate.	13. Retractor bush.	18. Retractor pin.
4. Bridge pipe.	9. Backing plate.	14. Plate.	19. Bleed screw.
5. Keep plate.	10. Dust seal.	15. Cap.	20. Bleed screw ball.

the backing plate on the piston spigot, and with the piston suitably supported press the backing plate fully home.

Insert the retractor bush into the bore of the piston. Lightly lubricate the piston seal with brake fluid (if there is any doubt about the condition of this component it should be renewed) and fit it to the piston face. Attach and secure the plate with the screws, and peen-lock the screws.

Check that the piston and the cylinder bore are thoroughly clean and show no signs of damage. Locate the piston assembly on the end of the retractor pin, and

with the aid of a hand press slowly apply an even pressure to the backing plate and press the assembly into the cylinder bore. Ensure that the piston assembly is in correct alignment in relation to the cylinder bore and that the piston seal does not become twisted or trapped as it enters the cylinder bore. Engage the lip of the dust seal with the lip on the cylinder block face.

Reassemble the cylinder blocks to the calliper and fit the bridge pipe, ensuring that it is correctly positioned (with the near-vertical part of the pipe farthest from the wheel). If the complete brake unit has been removed it should be replaced as detailed in Section M.10.

Remove the blank, replace the supply pipe, and fit the friction pads as described in Section MMM.12.

Finally, bleed the system (Section MMM.2) and check for leaks with the brakes fully applied.

Section MMM.10

REPLACING A BRAKE UNIT

The replacement procedure is a reversal of the instructions given in Section MMM.7, with the exception of the following details. Replace the brake pads as detailed in Section MMM.12.

Check the gap between each side of the calliper and the disc. The difference should not exceed ·010 in. (·254 mm.) and the shims may be altered to obtain this figure. Bleed the system as detailed in Section MMM.2 and, finally, check for leaks with the brakes fully applied.

Section MMM.11

REMOVING THE FRICTION PADS

Remove the nut, washer, and bolt securing the keep plate and withdraw the plate. Withdraw the pad assemblies with a suitable hooked implement engaged in the hole in the lug of the securing plate.

Thoroughly clean the backing plate, dust seal, and the surrounding area of the calliper.

Section MMM.12

REPLACING THE FRICTION PADS

Where the original friction pads are to be refitted it is only necessary to reverse the instructions given in Section MMM.11.

If wear has reduced the pads to the minimum permissible thickness of ·025 in. (·635 mm.) the pads must be renewed. Press the piston assemblies with a suitable lever to the base of the cylinder bores against the resistance offered by the retractor pin and bush. Insert the new friction pad assemblies, replace the keep plate, and secure it with the bolt, washer, and nut.

Section MMM.13

RELINING THE HAND BRAKE

Unscrew and remove the adjuster bolt and locknut and swing the pad carriers away from the disc. Extract the split pin and withdraw the lever pivot pin (see Fig. MMM.7).

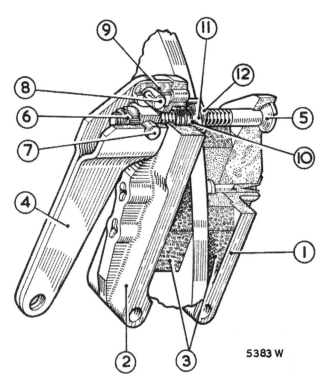

Fig. MMM.7

The hand brake carrier components

1.	Pad carrier (outer).	7.	Trunnion.
2.	Pad carrier (inner).	8.	Pivot pin.
3.	Pads.	9.	Pivot seat.
4.	Lever.	10.	Spring.
5.	Adjuster bolt.	11.	Spring retaining nut.
6.	Locknut.	12.	Spring plate.

Remove the bifurcated rivets from both carriers and prise off the worn linings. Place the new linings in position and secure them with new bifurcated rivets.

Place the lever in the position indicated in Fig. MMM.8. Hold the locknut firmly against the outer face of the trunnion and screw in the adjuster bolt until three or four threads engage in the locknut. Align the holes in the lever and pivot seat, fit the pivot pin, and lock it with a split pin.

Reset the clearance as detailed in Section MMM.14.

Section MMM.14

ADJUSTING THE HAND BRAKE

Adjustment to compensate for pad wear must be made at the hand brake units and not on the relay lever adjuster. The adjustments should be made in the following manner when the travel of the hand brake lever becomes excessive.

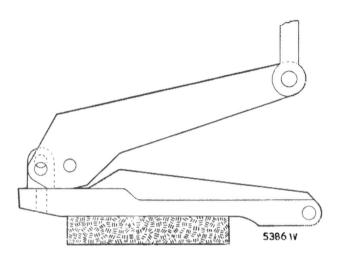

Fig. MMM.8

When replacing a hand brake friction pad place the lever against the inner carrier in the position shown. Hold the locknut against the trunnion and screw in the adjuster bolt three or four threads

Raise the rear of the car—making certain that the front wheels are suitably blocked to prevent the car running forward—and remove both rear wheels. The hand brake lever should be in the fully off position whilst the adjustments are made.

Slacken the brass adjuster nut fitted to the relay lever (located beside the front universal joint on the propeller shaft) so that the operating cable hangs loosely.

Tighten each adjuster bolt until the pads 'nip' the brake disc. Screw up the brass adjuster nut on the relay lever until all slackness is taken up, ensuring that there is no preload on the linkage.

Set the clearance between the pads and the brake disc by unscrewing each adjuster bolt approximately one-third of a turn. Make sure that the discs rotate freely.

Section MMM.15

REMOVING THE HAND BRAKE CABLE

Unscrew and remove the adjuster nut; withdraw the end of the cable from the lower end of the lever and remove the spring.

Disconnect the clips securing the cable assembly to the body.

Remove the clevis pins from the levers on the wheel brake units.

Unscrew the two nuts on the axle balance lever; separate the two halves of the lever and remove the cable and trunnion.

MMM.8

Section MMM.16

REMOVING A BRAKE DISC

Remove the brake unit as detailed in Section MMM.7 and withdraw the hub by the method described in Section KKK.1 (front) or Section HHH.2 (rear).

The rear disc is separated from the hub by removing the four securing nuts and washers. The front disc is secured to the hub by four set bolts with spring washers, and after removing these the two components may be separated.

Section MMM.17

REPLACING A BRAKE DISC

Assemble the brake disc to the hub by a reversal of the instructions given in Section MMM.16 and fit the assembly to the vehicle.

Check the disc for true rotation by clamping a dial indicator to a suitable fixed point on the vehicle with the needle pad bearing on the face of the disc. Run-out must not exceed ·006 in. (·152 mm.), and in the event of the value being exceeded the components should be examined for damage and, if necessary, renewed.

Replace the brake unit as detailed in Section MMM.10.

Section MMM.18

FLEXIBLE HOSES

The flexible pipes must show no signs of deterioration or damage and the bores should be cleared with a jet of

Fig. MMM.9

The arrow indicates the brass adjuster nut fitted to the cable relay lever

M.G. 'MGA'. Issue 1. 29723

compressed air. No attempt should be made to clear a blockage by probing as this may result in damage to the lining and serious restriction to fluid flow. Partially or totally blocked flexible pipes should always be renewed. When removing or refitting a flexible pipe the end sleeve hexagon should be held with the appropriate spanner to prevent the pipe from twisting. A twisted pipe will prove detrimental to efficient brake operation.

Removing a rear hose

The front end of the rear flexible hose is held in a bracket mounted on the right-hand battery box. Unscrew the metal pipe union nut and release the pipe. Hold the hexagon on the flexible hose with a spanner and remove the large retaining nut and its shakeproof washer from the under side of the support bracket. The pipe may now be unscrewed at its rear end from the three-way piece on the rear axle.

Removing a front hose

Unscrew the metal pipe union nuts at each end of the front hose. Hold the hexagon on the flexible hose and remove the nut and shakeproof washer on the under side of the mounting brackets.

Section MMM.19

REFACING BRAKE DISCS

Scoring of the brake discs is not detrimental, provided that the scoring is concentric, even, and is not excessive.

*Fig. MMM.*10

The arrow indicates the hand brake carrier pad adjusting bolt

The disc faces may be ground provided the following conditions are observed:

(1) The final thickness of the disc must not be less than between ·340 and ·330 in. (8·64 and 8·38 mm.).

(2) The disc faces must run true to within ·003 in. (·076 mm.).

(3) The faces must be parallel to within ·001 in. (·025 mm.).

Notes

SECTION N

ELECTRICAL EQUIPMENT

General description.

Section No. N.1 Battery maintenance.

Section No. N.2 Dynamo.

Section No. N.3 Removing and replacing the dynamo.

Section No. N.4 Dismantling the dynamo.

Section No. N.5 Servicing the dynamo.

Section No. N.6 Starter.

Section No. N.7 Removing and replacing the starter.

Section No. N.8 Servicing the starter.

Section No. N.9 Control box.

Section No. N.10 Fuses.

Section No. N.11 Electric horn.

Section No. N.12 Flashing direction indicators.

Section No. N.13 Windshield wiper.

Sections Nos. N.14 to N.24 Lamps.

Section No. N.25 Replacement bulbs.

Section No. N.26 Fitting a fog lamp.

Section No. N.27 Location and remedy of faults.

Section No. N.28 Windowless yoke dynamo.

Section No. N.29 Modified control box.

Section No. N.30 Headlamp beam setting.

Section No. N.31 Dry-charged batteries.

Section No. N.32 Modified European light unit

M.G. 'MGA'. Issue 6. 48539   N.1

207

GENERAL DESCRIPTION

The 12-volt electrical equipment incorporates compensated voltage control for the charging circuit. The positive earth system of wiring is employed.

The two 6-volt batteries, mounted to the rear of the seats, are accessible for examination and maintenance attention.

The dynamo is mounted on the right of the cylinder block and driven by endless belt from the engine crankshaft. A rotatable mounting enables the belt tension to be adjusted.

The control box is sealed and should not normally need attention. The fuses and spare fuses are carried in external holders.

The starter motor is mounted on the flywheel housing on the right-hand side of the engine unit and operates on the flywheel through the usual sliding pinion device.

The headlamps employ the double-filament dipping system. Both lamps dip according to the regulations existing in the country concerned.

Section N.1

BATTERY MAINTENANCE

In order to keep the batteries in good condition, a periodical inspection must be made.

Unscrew the five quick-release fasteners securing the panel immediately behind the seats and lift the panel away to obtain access to the batteries.

Topping up

Weekly, remove the filler plug from each cell and examine the level of the electrolyte. Add distilled water to bring the level of the electrolyte just above the separators.

NOTE.—Do not use tap-water and do not use a naked light when examining the condition of the cells. Wipe away all dirt and moisture from the top of the battery.

Testing the condition of the battery

Every 6,000 miles (10000 km.) examine the condition of the batteries by taking hydrometer readings. The hydrometer contains a graduated float on which is indicated the specific gravity of the acid in the cell from which the same is taken.

The specific gravity readings and their indications are as follows:

Climates below 27° C. (80° F.)

1·270 to 1·290	Cell fully charged.
1·190 to 1·210	Cell about half-discharged.
1·110 to 1·130	Cell fully discharged.

Climates frequently above 27° C. (80° F.)

1·210 to 1·230	Cell fully charged.
1·130 to 1·150	Cell about half-discharged.
1·050 to 1·070	Cell fully discharged.

These figures are given assuming an electrolyte temperature of 16° C. (60° F.). If the temperature of the electrolyte exceeds this, ·002 must be added to hydrometer readings for each 3° C. (5° F.) rise to give the true specific gravity. Similarly ·002 must be subtracted from hydrometer readings for every 3° C. (5° F.) below 16° C. (60° F.).

The readings of all the cells should be approximately the same. If one cell gives a reading very different from the rest it may be that the electrolyte has been spilled or has leaked from the cell or there may be an internal fault. Should a battery be in a low state of charge, it should be recharged by taking the car for a long daytime run or by charging from an external source of D.C. supply at a current rate of 5 amperes until the cells are gassing freely.

After examining the battery, check the vent plugs, making sure that the air passages are clear.

Storage

If a battery is to be out of use for any length of time, it should first be fully charged and then given a freshening charge about every fortnight.

A battery must never remain in a discharged condition, as the plates will become sulphated.

Initial filling and charging

When a new battery has been supplied dry it is necessary to fill the cells with electrolyte of the correct specific gravity.

All batteries, including those having type suffix letter 'Z' (e.g. SGZ, etc.) and those having no additional suffix letter (e.g. SG, BT, etc.), are assembled with dry separators. The specific gravity of the filling-in solution depends upon the climate in which the battery is to be used (i.e. 1·260 for climates below 27° C. [80° F.] and 1·210 for climates frequently above 27° C. [80° F.]). For more details of the requirements of 'dry-charged' batteries see Section N.31.

The electrolyte is prepared by mixing distilled water and concentrated sulphuric acid 1·835 S.G. The mixing must be carried out in a lead-lined tank or a suitable glass or earthenware vessel. Steel or iron containers must **not** be used. The acid must be added slowly to the water, while the mixture is stirred with a glass rod. **Never add the water to the acid,** as the severity of the resulting chemical reaction may have dangerous consequences.

Heat is produced by the mixture of acid and water, and it should, therefore, be allowed to cool before it is poured into the battery, otherwise the plates, separators and moulded container may be damaged.

The temperature of the filling-in acid, battery and charging room should be above 0° C. (32° F.).

To produce electrolyte of the correct specific gravity:

To obtain specific gravity (corrected to 60° F.[16°C.])	Add 1 part by volume of 1·835 S.G. acid to distilled water by volume as below
1·260	3 parts
1·210	4 parts

Carefully break the seals in the filling holes and half-fill each cell in the battery with dilute sulphuric acid solution of the appropriate specific gravity (according to temperature). The quantity of electrolyte required to half-fill a two-volt cell is ½ pint (·28 litre). Allow to stand for at least six hours, then complete the filling of the cells by the addition of more diluted acid of the same specific gravity as before until the level reaches the bottom of the filling holes, and allow the battery to stand for at least another two hours before commencing the first charge.

Charge at a constant current of 3·5 amps. until the voltage and temperature-corrected specific gravity readings show no increase over five successive hourly readings. This period is dependent upon the length of time the battery has been stored since manufacture, and will be from 40 to 80 hours, but usually not more than 60.

Throughout the charge the acid must be kept level with the tops of the separators in each cell by the addition of electrolyte of the same specific gravity as the original filling-in acid.

If, during charge, the temperature of the acid in any cell of the battery reaches the maximum permissible temperature of 38° C. (100° F.) in a climate below 80° F. (27° C.) or 49° C. (120° F.) in a climate frequently above 80° F. (27° C.), the charge must be interrupted and the battery temperature allowed to fall at least 5·5°C. (10° F.) before charging is resumed.

At the end of the first charge, i.e. when specific gravity and voltage measurements remain constant, carefully check the specific gravity in each cell to ensure that it lies within the limits specified. If any cell requires adjustment, the electrolyte above the plates must be siphoned off and replaced either with acid of the strength used for the original filling in, or distilled water, according to whether the specific gravity is too low or too high respectively. After such adjustment, the gassing charge should be continued for one or two hours to ensure adequate mixing of the electrolyte. Re-check, if necessary, repeating the procedure until the desired result is obtained.

Section N.2

DYNAMO

To test on vehicle when dynamo is not charging

(1) Make sure that belt slip is not the cause of the trouble. It should be possible to deflect the belt approximately ½ in. (13 mm.) at the centre of its longest run between two pulleys with moderate hand pressure. If the belt is too slack, loosen the two dynamo suspension bolts and then the bolt of the slotted adjustment link. A gentle pull on the dynamo outwards will enable the correct tension

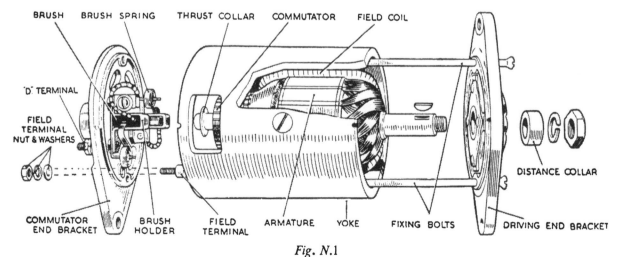

Fig. N.1

An exploded view of the dynamo

to be applied to the belt and all three bolts should then be tightened firmly.

(2) Check that the dynamo and control box are connected correctly. The dynamo terminal 'D' should be connected to the control box terminal 'D' and the dynamo terminal 'F' connected to the control box terminal 'F'.

(3) After switching off all lights and accessories, disconnect the cables from the dynamo terminals marked 'D' and 'F' respectively.

(4) Connect the two terminals with a length of wire.

(5) Start the engine and set to run at idling speed.

(6) Clip the negative lead of a moving-coil-type voltmeter, calibrated 0–20 volts, to one dynamo terminal and the other lead to a good earthing point on the dynamo yoke.

(7) Gradually increase the engine speed, when the voltmeter reading should rise rapidly and without fluctuation. Do not allow the voltmeter reading to reach 20 volts. Do not race the engine in an attempt to increase the voltage. It is sufficient to run the dynamo up to a speed of 1,000 r.p.m.

If there is no reading, check the brush gear.

If the reading is low (approximately 1 volt), the field winding may be faulty.

If the reading is approximately 5 volts, the armature winding may be faulty.

(8) Remove the dynamo cover band and examine the brushes and commutator. Hold back each of the brush springs and move the brush by pulling gently on its flexible connector. If the movement is sluggish, remove the brush from its holder and ease the sides by lightly polishing on a smooth file. Always replace brushes in their original positions. If the brushes are worn so that they no longer bear on the commutator, or if the brush flexible lead has become exposed on the running face, new brushes must be fitted. If the commutator is blackened or dirty, clean it by holding a petrol-moistened cloth against it while the engine is turned slowly by hand-cranking. Re-test the dynamo; if there is still no reading on the voltmeter there is an internal fault and the complete unit should be renewed.

If the dynamo is in good order, leave the temporary link in position between the terminals and restore the original connections, taking care to connect the dynamo terminal 'D' to the control box terminal 'D' and the dynamo terminal 'F' to the control box terminal 'F'. Remove the lead from the 'D' terminal on the control box and connect the voltmeter between this cable and a good earthing point on the vehicle. Run the engine

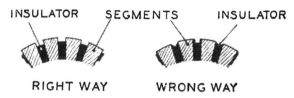

Fig. N.2

The correct method of undercutting the dynamo commutator

as before. The reading should be the same as that measured directly at the dynamo. No reading on the voltmeter indicates a break in the cable to the dynamo. Carry out the same procedure for the 'F' terminal, connecting the voltmeter between cable and earth. Finally remove the link from the dynamo. If the reading is correct test the control box (Section N.9).

Section N.3

REMOVING AND REPLACING THE DYNAMO

To remove the dynamo, disconnect the dynamo leads from the dynamo terminals.

Slacken all four attachment bolts and pivot the dynamo towards the cylinder block to enable the fan belt to be removed from the dynamo pulley. The dynamo can then be removed by completely removing the two upper and one lower attachment bolts.

Replacement of the dynamo is an exact reversal of this procedure.

Section N.4

DISMANTLING THE DYNAMO

Take off the dynamo pulley.

Remove the cover band, hold back the brush springs and remove the brushes from their holders.

Unscrew the locknuts from the through-bolts at the commutator end. Withdraw the two through-bolts from the driving end.

Remove the nut, spring washer and flat washer from the smaller terminal (i.e. field terminal) on the commutator end bracket and remove the bracket from the dynamo yoke.

The driving end bracket, together with the armature, can now be lifted out of the yoke.

The driving end bracket which, on removal from the yoke, has withdrawn with it the armature and armature shaft ball bearing, need not be separated from the shaft unless the bearing is suspected and requires examination, in which event the armature should be removed from the end bracket by means of a hand press.

Section N.5

SERVICING THE DYNAMO

Brushes

Test if the brushes are sticking. Clean them with petrol and, if necessary, ease the sides by lightly polishing with a smooth file. Replace the brushes in their original positions.

Test the brush spring tension with a spring scale if available. The correct tension is 20–5 oz. (567–709 gm.). Fit a new spring if the tension is low.

If the brushes are worn so that the flexible lead is exposed on the running face, new brushes *must* be fitted. Brushes are pre-formed so that bedding to the commutator is unnecessary.

Commutator

A commutator in good condition will be smooth and free from pits or burned spots. Clean the commutator with a petrol-moistened cloth. If this is ineffective, carefully polish with a strip of fine glass-paper while rotating the armature. To remedy a badly worn commutator, mount the armature (with or without the drive end bracket) in a lathe, rotate at high speed and take a light cut with a very sharp tool. Do not remove more metal than is necessary. Polish the commutator with very fine glass-paper. Undercut the mica insulation between the segments to a depth of $\frac{1}{32}$ in. (·8 mm.) with a hacksaw blade ground down to the thickness of the mica.

Field coils

Test the field coils, without removing them from the dynamo yoke, by means of an ohmmeter. The reading on the ohmmeter should be between 6·0 and 6·3 ohms. If this is not available, connect a 12-volt D.C. supply with an ammeter in series between the field terminal and the dynamo yoke. The ammeter reading should be approximately 2 amps. If no reading is indicated the field coils are open-circuited and must be renewed. To test for earthed field coils, unsolder the end of the field winding from the earth terminal on the dynamo yoke and, with a test lamp connected from supply mains, test across the field terminal and earth. If the lamp lights, the field coils are earthed and must be renewed.

When fitting field coils, carry out the procedure outlined below, using an expander and wheel-operated screwdriver:

(a) Remove the insulation piece which is provided to prevent the junction of the field coils from contacting the yoke.

(b) Mark the yoke and pole-shoes in order that they can be refitted in their original positions.

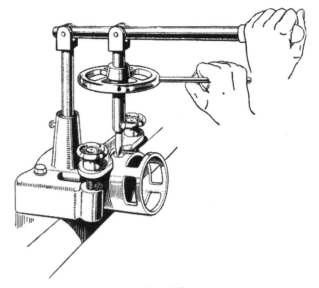

Fig. N.3

Showing the use of a wheel-operated screwdriver to remove the pole-shoe attachment screws

(c) Unscrew the two pole-shoe retaining screws by means of the wheel-operated screwdriver.

(d) Draw the pole-shoes and coils out of the dynamo yoke and lift off the coils.

(e) Fit the new field coils over the pole-shoes and place them in position inside the yoke. Take care to ensure that the taping of the field coils is not trapped between the pole-shoes and the yoke.

(f) Locate the pole-shoes and field coils by lightly tightening the fixing screw.

(g) Insert the pole-shoe expander, open it to the fullest extent and tighten the screws.

(h) Finally tighten the screws by means of the wheel-operated screwdriver and lock them by caulking.

(j) Replace the insulation piece between the field coil connections and the yoke.

Armature

The testing of the armature winding requires the use of a voltage drop test and growler. If these are not available, the armature should be checked by substitution. No attempt should be made to machine the armature core or to true a distorted armature shaft.

Bearings

Bearings which are worn to such an extent that they will allow side movement of the armature shaft must be replaced by new ones.

To fit a new bearing at the commutator end of the dynamo proceed as follows:

(a) Press the bearing bush out of the commutator end bracket.

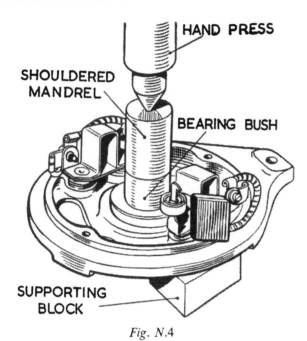

Fig. N.4

The method of pressing out the commutator end bracket bush is shown in this illustration

(b) Press the new bearing bush into the end bracket, using a shouldered mandrel of the same diameter as the shaft which is to fit in the bearing.

Before fitting the new bearing bush allow it to stand completely immersed in thin engine oil for 24 hours, to fill the pores of the bush with lubricant.

The ball bearing at the driving end is renewed as follows:

(a) Knock out the rivets which secure the bearing retaining plate to the end bracket and remove the plate.

(b) Press the bearing out of the end bracket and remove the corrugated washer, felt washer and oil retaining washer.

(c) Before fitting the replacement bearing see that it is clean and pack it with a high-melting-point grease.

(d) Place the oil retaining washer, felt washer and corrugated washer in the bearing housing in the end bracket.

(e) Locate the bearing in the housing and press it home by means of a hand press.

(f) Fit the bearing retaining plate. Insert the new rivets from the inside of the end bracket and open the rivets by means of a punch to secure the plate rigidly in position.

Reassembly

The reassembly of the dynamo is a reversal of the operations described in Section N.4

If the end bracket has been removed from the armature in dismantling, press the bearing end bracket onto the armature shaft, taking care to avoid damaging the end plate and armature winding.

Add a few drops of oil through the hole in the armature end cover.

Section N.6

THE STARTER

To test on vehicle

Switch on the lamps and operate the starter control. If the lights go dim, but the starter is not heard to operate, an indication is given that current is flowing through the starter windings but that the starter pinion is meshed permanently with the geared ring on the flywheel. This was probably caused by the starter being operated while the engine was still running. In this case the starter must be removed from the engine for examination.

Should the lamps retain their full brilliance when the starter switch is operated, check that the switch is functioning. If the switch is in order, examine the connections at the battery, starter switch and starter, and also check the wiring between these units. Continued failure of the starter to operate indicates an internal fault, and the starter must be removed from the engine for examination.

Sluggish or slow action of the starter is usually caused by a poor connection in the wiring which produces a high resistance in the starter circuit. Check as described above.

Damage to the starter drive is indicated if the starter is heard to operate but does not crank the engine.

Section N.7

REMOVING AND REPLACING THE STARTER

Release the starter cable from the terminal and unscrew the two starter securing bolts. Manœuvre the starter forwards below the oil filter, then rearwards and upwards.

Section N.8

SERVICING THE STARTER

Examination of commutator and brush gear

Remove the starter cover band (A) (Fig. N.5) and examine the brushes (C) (Fig. N.5) and the commutator. Hold back each of the brush springs (B) (Fig. N.5) and move the brush by pulling gently on its flexible connector. If the movement is sluggish remove the brush from its holder and ease the sides by lightly polishing with a smooth file. Always replace brushes in their original positions. If the brushes are worn so that they no longer bear on the commutator, or if the brush flexible lead has become exposed on the running face, they must be renewed.

If the commutator is blackened or dirty, clean it by holding a petrol-moistened cloth against it while the armature is rotated.

Secure the body of the starter in a vice and test by connecting it with heavy-gauge cables to a battery of the correct voltage. One cable must be connected to the starter terminal and the other held against the starter body or end bracket. Under these light load conditions the starter should run at a very high speed.

If the operation of the starter is still unsatisfactory, the starter should be dismantled for detailed inspection and testing.

Dismantling

Take off the cover band " A " (Fig. N.5) at the commutator end, hold back the brush springs " B " (Fig. N.5) and take out the brushes " C " (Fig. N.5) from their holders.

Withdraw the jump ring and shims from the armature shaft at the commutator end and remove the armature complete with drive from the commutator end bracket and starter frame.

Remove the terminal nuts " E " and washers " F " from the terminal post " G " at the commutator end bracket and also withdraw the two through bolts. Remove the commutator end bracket and the attachment bracket from the starter frame.

Brushes

(a) Test the brush springs with a spring scale. The correct tension is 30–40 oz. (850–1134 gm.). Fit a new spring if the tension is low.

(b) If the brushes are worn so that they no longer bear on the commutator, or if the flexible connector has become exposed on the running face, they must be renewed. Two of the brushes are connected to terminals eyelets attached to the brush boxes on the commutator end bracket. The other two brushes (Fig. N.5) are connected to tappings on the field coils.

The flexible connectors must be removed by un-soldering and the connectors of the new brushes secured in place by soldering. The brushes are pre-formed, so that bedding of the working face to the commutator is unnecessary.

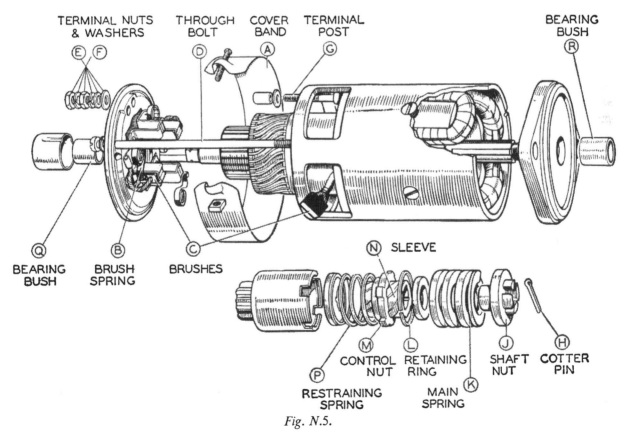

Fig. N.5.

An exploded view of the starter and drive.

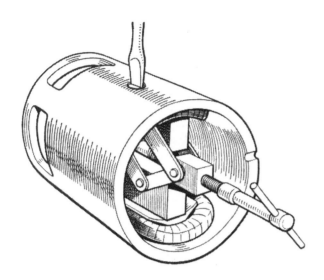

Fig. N.6.
An expander in use for fitting pole shoes.

Drive

If the pinion is tight on the sleeve, wash in paraffin; replace any worn or damaged parts.

To dismantle the drive, extract the split pin and remove the shaft nut " J " (Fig. N.5); withdraw the main spring and collar.

Rotate the barrel to push out the sleeve; remove the barrel and pinion.

The barrel and pinion are supplied as an assembly but the parts may be separated by extracting the retaining ring " L."

Note.—Should either the control nut or screwed sleeve be damaged, a replacement assembly, consisting of a screwed sleeve and control nut, must be fitted. These components must not be fitted individually.

Commutator

A commutator in good condition will be smooth and free from pits and burned spots. Clean the commutator with a cloth moistened with petrol. If this is ineffective, carefully polish with a strip of fine glass-paper, while rotating the armature. To remedy a badly worn commutator, dismantle the starter drive as described above and remove the armature from the end bracket. Now mount the armature in a lathe, rotate it at a high speed and take a light cut with a very sharp tool. Do not remove any more metal than is absolutely necessary, and finally polish with very fine glass-paper.

The mica on the **starter** commutator **must not be undercut.**

Field coils

The field coils can be tested for an open circuit by connecting a 12-volt battery, having a 12-volt bulb in one of the leads, to the tapping point of the field coils to which the brushes are connected, and the field terminal post. If the lamp does not light, there is an open circuit in the wiring of the field coils.

Lighting of the lamp does not necessarily mean that the field coils are in order, as it is possible that one of them may be earthed to a pole shoe or to the yoke. This may be checked by removing the lead from the brush connector and holding it on a clean part of the starter yoke. Should the bulb now light it indicates that the field coils are earthed.

Should the above tests indicate that the fault lies in the field coils, they must be renewed. When renewing field coils carry out the procedure detailed in the Dynamo Section N.5.

Armature

Examination of the armature will in many cases reveal the cause of failure, e.g. conductors lifted from the commutator due to the starter being engaged while the engine is running and causing the armature to be rotated at an excessive speed. A damaged armature must in all cases be renewed—no attempt should be made to machine the armature core or to true a distorted armature shaft.

Bearings (commutator end)

Bearings which are worn to such an extent that they will allow excessive sideplay of the armature shaft must be renewed. To renew the bearing bush, proceed as follows:—

Press the new bearing bush into the end bracket, using a shouldered mandrel of the same diameter as the shaft which is to fit in the bearing.

The bearing bush is of the porous phosphor-bronze type, and before fitting, **new bushes should be allowed to stand completely immersed for twenty-four hours in thin engine oil in order to fill the pores of the bush with lubricant.**

Reassembly

The reassembly of the starter is a reversal of the operations described in this section.

Section N.9

THE CONTROL BOX

Regulator adjustment

The regulator is carefully set before leaving the Works to suit the normal requirements of the standard equipment, and in general it should not be necessary to alter it. If, however, the battery does not keep in a charged condition, or if the dynamo output does not fall when the battery is fully charged, it may be advisable to check the setting and, if necessary, to readjust it.

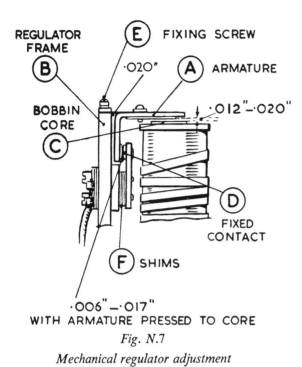

REGULATOR FRAME **B**
E FIXING SCREW
·020"
A ARMATURE
BOBBIN CORE **C**
·012"–·020"
D
FIXED CONTACT
F SHIMS
·006"–·017"
WITH ARMATURE PRESSED TO CORE

Fig. N.7

Mechanical regulator adjustment

It is important, before altering the regulator setting, when the battery is in a low state of charge, to check that its condition is not due to a battery defect or to the dynamo belt slipping.

How to check and adjust electrical setting

The regulator setting can be checked without removing the cover of the control box.

Withdraw the cables from the terminals marked 'A' and 'A1' at the control box and join them together. Connect the negative lead of a moving-coil voltmeter (0–20 volts full-scale reading) to the 'D' terminal on the dynamo and connect the other lead from the meter to a convenient chassis earth.

Slowly increase the speed of the engine until the voltmeter needle flicks and then steadies; this should occur at a voltmeter reading between the limits given below for the appropriate temperature of the regulator.

Setting at 10° C. (50° F.)	16·1–16·7 volts
„ „ 20° C. (68° F.)	15·8–16·4 „
„ „ 30° C. (86° F.)	15·6–16·2 „
„ „ 40° C. (104° F.)	15·3–15·9 „

If the voltage at which the reading becomes steady occurs outside these limits, the regulator must be adjusted.

Shut off the engine, remove the control box cover, release the locknut (A) (Fig. N.8) holding the adjusting screw (B) and turn the screw in a clockwise direction to raise the setting or in an anti-clockwise direction to lower the setting. Turn the adjusting screw a fraction of a turn and then tighten the locknut.

When the dynamo is run at a high speed on open circuit, it builds up a high voltage. When adjusting the regulator, do not run the engine up to more than 3,000 r.p.m. or a false voltmeter reading will be obtained.

Mechanical setting

The mechanical setting of the regulator is accurately adjusted before leaving the Works, and provided that the armature carrying the moving contact is not removed, the regulator will not require mechanical adjustment. If, however, the armature has been removed from the regulator for any reason, the contacts will have to be reset. To do this, proceed as follows:

(1) Slacken the two armature fixing screws (E) (Fig. N.7). Insert a ·020 in. (·51 mm.) feeler gauge between the back of the armature (A) and the regulator frame.

(2) Press back the armature against the regulator frame and down onto the top of the bobbin core with the gauge in position and lock the armature by tightening the two fixing screws.

(3) Check the gap between the under side of the arm and the top of the bobbin core. This must be ·012 to ·020 in. (·30 to ·51 mm.). If the gap is outside these limits correct by adding or removing shims (F) (Fig. N.7) at the back of the fixed contact (D) or, in later types, by carefully bending the fixed contact bracket.

(4) Remove the gauge and press the armature down, when the gap between the contacts should be between ·006 in. (·15 mm.) and ·017 in. (·43 mm.).

Cleaning contacts

To render the regulator contacts accessible for cleaning, slacken the screws securing the plate carrying the fixed contact. It will be necessary to slacken the upper screw (C) (Fig. N.8) a little more than the lower screw (D), so that the contact plate can be swung outwards. Clean the contacts by means of fine carborundum stone or fine emery-cloth. Carefully wipe away all traces of dirt or other foreign matter. Finally tighten the securing screws.

CUT-OUT

Adjustment

If it is suspected that the cutting-in speed of the dynamo is too high, connect a voltmeter between the terminals marked 'D' and 'E' at the control box and slowly raise the engine speed. When the voltmeter reading rises to between 12·7 and 13·3 volts the cut-out contacts should close.

If the cut-out has become out of adjustment and operates at a voltage outside these limits it must be reset. To make the adjustment, slacken the locknut (E) (Fig. N.8) and turn the adjusting screw (F) a fraction of a turn in a clockwise direction to raise the operating voltage

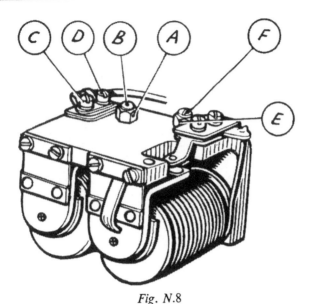

Fig. N.8

The cut-out and regulator assembly

or in an anti-clockwise direction to lower the voltage. Tighten the locknut after making the adjustment.

Cleaning

To clean the contacts remove the cover, place a strip of fine glass-paper between the contacts and then, closing the contacts by hand, draw the paper through. This should be done two or three times, with the rough side towards each contact.

Radio suppression

When it is desired to fit suppressors for radio equipment, make sure that this is done only in accordance with recommended practice. Suppressors and capacitors wrongly fitted may cause damage to the electrical equipment.

Section N.10

FUSES

The fuses are mounted in a separate fusebox and are therefore accessible without removing the control box cover.

Units protected

The units which are protected by each fuse can readily be identified by referring to the wiring diagram on page N.14.

Blown fuses

A blown fuse is indicated by the failure of all the units protected by it, and is confirmed by examination of the fuse, which can easily be withdrawn from the spring clips.

If it has blown, the fused state of the wire will be visible inside the glass tube. Before renewing a blown fuse, inspect the wiring of the units that have failed for evidence of a short circuit or other faults which may have caused the fuse to blow, and remedy the cause of the trouble.

Section N.11

THE ELECTRIC HORN

If the horn fails or becomes uncertain in its action, it does not follow that the horn has broken down. First ascertain that the trouble is not due to a loose or broken connection in the wiring of the horn. If the fuse has blown, examine the wiring for the fault and replace with the spare fuse provided.

The performance of a horn may be upset by a loose fixing bolt, or by some component near the horn being loose. If after carrying out the above examination the trouble is not rectified, the horn may need adjustment.

Adjustment does not alter the pitch of the note: it merely takes up wear of moving parts. When adjusting the horn, short-circuit the fuse, otherwise it is liable to blow. Again, if the horn will not sound on adjustment, release the push instantly.

Adjustment

Remove the fixing screw from the top of the horn and take off the cover. Detach the cover securing bracket by springing it out of its location.

Slacken the locknut on the fixed contact and rotate the adjusting nut until the contacts are just separated (indicated by the horn failing to sound). Turn the adjusting nut half a turn in the opposite direction and secure it in this position by tightening the locknut.

Section N.12

FLASHING DIRECTION INDICATORS

The flashing direction indicators are operated by a pneumatic time switch through a flasher unit and a relay to the dual-filament bulbs in the side and tail lamps. In the event of failure, carry out the following procedure:
  (1) Check bulbs for broken filaments.
  (2) Refer to the wiring diagram and check over flasher circuit connections.
  (3) Switch on the ignition and check that terminal 'B' on the flasher is at 12 volts with respect to earth.
  (4) Connect together terminals 'B' and 'L' at the flasher unit and operate the direction indicator switch.

If the flasher lights now work, the flasher unit is defective and must be renewed.

If the lights do not work the relay is defective and must be renewed.

The length of time the flasher is operating can be altered by screwing up the adjusting screw located in the small boss at the back of the time switch. Screw in to lengthen the time of operation and out to shorten the period.

Section N.13

THE WINDSHIELD WIPER

Normally the windshield wiper will not require any servicing apart from the occasional renewal of the rubber blades.

Should any trouble be experienced, first check for loose connections, worn insulation, etc., before dismantling the motor.

1. To detach the cable rack from the motor and gearbox
Unscrew the pipe union nut.
Remove the gearbox cover.
Remove the split pin and washer from the crank-pin and final gear wheel.
Lift off the connecting link.

2. Commutator dirty
Remove the connecting leads to the terminals, and withdraw the three screws securing the cover at the commutator end. Lift off the cover. Clean the commutator with a cloth moistened with petrol (gasoline) and carefully remove any carbon dust from between the commutator segments.

3. Brush lever stiff or brushes not bearing on commutator
Check that the brushes bear freely on the commutator. If they are loose and do not make contact, a replacement

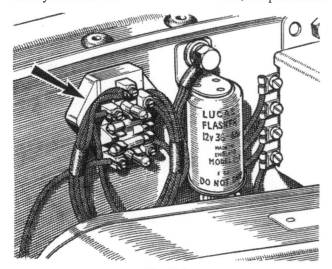

Fig. N.9
The fuses and flasher unit

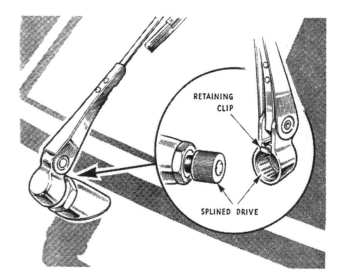

Fig. N.10
A windshield wiper arm and drive

tension spring is necessary. The brush levers must be free on their pivots. If they are stiff they should be freed by working them backwards and forwards by hand and by applying a trace of thin machine oil. Packing shims are fitted beneath the legs of the brush to ensure that the brushes are central and that there is no possibility of the brush boxes fouling the commutator. If the brushes are considerably worn they must be replaced by new ones.

4. Motor operates but does not transmit motion to spindles
Remove the cover of the gearbox. A push-pull motion should be transmitted to the inner cable of the flexible rack. If the cross-head moves sluggishly between the guides, lightly smear a small amount of medium grade engine oil in the groove formed in the die-cast housing. When overhauling, the gear must be lubricated by lightly packing the gearbox with a grease to Ref. D (page P.2).

5. Thrust screw adjustments
The thrust screw is located on the top of the cross-head housing. To adjust, slacken the locknut, screw down the thrust screw until it contacts the armature and then turn back a fraction of a turn. Hold the thrust screw with a screwdriver and tighten the locknut.

6. To remove the motor
Detach the cable rack from the motor and gearbox as detailed above. Disconnect the lead. Remove the two screws securing the mounting bracket to remove the motor.

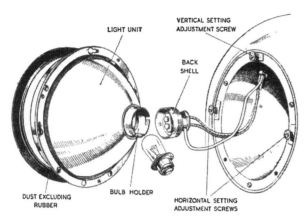

Fig. N.11
A light unit

Section N.14

THE HEADLAMPS

The headlamps are built into the wings and are fitted with double-filament bulbs. The design is such that the bulb is correctly positioned in relation to the reflector, and no focusing is required when a replacement bulb is fitted.

The anti-dazzle device

The double-filament bulbs are controlled by a foot-operated dipping switch deflecting both headlamp beams downwards to avoid dazzle.

Certain countries have lighting regulations to which the foregoing arrangements do not conform, and cars exported to such countries have suitably modified lighting equipment.

Section N.15

THE LIGHT UNITS

The light units consist of a lamp glass, reflector, and a back shell. The light unit is located to the front wing by three spring-loaded attachment screws in a domed shield attached to the wing. The back of the lamp is therefore sealed to give complete protection.

A dust- and weather-excluding rubber is fitted in the recess of the rim of the light unit and a plated rim is fitted over this to complete the weather-sealing.

Section N.16

REMOVING THE LIGHT UNITS

To remove the light unit for bulb replacement, unscrew the retaining screw at the bottom of the plated lamp rim and lift the rim away from the dust-excluding rubber.

Remove the dust-excluding rubber, which will reveal the three spring-loaded screws. Press the light unit inwards against the tension of the springs and turn it in an anti-clockwise direction until the heads of the screws can pass through the enlarged ends of the keyhole slots in the lamp rim.

This will enable you to withdraw the light unit sufficiently to give attention to the wiring and bulbs.

Section N.17

SETTING THE HEADLAMPS

The lamps should be set so that the main driving beams are parallel with the road surface or in accordance with your local regulations.

If adjustment is required, this is achieved by removing the plated rim and dust-excluding rubber as indicated in Section N.16.

Vertical adjustment can then be made by turning the screws at the top of the lamp in the necessary direction.

Horizontal adjustment can be effected by using the adjustment screws on each side of the light unit. (See Fig. N.12.)

Section N.18

REPLACING HEADLAMP BULBS

Twist the back shell anti-clockwise and pull it off. Withdraw the bulb from the holder.

Insert the replacement bulb in the holder, making sure that the slot in the periphery of the bulb flange engages the projection in the holder.

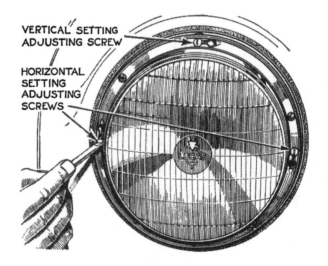

Fig. N.12
The headlamp setting screws

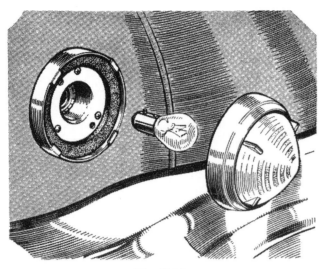

Fig. N.13.
Sidelamp bulb removal.

Engage the projections on the back shell with the slots of the holder, press it on and twist it clockwise until it engages with its catch.

Section N.19

REPLACING THE LIGHT UNITS

Position the light unit so that the heads of the adjusting screws coincide with the enlarged ends of the attachment slots. Push the light unit towards the wing to compress the springs and turn the unit to the right as far as it will go, that is, approximately $\frac{1}{2}$ in. (13 mm.).

Replace the dust-excluding rubber on the light rim with its flanged face forward and refit the plated rim.

Section N.20

THE TAIL-LAMPS AND STOP-LIGHTS

The tail-lamps are of the double-filament type, the second filament giving a marked increase in brilliance when the brakes are applied.

To obtain access to the bulbs, remove the glass by withdrawing the two screws. The bulbs are held in bayonet-type holders with offset pins to ensure correct fitting.

Section N.21

CLEANING THE LAMPS

Care must be taken when handling headlamp reflectors to prevent them from becoming finger-marked. If they do become marked a transparent and colourless protective covering enables any finger-marks to be removed

by polishing with a chamois-leather or a very soft dry cloth. **Do not use metal polish on reflectors.**

Chromium-plated surfaces such as lamp rims should be washed with plenty of water, and when the dirt is completely removed they may be polished with a chamois-leather or soft dry cloth. **Do not use metal polishes on chromium plating.**

Section N.22

THE SIDELAMPS

To obtain access to the bulb press the lamp front inwards and turn it anti-clockwise until it is free to be withdrawn. Reverse this movement to replace the front.

The locating pins on the bulbs are offset to ensure that it is fitted correctly to give increased brilliance when the flashing equipment is operating.

Section N.23

THE NUMBER-PLATE ILLUMINATION

The number-plate is illuminated by a separate lamp and the domed cover is removed for bulb replacement by unscrewing the slotted screw and withdrawing the cover.

Section N.24

THE PANEL AND WARNING LIGHTS

The locations of the lamps illuminating the instruments and the warning lights are shown by arrows on the illustrations on page N.15.

The bulbs are accessible from below the instrument panel.

Fig. N.14.
The number-plate lamp.

N.13

WIRING DIAGRAM

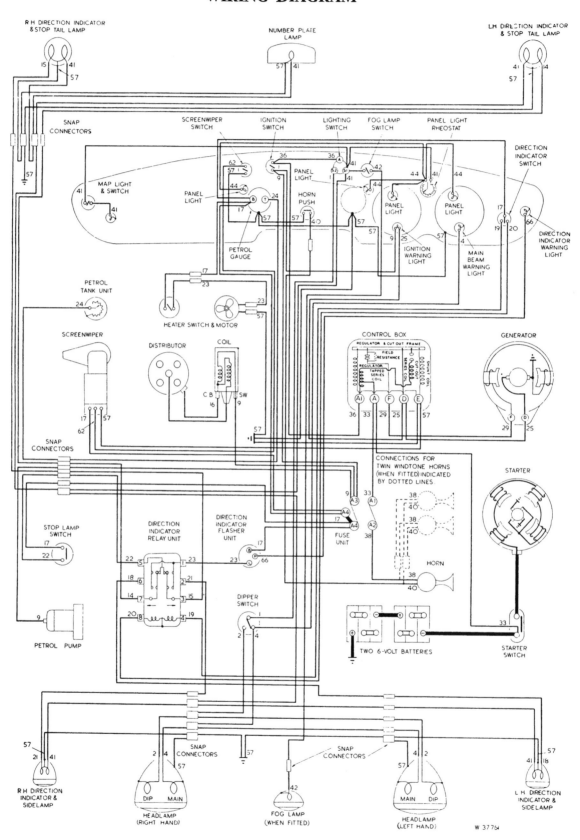

For index to **cable** colour **code** numbers see page N.15.

Section N.25

REPLACEMENT BULBS

	B.M.C. Part No.				Watts	Volts
Headlamps (Home and Export R.H.D.—dip left)	13H140	..	..	..	50/40	12
Headlamps (Europe and U.S.A. L.H.D.—dip right)	13H141	..	..	..	50/40	12
Headlamps (Europe except France—vertical dip)	3H921	..	..	..	45/40	12
Headlamps (Europe except France—vertical dip) from Car No. 58918	13H138	..	..	..	45/40	12
Headlamps (France—vertical dip) from Car No. 60340 ..	13H139	..	..	..	45/40	12
Sidelamp and stop/tail lamp	1F9026	..	..	..	6/21	12
Number-plate illumination lamp	2H4817	..	..	..	6	12
Panel lamps	2H4732	..	..	..	2·2	12

Section N.26

FITTING A FOG LAMP

A fog lamp is not fitted as standard equipment, but can be supplied as an optional extra. The necessary wiring together with the switch (marked 'F' on the instrument panel) is already provided to accommodate the fitment.

To fit a fog lamp bracket remove the over-rider and place the bracket in position. Mark off and drill a further hole through the bumper to accommodate an additional $\frac{5}{16}$ in. screw.

When mounted the lamp is connected up to the spare red and yellow lead located behind the radiator grille to the right-hand side.

The necessary parts are shown below, together with their part reference numbers.

ADH785	Fog lamp	2
AHH5454	Lead—fog lamp	2
AHH5521	Bracket—right-hand	1
AHH5520	Bracket—left-hand	1
RG103	Grommet—lead..	4
HZS0506	Screw—bracket to blade—lower	2
FNZ105	Nut	2
LWZ205	Spring washer	2
PMP0518	Screw	2
PMP105	Washer	2

Section N.27

LOCATION AND REMEDY OF FAULTS

Although every precaution is taken to eliminate possible causes of trouble, failure may occasionally develop through lack of attention to the equipment, or

Fig. N.15

The panel and warning lights

KEY TO CABLE COLOURS

1 Blue	15 White *with* Brown	28 Yellow *with* White	41 Red	54 Purple *with* Green
2 Blue *with* Red	16 White *with* Black	29 Yellow *with* Green	42 Red *with* Yellow	55 Purple *with* Brown
3 Blue *with* Yellow	17 Green	30 Yellow *with* Purple	43 Red *with* Blue	56 Purple *with* Black
4 Blue *with* White	18 Green *with* Red	31 Yellow *with* Brown	44 Red *with* White	57 Black
5 Blue *with* Green	19 Green *with* Yellow	32 Yellow *with* Black	45 Red *with* Green	58 Black *with* Red
6 Blue *with* Purple	20 Green *with* Blue	33 Brown	46 Red *with* Purple	59 Black *with* Yellow
7 Blue *with* Brown	21 Green *with* White	34 Brown *with* Red	47 Red *with* Brown	60 Black *with* Blue
8 Blue *with* Black	22 Green *with* Purple	35 Brown *with* Yellow	48 Red *with* Black	61 Black *with* White
9 White	23 Green *with* Brown	36 Brown *with* Blue	49 Purple	62 Black *with* Green
10 White *with* Red	24 Green *with* Black	37 Brown *with* White	50 Purple *with* Red	63 Black *with* Purple
11 White *with* Yellow	25 Yellow	38 Brown *with* Green	51 Purple *with* Yellow	64 Black *with* Brown
12 White *with* Blue	26 Yellow *with* Red	39 Brown *with* Purple	52 Purple *with* Blue	65 Dark Green
13 White *with* Green	27 Yellow *with* Blue	40 Brown *with* Black	53 Purple *with* White	66 Light Green
14 White *with* Purple				

damage to the wiring. The following pages set out the recommended procedure for a systematic examination to locate and remedy the causes of some of the more usual faults encountered.

The sources of trouble are by no means always obvious, and in some cases a considerable amount of deduction from the symptoms is needed before the cause is disclosed.

For instance, the engine might not respond to the starter switch; a hasty inference would be that the starter motor is at fault. However, as the motor is dependent on the batteries, it may be that the batteries are exhausted.

This, in turn, may be due to the dynamo failing to charge the batteries, and the final cause of the trouble may be, perhaps, a loose connection in some part of the charging circuit.

If, after carrying out an examination, the cause of the trouble is not found, the equipment should be checked by the nearest Lucas Service Depot or Agent.

CHARGING CIRCUIT

1. Batteries in low state of charge

(a) This state will be shown by lack of power when starting, poor light from the lamps, and hydrometer readings below 1·200. It may be due to the dynamo not charging or giving low or intermittent output. The ignition warning light will not go out if the dynamo fails to charge, or will flicker on and off in the event of intermittent output.

(b) Examine the charging and field circuit wiring, tightening any loose connections or replacing broken cables. Pay particular attention to the battery connections.

(c) Examine the dynamo driving belt; take up any undue slackness by swinging the dynamo outwards on its mounting after slackening the attachment bolts.

(d) Check the regulator setting and adjust if necessary.

(e) If, after carrying out the above, the trouble is still not cured, have the equipment examined by a Lucas Service Depot or Agent.

2. Batteries overcharged

This will be indicated by burnt-out bulbs, very frequent need for topping up the batteries, and high hydrometer readings. Check the charge reading with an ammeter when the car is running. It should be of the order of only 3–4 amperes.

If the ammeter reading is in excess of this value, it is advisable to check the regulator setting and adjust if necessary.

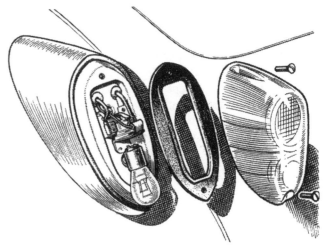

Fig. N.16
Removing a tail lamp bulb

STARTER MOTOR

1. Starter motor lacks power or fails to turn engine

(a) See if the engine can be turned over by hand. If not, the cause of the stiffness in the engine must be located and remedied.

(b) If the engine can be turned by hand, first check that the trouble is not due to a discharged battery.

(c) Examine the connections to the batteries, starter and starter switch, making sure that they are tight and that the cables connecting these units are not damaged.

(d) It is also possible that the starter pinion may have jammed in mesh with the flywheel, although this is by no means a common occurrence. To disengage the pinion, rotate the squared end of the starter shaft by means of a spanner.

2. Starter operates but does not crank engine

This fault will occur if the pinion of the starter drive is not allowed to move along the screwed sleeve into engagement with the flywheel, due to dirt having collected on the screwed sleeve. Remove the starter and clean the sleeve carefully with paraffin (kerosene).

3. Starter pinion will not disengage from flywheel when engine is running

Stop the engine and see if the starter pinion is jammed in mesh with the flywheel, releasing it if necessary by rotation of the squared end of the starter shaft. If the pinion persists in sticking in mesh, have the equipment examined at a Service Depot. Serious damage may result to the starter if it is driven by the flywheel.

M.G. 'MGA'. Issue 2. 22142

N.16

LIGHTING CIRCUITS

1. Lamps give insufficient illumination

(a) Test the state of charge of the battery, recharging it if necessary from an independent electrical supply.

(b) Check the setting of the lamps.

(c) If the bulbs are discoloured as the result of long service, they should be renewed.

2. Lamps light when switched on but gradually fade out

As paragraph 1 (a).

3. Brilliance varies with speed of car

(a) As paragraph 1 (a).

(b) Examine the battery connections, making sure that they are tight, and renew any faulty cables.

Section N.28

WINDOWLESS YOKE DYNAMO

Engines numbered from 487 are fitted with a new dynamo (Part No. 11G220) without brush gear inspection windows (see Fig. N.17). Access to the brush gear in these dynamos is gained by undoing the two through-bolts and withdrawing the commutator end bracket. Every 12,000 miles (19200 km.) the unit should be partially dismantled for the inspection of brush gear and commutator.

To check the brush spring tension, the yoke should be completely withdrawn from the armature and the commutator end bracket refitted to the shaft.

When reassembling a windowless yoke dynamo the brushes must first be held clear of the commutator in the usual way, i.e. by partially withdrawing the brushes from their boxes until each brush is trapped in position by the side pressure of its spring. The brushes can be released onto the commutator with a small screwdriver or similar tool when the end bracket is assembled to within about $\frac{1}{2}$ in. (13 mm.) of the yoke. Before closing the gap between the end bracket and yoke, see that the springs are in correct contact with the brushes.

Coil steady plate, Part No. 11G221, should always be used with dynamo, Part No. 11G220.

Section N.29

MODIFIED CONTROL BOX

A modified C.V.C. control box, model RB106/2, with revised settings (Part No. AHH5356) is introduced on later cars. Servicing instructions remain as before (see Section N.9), but adjustments must be made within 30 seconds, otherwise heating of the shunt winding will cause false settings to be made.

The voltmeter readings should be within the limits given below at approximately 1,500 dynamo r.p.m. and according to the ambient temperature:

Setting at 10° C. (50° F.) .. 15·9 to 16·5 volts
Setting at 20° C. (68° F.) .. 15·6 to 16·2 volts
Setting at 30° C. (86° F.) .. 15·4 to 16·0 volts
Setting at 40° C. (104° F.) .. 15·1 to 15·7 volts

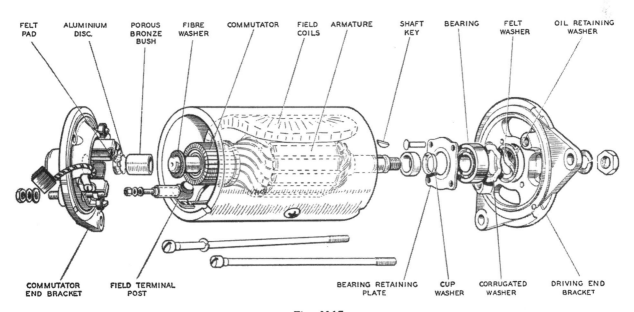

*Fig. N.*17

The windowless yoke dynamo

Section N.30

HEADLAMP BEAM SETTING

Refer to Section N.17 for details of the headlamp adjustment screws.

In the absence of specialized proprietary equipment the setting of the lamps can be carried out by placing the vehicle squarely in front of a blank wall at a distance of 25 ft. (7·6 m.) or more, taking care that the surface on which the car is standing is level and not sloping in relation to the wall. The vehicle should be loaded. It will be found an advantage to cover one lamp while setting the other.

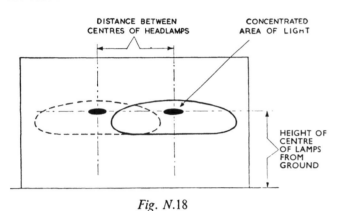

Fig. N.18
Headlamp—alignment

Section N.31

DRY-CHARGED BATTERIES

'Dry-charged' batteries are supplied without electrolyte but with the plates in a charged condition. This ensures that there is no deterioration of the battery if it is stored for a period before use. These batteries have the type suffix letter 'Z' (e.g. SGZ, etc.).

Filling the cells with electrolyte of the correct specific gravity (see **'Initial filling and charging'** of Section N.1) **in one operation** renders the battery capable of giving a starting discharge **one hour after filling.** The temperature of the filling-in solution, battery, and filling room should be maintained between 60° F. (16° C.) and 100° F. (38° C.). If the battery has been stored in a cool place, it should be allowed to warm up to room temperature before filling.

When time permits, a freshening charge at the normal recharge rate of the battery will ensure that the battery is fully charged. During the charge keep the electrolyte level with the top edge of the separators by the addition of distilled water. Check the electrolyte specific gravity

at the end of the charge: if 1·270 acid was used to fill the battery, the specific gravity should now be between 1·270 and 1·290; if 1·210 acid was used, the specific gravity should be between 1·210 and 1·230.

Section N.32

MODIFIED EUROPEAN LIGHT UNIT

Cars exported to Europe are now fitted with the new European-type headlamps. These lamp units are fitted with special bulbs and front lenses giving an asymmetrical beam to the right-hand side. This modification was introduced on the following cars:

From Car No. 58918 (Europe except France).
From Car No. 60340 (France).

Access to the bulb is gained in the same way as described in Section N.16. The bulb, however, is released from the reflector by withdrawing the three-pin socket and pinching the two ends of the wire retaining clip to clear the bulb flange (see Fig. N.18).

When replacing the bulb care must be taken to see that the rectangular pip on the bulb flange engages the slot in the reflector seating for the bulb.

Replace the spring clip with its coils resting in the base of the bulb flange and engaging in the two retaining lugs on the reflector seating.

The appropriate replacement bulbs are listed in Section N.25. They are not interchangeable with those used in conjunction with the Continental-type headlamps previously fitted.

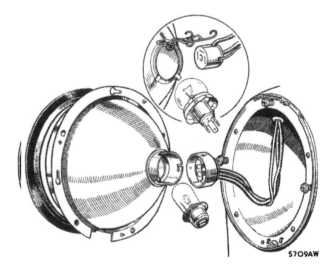

Fig. N.19
The headlamp light unit, with the European-type lamp bulb arrangement inset

SECTION NN

THE ELECTRICAL EQUIPMENT
(MGA 1600 and MGA 1600 [Mk. II])

WIRING DIAGRAM

(MGA 1600)

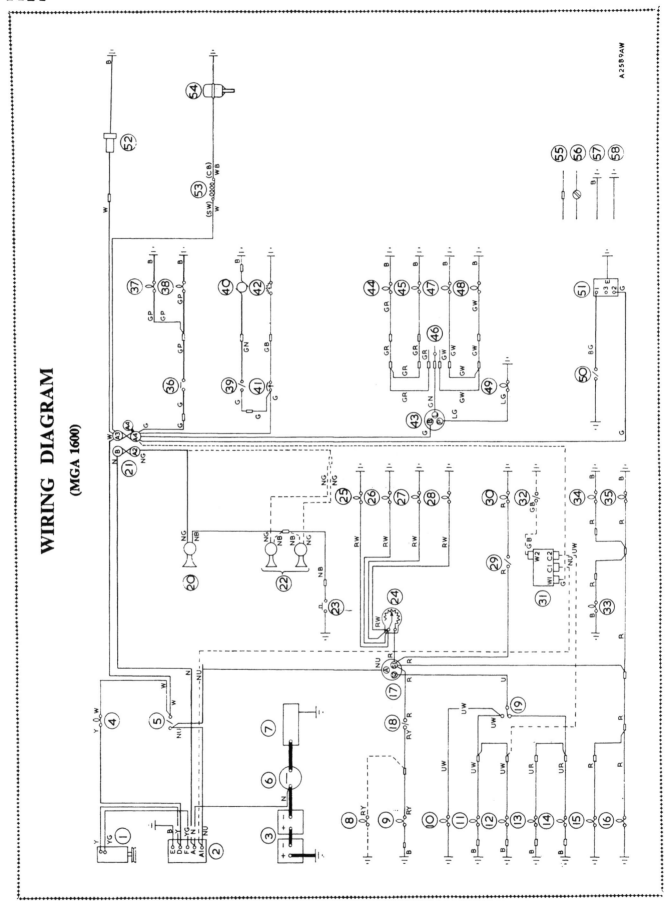

A2589AW

KEY TO WIRING DIAGRAM (R.H.D. AND L.H.D.)

No.	Description	No.	Description	No.	Description
1.	Generator.	21.	Fuse unit.	40.	Heater motor.
2.	Control box.	22.	Twin windtone horns (if fitted).	41.	Fuel gauge.
3.	Two 6-volt batteries.	23.	Horn-push.	42.	Fuel tank unit.
4.	Ignition warning light.	24.	Panel lamp rheostat.	43.	Flasher unit.
5.	Ignition switch.	25.	Panel lamp.	44.	L.H. rear flasher.
6.	Starter switch.	26.	Panel lamp.	45.	L.H. front flasher.
7.	Starter motor.	27.	Panel lamp.	46.	Flasher switch.
8.	R.H. fog lamp (if fitted).	28.	Panel lamp.	47.	R.H. front flasher.
9.	L.H. fog lamp.	29.	Map lamp switch.	48.	R.H. rear flasher.
10.	Main beam warning light.	30.	Map lamp.	49.	Flasher warning light.
11.	R.H. headlamp main beam.	31.	Headlamp flick relay.	50.	Windshield wiper switch.
12.	L.H. headlamp main beam.	32.	Headlamp flick switch.	51.	Windshield wiper motor.
13.	L.H. headlamp dip beam.	33.	L.H. tail lamp.	52.	Fuel pump.
14.	R.H. headlamp dip beam.	34.	Number-plate lamp.	53.	Ignition coil.
15.	L.H. pilot lamp.	35.	R.H. tail lamp.	54.	Distributor.
16.	R.H. pilot lamp.	36.	Stop lamp switch.	55.	Snap connectors.
17.	Lighting switch.	37.	L.H. stop lamp.	56.	Terminal blocks or junction box.
18.	Fog lamp switch.	38.	R.H. stop lamp.	57.	Earth connections made via cable.
19.	Dipper switch.	39.	Heater switch (when fitted).	58.	Earth connections made via fixing bolts.
20.	Horn.				

CABLE COLOUR CODE

B	Black	P	Purple	Y	Yellow
U	Blue	R	Red	D	Dark
N	Brown	S	Slate	L	Light
G	Green	W	White	M	Medium
K	Pink				

When a cable has two colour code letters the first denotes the main colour and the second denotes the tracer colour

Section NN.1

FRONT PILOT AND FLASHING INDICATOR LAMPS

To gain access to the front pilot and flashing indicator bulbs press the front of the lamp inwards and turn it in a clockwise direction.

Both bulbs have single filaments and may be replaced either way round.

Section NN.2

REAR FLASHING INDICATOR LAMPS

Fold back the rubber lip surrounding the lamp rim and withdraw the rim and lens.

The bulb has a single filament and may be replaced either way round.

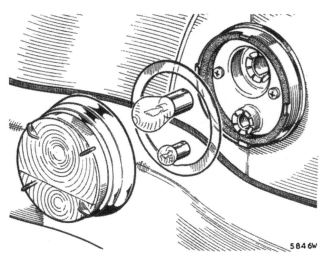

Fig. NN.1

A front pilot and flashing indicator bulb with the lens and rim removed

Section NN.3

REPLACEMENT BULBS

	B.M.C. Part No.				Watts	Volts
Headlamps (Home and Export R.H.D.—dip left)	13H140	..	..	..	50/40	12
Headlamps (Export and U.S.A. L.H.D.—dip right)	13H141	..	..	..	50/40	12
Headlamps (Europe except France—dip vertical)	13H138	..	..	..	45/40	12
Headlamps (France L.H.D.—dip vertical)	13H139	..	..	..	45/40	12
Headlamps (Sweden R.H.D.—dip vertical)	3H921	..	..	..	45/40	12
Headlamps (Sweden R.H.D.—left dip, from Car No. 72040) ..	13H138	..	..	..	45/40	12
Stop/tail lamps	1F9026	..	..	..	6/21	12
Number-plate illumination lamp	2H4817	..	..	..	6	12
Flashing indicator lamps	1F9012	..	..	..	21	12
Panel lamps	2H4732	..	..	..	2·2	12

Section NN.4

HEADLAMPS

From Car No. 70222 Mk. VIII headlamps with sealed-beam units (Part No. BHA4144) were fitted to cars exported to U.S.A.

From Car No. 72040 cars exported to Sweden have special headlamps with asymmetrical left dip (Part No. BMK391). These are interchangeable with earlier types in pairs only.

Section NN.5

LUCAS C40/1 DYNAMO

From Engine No. 16GA6272 a later type of dynamo with increased output, Lucas type C40/1, was introduced, complete with Lucar connectors, modified Lucas type RB106/2 control box, and a new-type ignition coil bracket.

The modified control box must not be fitted with the earlier-type dynamo.

Section NN.6

DISMANTLING A C40/1 DYNAMO

The instructions for dismantling the dynamo are basically the same as those given in Section N.4. The C40/1 dynamo has a windowless yoke, and is therefore not fitted with a cover band.

Access to the brushes is obtained by removing the commutator end bracket.

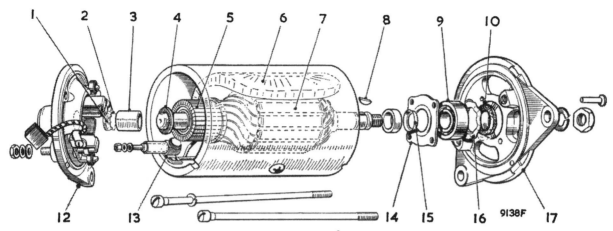

Fig. NN.2

The C40/1 dynamo

1. Felt pad.
2. Aluminium disc.
3. Bronze bush.
4. Fibre washer.
5. Commutator.
6. Field coils.
7. Armature.
8. Shaft key.
9. Bearing.
10. Felt washer.
12. Commutator end bracket.
13. Field terminal post.
14. Bearing retaining plate.
15. Cup washer.
16. Corrugated washer.
17. Driving end bracket.

Section NN.7

SERVICING A C40/1 DYNAMO

The instructions for servicing the dynamos are generally the same as given in Section N.5, with the following exceptions.

Brushes

The minimum permissible length of a worn brush is $\frac{9}{32}$ in. (7·14 mm.). Badly worn brushes must be renewed and the new brushes bedded to the commutator. The correct spring tension is 30 oz. (·85 kg.), maximum, on a new brush, and 13 oz. (·37 kg.), minimum, on a brush worn to $\frac{9}{32}$ in. (7·14 mm.).

Commutator

The later type of commutators are moulded, and may be reskimmed to a minimum diameter of 1·450 in. (36·8 mm.). The undercut must conform to the following dimensions:

Width .. ·040 in. (1·016 mm.)

Depth .. ·020 to ·035 in. (·508 to ·889 mm.)

It is important that the side of the undercut should clear the moulding material by a minimum of ·015 in. (·381 mm.).

Field coil

The resistance of the field coil is 6·0 ohms.

Bearings

To remove the bearing bush in the commutator end

plate screw a $\frac{5}{8}$ in. tap squarely into the bush and withdraw the bush: then remove the felt ring and its retainer.

When refitting the bearing plate to the front bracket insert the rivets from the outer face of the bracket.

The part numbers of the new components are as follows:

Dynamo, Lucas type C40/1 ..	..	13H219
Ignition coil bracket ..	.. ..	12H51
Control box, Lucas, Type RB106/2 ..		3H1836

Section NN.8

MODIFIED RB106/2 CONTROL BOX

The instructions for adjusting the modified Lucas Type RB106/2 control box are as follows.

Regulator adjustment

The electrical setting of the control unit can be checked without removing the cover. Use a good-quality moving-coil voltmeter (0 to 20 volts).

Remove the cables from the control box terminals 'A' and 'A1' and connect the cables together.

Connect the negative lead of the voltmeter to the control box terminal 'D' and connect the other lead to the terminal 'E'.

Run the dynamo at 3,000 r.p.m., and watch the voltmeter reading, which should be between the limits given below, according to the ambient temperatures.

Ambient temperature				Open-circuit voltage
50° F. (10° C.)	..	..	..	16·1 to 16·7
68° F. (20° C.)	..	..	..	16·0 to 16·6
86° F. (30° C.)	..	..	..	15·9 to 16·5
104° F. (40° C.)	..	..	..	15·8 to 16·4

An unsteady voltmeter reading may be due to dirty contacts, but if the reading is outside the appropriate limits the regulator must be adjusted.

Switch off the engine, remove the control box cover, restart the engine, and run the dynamo at 3,000 r.p.m. Turn the regulator adjusting screw (1, Fig. NN.3) in a clockwise direction to raise the setting, or in an anti-clockwise direction to lower the setting.

NOTE.—The operations of checking and adjusting the regulator should be completed within 30 seconds, otherwise false readings and settings, due to the heating of the shunt coil, will be made.

After adjustment a further check of the setting should be made by switching off and restarting the engine and then raising the dynamo speed to 3,000 r.p.m., when the open-circuit voltage must conform to the figures stated.

Refit the control box cover and restore the original connections.

Cut-out adjustments

To check the voltage at which the cut-out operates remove the control box cover and connect the voltmeter between terminals 'D' and 'E'.

Start the engine and slowly increase speed until the cut-out contacts are seen to close, noting the voltage at which this occurs. It should be 12·7 to 13·3 volts.

An alternative method of determining the exact point of contact closure is to switch on an electrical load, such as a pair of headlamps, when the instant of contact closure will be indicated by a slight flick of the voltmeter pointer.

If the cut-out operates outside the above limits it will be necessary to adjust it to within the limits. To do this turn the adjusting screw (2, Fig. NN.3) in a clockwise direction to raise the setting or in an anti-clockwise direction to reduce the setting.

Turn the screw only a fraction of a turn at a time and test the setting after each adjustment by increasing the engine speed from zero and noting the voltmeter reading at the instant of contact closure.

NOTE.—Like the regulator, the setting of the cut-out must be carried out as quickly as possible to avoid errors due to the heating of the shunt coil.

Having set the cut-in voltage correctly, the 'drop-off' setting should now be checked, and adjusted if necessary so that the instant of contact opening occurs between 8·5 and 11·0 volts.

To check the voltage at which the contacts open remove the control box cover, disconnect the cables from the control box terminals 'A' and 'A1', and join these cables together. Connect the voltmeter between terminal 'A1' and earth; start the engine and run up to speed.

Decelerate the engine slowly and watch the voltmeter pointer, which will return to zero immediately the contact points open. The opening of the contacts should occur between 8·5 and 11·0 volts.

Should the opening of the contacts occur outside these limits, the setting of the fixed contact must be adjusted.

Using a pair of thin-nosed pliers, carefully bend the fixed contact blade towards the bobbin to reduce the drop-off voltage, or away from the bobbin to increase the drop-off voltage. After each adjustment, which should be very small, test the setting, as previously described, and readjust as necessary.

Restore the original connections and refit the cover.

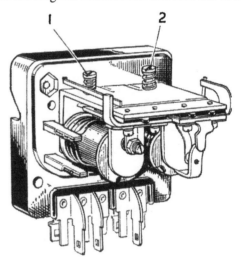

Fig. NN.3

The control box with cover removed

1. Regulator adjusting screw.
2. Cut-out adjusting screw.

Section NN.9

Mk. X SEALED-BEAM HEADLAMPS

Commencing at Chassis No. 103857, Mk. X sealed-beam light units are fitted to all cars exported to the U.S.A.

The lamp housing is secured to the wing by four screws and the back-shell is retained in the housing by a single coil spring. Two screws, each having a flange beneath its head, engage in slotted lugs on the rim of the back-shell and bear against the lamp housing; the screws are used to adjust the vertical and horizontal alignment. The lamp wiring passes through the housing and terminates in a three-hole socket.

Three pins at the back of the sealed-beam unit engage the holes in the socket, and the unit is retained in the back-shell by a lamp retaining plate that is secured to the back-shell by three screws. The lamp rim engages two lugs at the top of the housing and is retained in position by a screw.

To gain access to the lamp unit remove the rim retaining screw and lift the rim off the locating lugs. Slacken the three lamp retaining plate screws, turn the plate anticlockwise, and remove it from the back-shell. Withdraw the lamp unit and disengage the three-pin plug.

To refit the lamp unit engage the three-pin plug and place the unit in the back-shell; ensure that the three lugs formed on the rear circumference of the unit engage the slots in the back-shell. Refit the lamp retaining plate, press it firmly, and rotate it in a clockwise direction to the full extent of the slotted holes. Tighten the retaining plate screws. Fit the rim over the locating lugs, press the rim downwards and inwards, and secure it with its retaining screw.

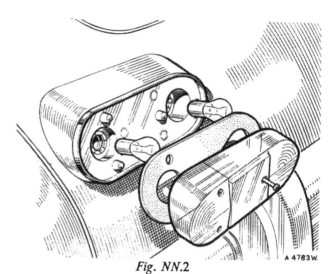

Fig. NN.2

To gain access to the flasher lamp bulb only remove the two outer screws and slide off the flasher lamp cover. To gain access to the tail/stop lamp bulb remove all three screws and the two-piece cover complete

Section NN.10

TAIL/STOP AND FLASHER LAMPS
(MGA 1600 [Mk. II])

Each tail/stop flasher lamp is secured to the tail plinth by two screws. The plinths incorporate two studs that pass through the body shell and are secured by washers and nuts.

The tail/stop and flasher lamp is a combined unit and is enclosed in a two-piece cover that incorporates a sealed reflector. The tail/stop lamp cover is secured by a single screw; the flasher cover slides onto one end of the tail/stop lamp cover and is secured by two screws.

A 24-watt single-filament bulb is used in the flasher lamp and a 21/6-watt double-filament bulb is used in the tail/stop lamp bulb. To ensure that the stop light gives the brighter light the tail/stop lamp bulb has offset pegs and can only be fitted in one position.

To gain access to the flasher lamp bulb only, remove the two outer screws and slide off the flasher lamp cover. To gain access to the tail/stop lamp bulb remove all three screws and the two-piece cover complete.

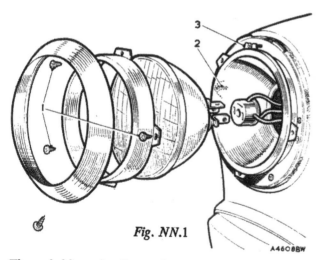

Fig. NN.1

The sealed-beam headlamp, showing the locating lugs and:

1. Retaining plate screw.   2. Horizontal adjustment screw.
3. Vertical adjustment screw.

SECTION O

THE WHEELS, TYRES, AND JACKING

THE WHEELS, TYRES, AND JACKING

O

GENERAL

The spare wheel is housed in the luggage compartment and clamped in position beneath the cover.

Remember that the spare wheel tyre pressure should be maintained at the correct running pressure for the rear wheels.

Pressures are given in 'GENERAL DATA'.

The jack and tyre pump are contained in the tool roll strapped above the spare wheel in the luggage compartment.

Disc wheels are fitted as standard and wire wheels are available as optional equipment.

TYRE MAINTENANCE

Even tyre wear is promoted by changing the positions of the tyres on the car at intervals of about 3,000 miles (4800 km.). The spare tyre should take its turn with the others (see Fig. O.8).

Attention should be paid to the following points, with a view to obtaining the maximum mileage from the tyre equipment of the vehicle.

Test the pressures of the tyres every 500 miles (800 km.) by means of a suitable gauge, and restore any air lost. It is not sufficient to make a visual examination of the tyre for correct inflation. Inflate the spare wheel to the correct rear wheel pressure.

Keep the treads free from grit and stones and carry out any necessary repairs. Clean the wheel rims and keep them free from rust. Paint the wheels if required.

Keep the brakes and clutch adjusted correctly and in good order. Fierceness or uneven action in either of these units has a destructive effect on the tyres.

Misalignment is a very costly error. Suspect it if rapid wear of the front tyres is noticed, and correct the fault at once. See Section J.10 for details of front wheel alignment.

Should the tyres get oily, petrol (gasoline) should be applied sparingly and wiped off at once.

Avoid under- and over-inflation.

Avoid kerbing and other causes of severe impact.

Have damage repaired immediately.

Remove tyres in time for remoulding.

Section O.1

JACKING UP THE CAR

When jacking a front wheel the jack pad should be engaged in the depression in the lower suspension arm spring seating.

At the rear the jack should be placed below the rear spring centre plate or under the spring as close to the axle as possible.

Always apply the hand brake and place blocks each side of the wheels remaining on the ground when the front or rear of the car is to be raised.

The car must not be jacked under the frame side-members.

Section O.2

REMOVING AND REPLACING ROAD WHEELS
Disc wheels

Remove the hub cover by inserting the flattened end of the wheel nut spanner in the recess provided adjacent to the retaining studs and giving it a sideways twist.

Remove the four nuts securing the road wheels to the hub. The wheel nuts have right-hand threads, i.e. turn clockwise to tighten and anti-clockwise to remove. Lift the road wheel from the studs.

Reverse this procedure when replacing the road wheel. Ensure that the brake adjustment hole seal is in position. Line up one of the holes in the wheel with that in the drum to enable the brakes to be adjusted without removing the wheel. Tighten the nuts to a torque wrench reading of 60 to 62·5 lb. ft. (8·3 to 8·65 kg. m.). **Do not overtighten.**

To refit the hub disc, the rim should be placed over two of the buttons on the wheel centre and the outer face given a sharp blow of the fist over the third button.

Wire wheels

Use the copper mallet provided in the tool kit to slacken the winged hub nut securing the wheel on the splines. The hub nuts on the left-hand side of the car have right-hand threads (unscrew anti-clockwise) and the nuts on the right-hand side of the car have left-hand threads (unscrew clockwise).

When replacing a wheel, ensure that the brake adjustment hole seals in the brake-drum are in position.

Section O.3

VALVES

Valve caps, in addition to preventing dirt from entering the valve, form a secondary air seal and should always be fitted. The valves may be tested for airtightness by rotating the wheel until the valve is at the top and inserting its end in an eggcup full of water. If bubbles appear the seating is faulty and should be removed and a new one fitted. It is advisable to change the valve interiors every 12 months.

Section O.4

TYRE REMOVAL

Remove all valve parts to deflate the tyre completely, and push both edges into the base of the rim at a point

O 2

M.G. 'MGA'. Issue 4. 40954

234

diametrically opposite the valve. Lever the cover edge, near the valve, over the rim of the wheel, using two levers at intervals of 6 in. (15 cm.) apart.

NOTE.—Do not attempt to stretch the edges of the tyre cover over the rim edge.

Force is entirely unnecessary and is detrimental, as it tends to damage the wired edges. Fitting or removing is quite easy if the tyre edges are carefully adjusted into the rim base; if found difficult, the operation is not being performed correctly.

Remove the tube carefully; do not pull on the valve. Stand the tyre and wheel upright, keeping the bead on the base of the rim. Lever the bead over the rim flange and at the same time push the wheel away from the cover with the other hand.

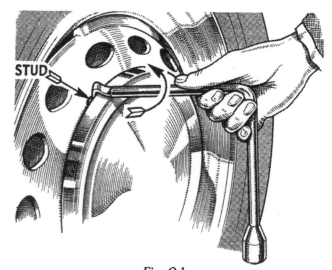

Fig. O.1
Removing a hub cover

Section O.5

THE IMPORTANCE OF BALANCE

In order to obtain good steering it is of importance to ensure that the wheels, with tyres fitted, are in good balance. To assist this, the tyre manufacturers are now marking their tyres with a white spot in the neighbourhood of the bead at the lightest point of the cover; similarly, they are marking the inner tubes with spots to indicate their heaviest point. When tyres are assembled care must therefore be taken to see that they are assembled with the spots on the cover coinciding with the spots on the tube.

It must be noted, in addition, that special balancing discs are fitted to the inside of the cover casing in some cases and that these should on no account be removed as the tyre balance will be upset if this is done. These balance discs are not repair patches and do not indicate any fault in the tyre.

Special balance weights, which cover a range of weights weighing from $\frac{1}{2}$ to $3\frac{1}{2}$ oz. in steps of $\frac{1}{2}$ oz., are supplied by the Dunlop Rubber Co. for attachment to the wheel rim under Part Nos. WBW/1 to 7.

Their use is advised to maintain the correct balance for the wheels, which must be within 8 to 12 in. oz. (·56 to ·85 cm. kg.).

The balance weights are fitted to the outside rim of the wheel.

Fig. O.2
Removing a disc wheel

Section O.6

FITTING TYRES AND TUBES

Inspect the inside of the cover carefully and remove all dirt. The wheel rim must be clean, free from rust and undamaged.

Dust the inside of the cover with french chalk. Inflate the tube until it begins to round out, then insert it in the cover.

Fig. O.3
Removing a wire wheel

Apply a frothy solution of soap and water generously around the centre base of the **tube,** extending upwards between the tyre beads and the tube itself for at least 2 in. (50·8 mm.) on both sides. Also apply the solution to the bottom and outside of the tyre beads. Do not allow the solution to run into the crown of the tyre. The solution must be strong enough to feel slippery when the fingers are wetted with the solution and rubbed together.

Mount the tyre on the rim immediately, whilst the soap solution is still wet.

Push one edge of the cover over the edge of the rim. It will go quite easily if the part first put on is fitted on the opposite side of the valve and is pushed right down into the rim base. Move it round so that its balance spots coincide with those of the inner tube when it is inserted with the valve passing through the hole in the rim. (Take care that the valve, fitted in the side of the tube, is on the correct side of the rim.)

Before inflating, be sure that the tyre beads are clear of the well of the rim all the way round and **push the valve into the tyre as far as possible in order to ensure the tube is not trapped between the bead and the rim, then pull it out again into its correct position.**

Inflate slowly until the beads are fully seated.

Remove the valve core to **deflate the tube completely.**

Reinflate to the correct working pressure (see page 9 in 'GENERAL DATA'). This procedure must be followed whenever a tube is fitted.

The object of the double inflation is to permit any stretched portions of the tube to readjust themselves in the cover and relieve any local strains in the tube.

In an emergency french chalk may be used as a substitute for the soap solution, provided it is evenly and generously applied. This practice, however, is not recommended.

Repairing tubes

Punctures or injuries must be vulcanized. Ordinary patches should only be used for emergencies and cannot be relied upon.

Section O.7

MAINTENANCE OF WIRE WHEELS

In wheel building the rim, hub shell, spokes, and nipples should be loosely assembled to bring the rim into as true a running position with the hub as practicable, while ensuring that the outside dish is maintained. (Outside dish is the distance from the edge of the rear rim flange to the flange of the hub shell.)

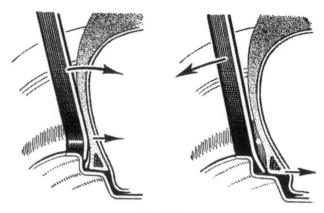

Fig. O.4

The use of tyre levers

Fig. O.5

Pushing the tyre bead into the well

Fig. O.6

Lifting the bead over the rim

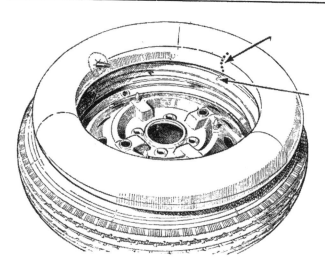

Fig. O.7

Balance marks on tyre and tube

by giving a slight additional tension to the appropriate spoke or sets of spokes.

It is important that as little additional tension as possible should be given when pulling the rim true in this manner, the desirable condition to aim at being that all spokes are as nearly as possible at the same tension. If excessive tension is required to bring the rim true the opposing spokes must be loosened slightly.

The experienced wheel builder will generally be able to gauge when the correct tension has been reached, either by the general feel of the spokes or by the ringing note which the spoke will give if lightly struck with a small spanner or similar metal object.

When building is completed the wheel should be examined carefully to ensure that no spoke ends protrude into the rim well. Any that do so should have their protruding ends carefully filed away.

When this condition is reached, and not before, the wheel should be mounted on a running hub. Each pair of spokes should then be tensioned carefully a small amount at a time, working from one pair of spokes and thence to the diametrically opposite pair of spokes, afterwards repeating the process on the opposed pairs, which are located at right-angles to the original pair of spokes tensioned, and so on.

At each stage of the tensioning the truth of the wheel should be checked carefully both for lateral and up-and-down movement, checking any tendency to out of truth

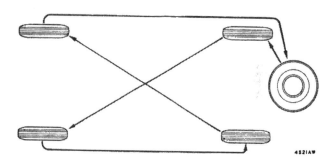

Fig. O.8

System of wheel changing to regularize tyre wear

Notes

SECTION OO

THE WHEELS, TYRES, AND JACKING

(MGA 1600 and MGA 1600 [Mk. II])

Section OO.1

HIGH-SPEED TYRE PRESSURE CONDITIONS

The new British motorways and current facilities for Continental touring give the motorist many opportunities of driving at high and sustained high speeds. In such conditions, and in competition work, the tyres are subjected to greater stresses than those produced during ordinary driving.

Many factors, some probably as important as the physical characteristics of the tyre itself, affect the speed at which it should be driven: road surface, air temperature, and in particular the duration of high-speed driving. However, a normal tyre in good condition and at the correct pressure can be relied upon to perform satisfactorily at speeds up to 80 m.p.h. (144 km./hr.) and intermittently in excess of this by 10 m.p.h. (18 km./hr.). If the car is to be driven consistently at speeds near the maximum of which it is capable special tyres should be fitted on the advice of the tyre manufacturers.

For the 'MGA 1600' fitted with 5·60—15 Gold Seal standard tyres we give the following recommendations:

(1) Normal motoring—standard tyre pressures as given in **'GENERAL DATA'** may be used.

(2) Fast motoring—pressures should be increased by 4 lb./sq. in. (·281 kg./cm.²).

(3) Sustained high speeds and competition work—pressures should be increased by 6 lb./sq. in. (·422 kg./cm.²).

These remarks do not apply to remoulded tyres since it is even more difficult to state with certainty what their maximum speed should be. Therefore when it is intended to indulge in high speeds or competition work we advise the use of first tread tyres.

Section OO.2

GOLD SEAL NYLON AND ROAD SPEED TYRES

Gold Seal nylon tyres for Home and Export (except West Germany) were introduced at Chassis No. 103192. Gold Seal (White Wall) and Road Speed (White Wall) tyres are available as optional extras. Standard Road Speed nylon tyres are fitted to cars exported to West Germany.

The tyre pressure recommendations for these tyres are as follows.

Dunlop Gold Seal nylon tyres

For all normal use, including motorways, etc., up to 100 m.p.h. (161 km.p.h.):

Front	.. ..	21 lb./sq. in. (1·47 kg./cm.²)
Rear	.. ..	24 lb./sq. in. (1·68 kg./cm.²)

Where maximum or near-maximum performance is required:

Front	.. ..	24 lb./sq. in. (1·68 kg./cm.²)
Rear	.. ..	27 lb./sq. in. (1·9 kg./cm.²)

Dunlop Road Speed RS.5 tyres

When the car is fitted with Dunlop Road Speed RS.5 tyres as an optional extra (recommended when the car is used predominantly at very high speeds):

Normal use:

Front	..	17 lb./sq. in. (1·1 kg./cm.²)
Rear	..	20 lb./sq. in. (1·4 kg./cm.²)

When maximum or near-maximum speeds are sustained for lengthy periods or for competition use:

Front	..	24 lb./sq. in. (1·68 kg./cm.²)
Rear	..	27 lb./sq. in. (1·9 kg./cm.²)

SECTION P

LUBRICATION

Correct lubrication of any piece of mechanism is of paramount importance, and in no instance is it of greater importance than in the correct choice of lubricant for a motor-car engine. Automobile engines have different characteristics, such as operating temperatures, oiling systems, size of oilways, clearances, and similar technicalities, and the use of the correct oil is therefore essential.

NOTE.—The letters given in brackets throughout the Manual refer to the appropriate section of the recommended lubricants table given on page P.2.

The following is a list of lubricants recommended :

A ENGINE AND AIR CLEANERS

Climatic conditions	Mobil	Shell	BP	Filtrate	Sternol	Duckham's	Castrol	Esso
Tropical and temperate down to 0° C. (32° F.)	Mobiloil A	Shell X—100 30	Energol S.A.E. 30	Filtrate Medium 30	Sternol W.W. 30	Duckham's NOL Thirty	Castrol X.L.	Esso Extra Motor Oil 20W/30
Extreme cold down to —12° C. (10° F.)	Mobiloil Arctic	Shell X—100 20W	Energol S.A.E. 20W	Filtrate Zero 20/20W	Sternol W.W. 20	Duckham's NOL Twenty	Castrolite	Esso Extra Motor Oil 20W/30
Arctic consistently below —12° C. (10° F.)	Mobiloil 10W	Shell Rotella 10W	Energol S.A.E. 10W	Filtrate Sub-Zero 10W	Sternol W.W. 10	Duckham's NOL Ten	Castrol Z	Esso Motor Oil 10

GEARBOX

	Mobil	Shell	BP	Filtrate	Sternol	Duckham's	Castrol	Esso
All conditions	Mobiloil A	Shell X—100 30	Energol S.A.E. 30	Filtrate Medium 30	Sternol W.W. 30	Duckham's NOL Thirty	Castrol X.L.	Esso Extra Motor Oil 20W/30

B REAR AXLE (HYPOID) AND STEERING GEAR

	Mobil	Shell	BP	Filtrate	Sternol	Duckham's	Castrol	Esso
All conditions down to —12° C. (10° F)	Mobilube G.X. 90	Shell Spirax 90 E.P.	Energol S.A.E. 90 E.P.	Filtrate Hypoid Gear 90	Ambroleum E.P. 90	Duckham's Hypoid 90	Castrol Hypoy	Esso Gear Oil G.P. 90
Arctic consistently below —12° C. (10° F.)	Mobilube G.X. 80	Shell Spirax 80 E.P.	Energol S.A.E. 80 E.P.	Filtrate Hypoid Gear 80	Ambroleum E.P. 80	Duckham's Hypoid 80	Castrol Hypoy Light	Esso Gear Oil G.P. 80

C WATER PUMP AND LUBRICATION NIPPLES

	Mobil	Shell	BP	Filtrate	Sternol	Duckham's	Castrol	Esso
All conditions	Mobilgrease M.P.	Shell Retinax A	Energrease L. 2	Filtrate Super Lithium Grease	Ambroline L.H.T.	Duckham's L.B. 10 Grease	Castrolease L.M.	Esso Multipurpose Grease H

D UTILITY LUBRICANT, S.U. CARBURETTER DAMPERS, OILCAN POINTS, ETC.

	Mobil	Shell	BP	Filtrate	Sternol	Duckham's	Castrol	Esso
All conditions	Mobiloil Arctic	Shell X—100 20W	Energol S.A.E. 20W	Filtrate Zero 20/20W	Sternol W.W. 20	Duckham's NOL Twenty	Castrolite	Esso Extra Motor Oil 20W/30

E UPPER CYLINDER LUBRICANT

	Mobil	Shell	BP	Filtrate	Sternol	Duckham's	Castrol	Esso
All conditions	Mobil Upperlube	Shell Upper Cylinder Lubricant	Energol U.C.L.	Filtrate Petroyle	Sternol Magikoyl	Duckham's Adcoid Liquid	Castrollo	Esso Upper Cylinder Lubricant

Section P.1

DAILY SERVICE

ENGINE (A)

Inspect the oil level in the engine and refill if necessary to the 'MAX' mark on the dipstick. The oil filler cap is on top of the rocker cover and is released by turning it anti-clockwise.

Section P.2

3,000 MILES (5000 Km.) SERVICE

ENGINE OIL CHANGE (A)

Drain the oil from the engine sump after 3,000 miles (5000 km.). The drain plug is on the right-hand side of the sump and should be removed after a journey, while the oil is still warm and will drain easily.

The sump capacity is 7½ pints (9 U.S. pints, 4·26 litres). Refill the engine with new oil.

AIR CLEANERS (A)

Wash the filter element in petrol and allow to dry. Re-oil the elements and allow to drain before reassembling.

It is only necessary to withdraw the two hexagon-headed screws and lift off the outer cover to release each corrugated element. Reassemble the front element with the corrugations clear of the breather spigot in the main filter case.

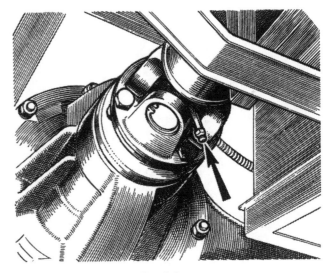

Fig. P.2
The nipple on the rear universal joint

CARBURETTER DAMPERS (D)

Unscrew the oil cap at the top of each suction chamber, pour in a small quantity of thin engine oil and replace the caps. Under no circumstances should a heavy-bodied lubricant be used. Failure to lubricate the piston dampers will cause the pistons to flutter and reduce acceleration.

STEERING GEAR (C)

Grease nipples are provided at the top and bottom of each swivel pin and on the steering tie-rods. The gun should be applied to the nipples and three or four strokes given.

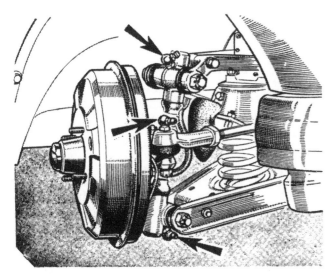

Fig. P.1
Nipples on the front suspension

Fig. P.3
Topping up the gearbox

Fig. P.4

The carburetter suction chamber damper oil must be replenished every 3,000 miles (5000 km.)

PROPELLER SHAFT (C)

The joint at each end of the propeller shaft is provided with a nipple. Later cars have a third propeller shaft nipple on the sliding spline joint.

GEARBOX (A)

Top up the oil level and ensure that the gearbox is not filled above the 'HIGH' mark on the dipstick. If the level is too high oil may get into the clutch case and cause clutch slip. The combined filler plug and dipstick is located beneath the rubber plug in the gearbox cover.

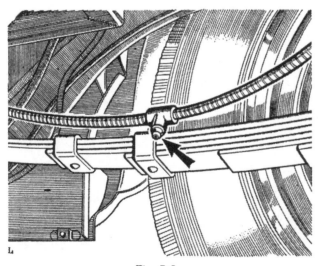

Fig. P.5

A hand brake cable greaser

REAR AXLE (B)

The combined filler and level plug is reached from below the rear of the car. The oil level should be replenished if necessary to the level of the filler plug hole. (See Fig. P.9)

NOTE.—It is essential that only Hypoid oil be used in the rear axle.

HAND BRAKE CABLE (C)

The grease nipple on the hand brake cable should be given three or four strokes with a grease gun.

Section P.3

6,000 MILES (10000 Km.) SERVICE

Carry out the instructions detailed in Section P.2 except those under 'GEARBOX' and 'REAR AXLE' (Section P.2), and continue with the following.

DISTRIBUTOR

Cam bearing (D)

Lift the rotor off the top of the spindle by pulling it squarely and add a few drops of oil to the cam bearing. Do not remove the screw which is exposed.

There is a clearance between the screw and the inner face of the spindle for the oil to pass.

Replace the rotor with its drive lug correctly engaging the spindle slot and push it onto the shaft as far as it will go.

Cam (C)

Lightly smear the cam with a very small amount of grease, or if this is not available, clean engine oil can be used.

Automatic timing control (D)

Carefully add a few drops of oil through the hole in the contact breaker base through which the cam passes. Do not allow the oil to get on or near the contacts. Do not over-oil.

Contact breaker pivot (D)

Add a spot of oil to the moving contact pivot pin.

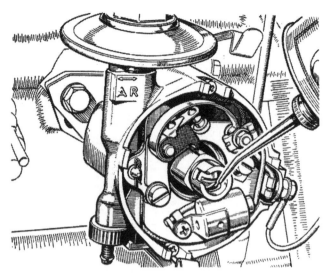

Fig. P.6

Distributor bearing lubrication

Fig. P.8

The external oil filter

GEARBOX OIL CHANGE (A)

Drain the gearbox oil.

When the gearbox has been drained completely, 4½ Imperial pints (5·6 U.S. pints, 2·56 litres) of oil are required to fill it. The oil should be poured in through the filler plug shown in Fig. P.3.

REAR AXLE OIL CHANGE (B)

Remove the drain plug and drain out the oil. Refill with Hypoid oil to the level of the filler plug hole.

Approximately 2¼ pints (2·7 U.S. pints, 1·28 litres) of oil are required to refill the axle.

ENGINE OIL FILTER (A)

Fit a new engine oil filter element. The filter is released by unscrewing the central bolt securing the filter body to the filter head. When fitting the new element, make sure that the seating washer for the filter body is in good condition and that the body is fitted securely to prevent oil leaks.

Ensure that the washers below the element inside the bowl are fitted correctly. The small felt washer must be positioned between the element pressure plate and the metal washer above the pressure spring. It is essential for correct oil filtration that the felt washer should be in good condition and be a snug fit on the centre-securing bolt.

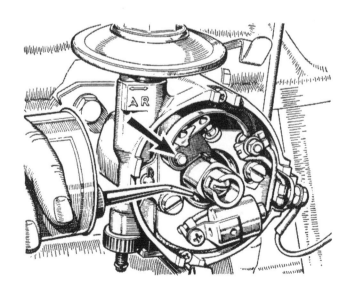

Fig. P.7

Lubricating the distributor advance mechanism

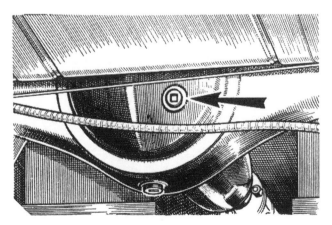

Fig. P.9

The rear axle level plug

DYNAMO (D)

Add two drops of engine oil to Ref. D, page P.2, in the lubrication hole in the centre of the rear end bearing plate.

Do not over-oil.

Section P.4

12,000 MILES (20000 Km.) SERVICE

Carry out the instructions detailed in Sections P.2 and P.3 in addition to the following.

FLUSHING THE ENGINE

Flush the engine with a flushing oil supplied by one of the recommended manufacturers (page P.2). This operation must be carried out prior to oil filter changing. Use approximately half the normal sump capacity and run the engine for 2½ to 3 minutes at a fast tick-over, after which special care must be taken to ensure complete drainage of the flushing oil.

It is recommended that at 24,000 miles (40000 km.) the sump and oil pump pick-up strainer should be removed for cleaning.

WATER PUMP (C)

Remove the water pump plug on the water pump casing and add a small quantity of grease. The lubri-

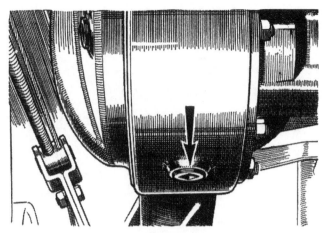

*Fig. P.*10
The rear axle drain plug

cation of the pump must be done very sparingly, otherwise grease will pass the bearings and reach the face of the carbon sealing ring, impairing its efficiency.

STEERING GEARBOX (B)

The two nipples for the steering gearbox and pinion are reached from under the bonnet.

Give the gearbox nipple 15 strokes **only,** and the pinion nipple five strokes **only** at the same time, with a gun filled with oil.

SECTION Q

SPECIAL TOOLS

Every Distributor servicing the M.G. (Series MGA) is recommended to maintain the special tools detailed in this list, as by their use damage to parts will be avoided and repairs generally will be greatly facilitated. For additional information refer to the Service Tool Catalogue (Part No. AKD770). When ordering service tools always quote the new part numbers.

Description	Original Part No.	New Part No.
Engine and clutch		
Exhaust valve seat finishing cutter	301079	18G25
Exhaust valve seat glaze breaker	—	18G25A
Exhaust valve seat top narrowing cutter	301080	18G25B
Inlet and exhaust valve seat bottom narrowing cutter	301081	18G25C
Valve seat cutter handle	301076	18G27A
Valve seat cutting tool fibre box	AJE5012	18G27B
Inlet valve seat top narrowing cutter	—	18G28B
Inlet valve seat bottom narrowing cutter	—	18G28C
Valve suction tool	—	18G29
Valve spring compressor	—	18G45
Piston ring compressor	—	18G55A
Water pump bearing remover and replacer	GT60	18G60
Oil pump relief valve grinding-in tool	—	18G69
Starting dog nut spanner	GT98	18G98
Clutch assembly gauging fixture	AJA5010	18G99A
Camshaft liner reamer (basic tool)	AJE5001	18G123A
Camshaft liner reamer cutter	—	18G123B
Camshaft liner reamer cutter	—	18G123E
Camshaft liner reamer cutter	—	18G123F
Camshaft liner reamer pilot	—	18G123L
Camshaft liner reamer pilot	—	18G123T
Camshaft liner reamer pilot	—	18G123AB
Camshaft liner reamer pilot	—	18G123AC
Camshaft liner reamer pilot	—	18G123AD
Fibre box	—	18G123AL
Camshaft liner remover and replacer (basic tool)	AJA5060	18G124A
Camshaft liner remover adaptor	—	18G124B
Camshaft liner remover adaptor	—	18G124C
Camshaft liner remover adaptor	—	18G124F
Camshaft liner remover adaptor	—	18G124H
Inlet valve seat finishing cutter	—	18G174
Inlet valve seat glaze breaker	—	18G174A
Valve seat cutter pilot	18G31	18G174D
Valve rocker bush remover and replacer	—	18G226
Clutch plate centralizer	67833	18G279
Impulse extractor—UNF.	301203	18G284
Main bearing cap removal adaptor	AJH5147	18G284A
Valve suction tool	66893	18G328

Description			Original Part No.	New Part No.
Gearbox				
Bearing and oil seal replacer adaptor handle (basic tool)	..	..	—	**18G134**
Gearbox extension oil seal replacer adaptor	..	.. .. ..	—	**18G134M**
Gearbox extension oil seal replacer adaptor	..	.. .. ..	—	**18G134N**
Gearbox front cover oil seal replacer	..	.. .. ..	—	**18G134Q**
Synchromesh unit assembly ring—second speed	..	.. ..	AJA5055	**18G222**
Synchromesh unit assembly ring—third speed and top	..	..	AJA5056	**18G223**
Dummy layshaft	..	..	AJA5057	**18G266**
Oil seal remover (basic tool)	..	.. ..	—	**18G389**
Oil seal remover adaptor	..	.. ..	—	**18G389B**
Gearbox front cover centralizer	..	.. ..	—	**18G598**
Front suspension and steering				
Front and rear hub plate remover (basic tool)	..	.. ..	AJA5019	**18G304**
Bolt adaptor—$\frac{7}{16}$ in. UNF.	..	.. ..	AJA5021	**18G304B**
Steering-wheel remover	..	.. ..	68827	**18G310**
Steering tie-rod pin spanner	..	.. ..	68965	**18G312**
Steering tie-rod 'C' spanner	..	.. ..	300813	**18G313**
Wire wheel hub remover..	..	.. ..	—	**18G363**
Rear axle and rear suspension				
Bevel pinion flange wrench	..	.. ..	AJA5062	**18G34A**
Differential bearing remover (basic tool)	..	.. ..	AJA5061	**18G47C**
Differential bearing remover adaptors ..	..	.. ..	—	**18G47T**
Bearing and oil seal replacer adaptor handle (basic tool)	..	..	—	**18G134**
Rear hub bearing remover, differential bearing replacer, and rear hub assembly replacer adaptor			—	**18G134P**
Bevel pinion and differential bearing setting gauge ..	..	..	—	**18G191B**
Bevel pinion bearing outer race remover (basic tool)	..	..	—	**18G264**
Bevel pinion bearing outer race remover adaptor	..	.. ..	—	**18G264E**
Bevel pinion bearing outer race remover adaptor	..	.. ..	—	**18G264F**
Partitioned fibre box	..	.. ..	—	**18G264K**
Rear hub nut spanner	..	.. ..	—	**18G152**
Bevel pinion bearing preload gauge . ..	..	.. ..	—	**18G207**
Bevel pinion inner race remover and replacer	..	.. ..	—	**18G285**
Front and rear hub plate remover (basic tool)	..	.. ..	AJA5019	**18G304**
Bolt adaptor—$\frac{7}{16}$ in. UNF. (2 off) ..	..	.. ..	AJA5021	**18G304B**
Hub remover thrust pad adaptor .. .	.	.. ..	AJA5146	**18G304J**
Wire wheel hub remover..	..	.. ..	—	**18G363**
Body				
Body jack and metal case	..	.. ..	—	**18G308B**
Miscellaneous				
Lockheed bleeder screw wrench ..	..	.. ..	46746	**18G353**
Torque wrench	..	..	AJE5010	**18G372**

18G25. Exhaust Valve Seat Finishing Cutter
Use with pilot 18G174D and handle 18G27A.

18G25

18G25A. Exhaust Valve Seat Glaze Breaker
Use with pilot 18G174D and handle 18G27A.

18G25A

18G25B. Exhaust Valve Seat Top Narrowing Cutter
Use with pilot 18G174D and handle 18G27A.

18G25B

18G25C. Inlet and Exhaust Valve Seat Bottom Narrowing Cutter
Use with pilot 18G174D and handle 18G27A.

18G25C

18G27A. Valve Seat Cutter Handle
A standardized type of handle for use with a wide range of cutters. A spanner can be used on the square end when cutting valve seats on models where the tommy bar will foul the engine bulkhead.

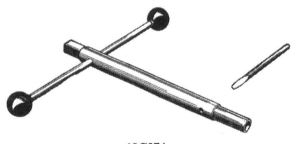

18G27A

18G28B. Inlet Valve Seat Top Narrowing Cutter
Use with pilot 18G174D and handle 18G27A.

18G28B

18G28C

18G28C. Inlet Valve Seat Bottom Narrowing Cutter
Use with pilot 18G174D and handle 18G27A.

18G29

18G29. Valve Suction Tool
A handle complete with a detachable suction pad 18G29A; a further suction pad 18G29B is available for use on smaller valves.

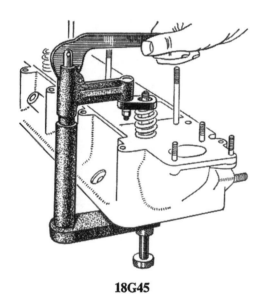

18G45

18G45. Valve Spring Compressor
This tool is designed for overhead-valve engines. It has a cam and lever action and screw adjustment. The adaptor ring is shaped to facilitate the fitting of cotters.

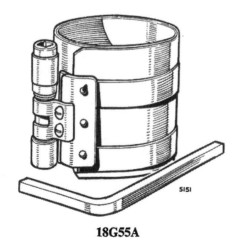

18G55A

18G55A. Piston Ring Compressor
Designed to cover a wide range of pistons, it is easy to operate and will compress the strongest piston ring, making assembly to the bore a quick and easy operation.

18G60. Water Pump Bearing Remover and Replacer

To safeguard against broken pump bodies this specially designed tool should be used when removing and replacing bearings. Comprising a drift, a dummy bearing, and a pilot, it aligns each bearing with its housing before the bearing is pressed into position.

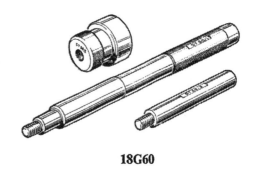

18G60

18G69. Oil Pump Relief Valve Grinding-in Tool

The small knurled knob at the end of this tool is turned to compress the rubber sleeve and increase its diameter until, when pressed into the valve, it will hold it securely while it is lapped to its seat.

18G69

18G98. Starting Dog Nut Spanner

A 'shock-type' spanner designed to enable the starting dog nut to be removed without locking the crankshaft.

18G98

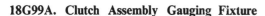

18G99A. Clutch Assembly Gauging Fixture

This tool provides for accurate adjustment of the levers and, additionally, affords a fixture upon which to dismantle and assemble the unit.

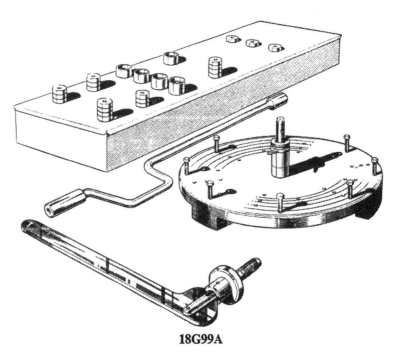

18G99A

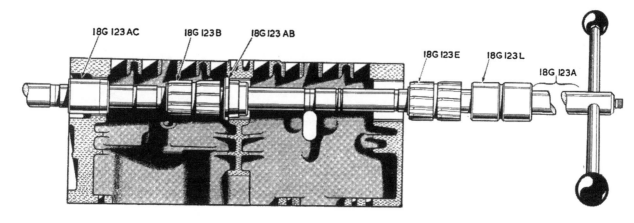

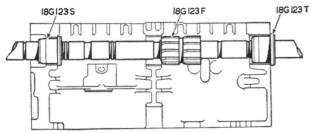

18G123A. Camshaft Liner Reamer (basic tool), Cutters, and Pilots

This equipment is essential when reconditioning cylinder blocks, otherwise camshaft liners cannot be reamed in line, and in consequence the clearance between the camshaft journal and liner will be incorrect. This basic tool must be used with the cutters and pilots supplied separately and shown on this page. Full instructions for using the equipment will be supplied with each basic tool.

18G123B. Camshaft Liner Reamer Cutter

18G123E. Camshaft Liner Reamer Cutter

18G123F. Camshaft Liner Reamer Cutter

18G123L. Camshaft Liner Reamer Pilot

18G123T. Camshaft Liner Reamer Pilot

18G123AB. Camshaft Liner Reamer Pilot

18G123AC. Camshaft Liner Reamer Pilot

18G123AD. Camshaft Liner Reamer Pilot

18G123AL. Camshaft Liner Reamer Component Fibre Box

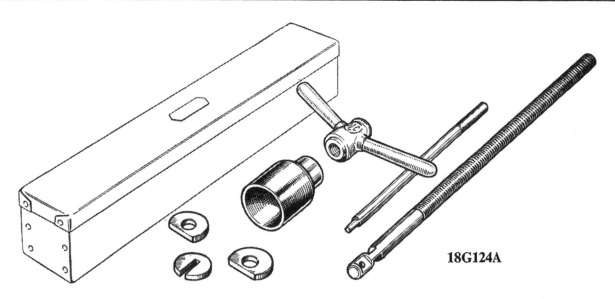

18G124A

18G124A. Camshaft Liner Remover and Replacer (basic tool)

The equipment consists of a basic tool 18G124A and various adaptors (supplied separately) for different types of engines and for this model shown below. Liners can be removed and replaced without the damage invariably associated with the use of improvised drifts. Full instructions for using the equipment will be supplied with each basic tool.

18G124B, 18G124C, 18G124F, and 18G124H. Camshaft Liner Remover Adaptors

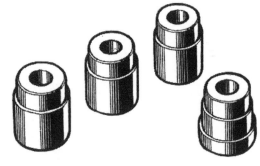

18G124B
18G124C

18G124F
18G124H

18G174. Inlet Valve Seat Finishing Cutter

Use with pilot 18G174D and handle 18G27A.

18G174

18G174A. Inlet Valve Seat Glaze Breaker

Use with pilot 18G174D and handle 18G27A.

18G174A

18G174D. Valve Seat Cutter Pilot

Use with the cutters listed above and handle 18G27A.

18G174D

18G226

18G226. Valve Rocker Bush Remover and Replacer

The flange of the driver is recessed to prevent the split bush opening when being driven into position; the anvil is also recessed to retain the rocker during the operation. Use of a light press is desirable when using this tool, alternatively a vice or copper-faced hammer may be used.

18G279

18G279. Clutch Plate Centralizer

When reassembling the single-plate clutch it is essential to use this tool to ensure that the clutch plate is concentric with the spigot bearing in the flywheel centre, otherwise it is impossible to assemble the gearbox to the engine. This tool will also fit the Wolseley Four-Fifty, Morris Oxford (Series MO), and Morris Six (Series MS) models.

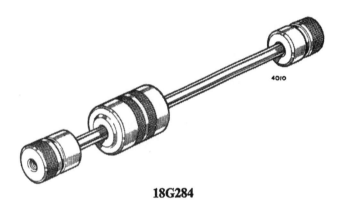

18G284

18G284. Impulse Extractor—UNF.

This universal impulse extractor when used with adaptor 18G284A will remove the main bearing cap.

18G284A

18G284A. Main Bearing Cap Removal Adaptor

The adaptor for use with impulse extractor 18G284.

18G134

18G134. Bearing and Oil Seal Replacer Adaptor Handle (basic tool)

For use with adaptor 18G134M or 18G134N to replace the gearbox extension oil seal.

18G134M and 18G134N. Oil Seal Replacer Adaptors

Used with handle 18G134 for the replacement of gearbox extension oil seals.

For later power units (from Engine No. 10990) use adaptor 18G134N.

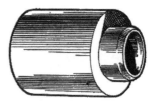

18G134M and 18G134N

Synchromesh Unit Assembly Rings
18G222—Second Speed
18G223—Third Speed and Top

This tool is specially designed to retain the balls and springs in the synchronizer hub while it is being pushed into the sliding sleeve.

18G222 and 18G223

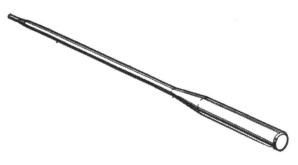

18G266. Dummy Layshaft

Reassembly of the layshaft into the laygear is greatly simplified by the use of this tool.

18G266

18G389. Oil Seal Remover (basic tool)

This basic tool together with the appropriate adaptor is essential for removing the gearbox extension oil seal easily and without damage to the extension.

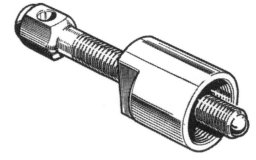

18G389

18G389B. Oil Seal Remover Adaptor

Use with basic tool 18G389.

18G389B

18G310

18G312

18G313

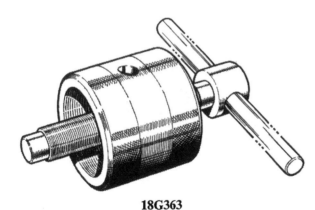

18G363

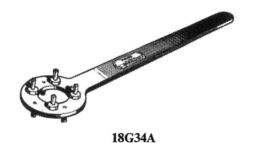

18G34A

18G310. Steering-wheel Remover

This extractor has been specially designed to remove the steering-wheel without damage. Dealers who already possess tool 55418 and the attachment (Part No. 56052) will find that this may also be used to withdraw the steering-wheel. It can also be used to extract Wolseley and Morris steering-wheels.

18G312. Steering Tie-rod Pin Spanner

This tool is designed for use in conjunction with the special spanner 18G313 for dismantling the steering tie-rod ball housing without damage.

In use it is clamped securely in a vice and the two holes of the ball housing are engaged with the two pins of the extractor, pushing the ball housing into the tool as far as it will go. Using the special spanner 18G313, engage the claws with the ball housing cap and unscrew it from the housing.

18G313. Steering Tie-rod 'C' Spanner

A tool with jaws designed to engage the shallow splines of the steering rack ball housing cap and remove it without damage.

18G363. Wire Wheel Hub Remover

Designed to withdraw left-hand and right-hand 'knock-on' hubs. The body is internally threaded with a left-hand thread in one end and a right-hand thread in the other.

18G34A. Bevel Pinion Flange Wrench

Used to hold the bevel pinion flange when unscrewing the flange retaining nut.

18G47C. Differential Bearing Remover

This basic tool simplifies the removal of the differential cage bearings without risk of damage. The adaptor for use on the 'B' type axle is shown under.

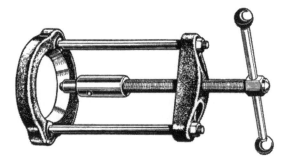

18G47C

18G47T. Differential Bearing Remover Adaptors

Designed for use with basic tool 18G47C (above), these adaptors must be used when removing the differential cage bearings.

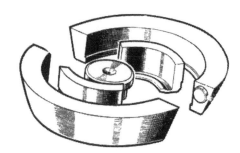

18G47T

18G134P. Rear Hub Bearing Remover, Differential Bearing Replacer, and Rear Hub Assembly Replacer Adaptor

Use with handle 18G134.

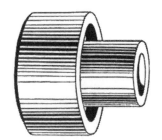

18G134P

18G191B. Bevel Pinion and Differential Bearing Setting Gauge

Indispensable in adjusting and setting 'A', 'B', and 'C' type standardized axles.

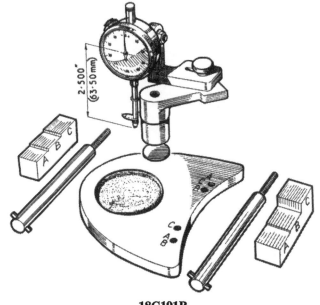

18G191B

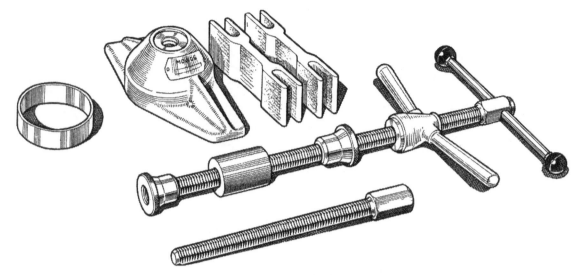

18G264. Bevel Pinion Bearing Outer Race Remover (basic tool)

Comprising a body, centre screw with extension and tommy-bar, wing nut, guide cone, and two distance pieces. A plain ring is also included to serve as a pilot when the rear bearing outer races are being replaced. Use with adaptors 18G264E and 18G264F.

18G264E, 18G264F

18G264E and 18G264F. Bevel Pinion Outer Race Remover Adaptors

Use with 18G264.

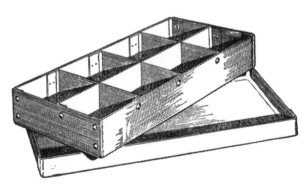

18G264K

18G264K. Partitioned Fibre Box

A strong fibre box for storing the bevel pinion bearing outer race remover adaptors.

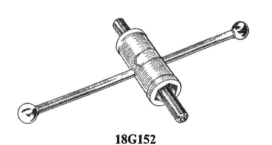

18G152

18G152. Rear Hub Nut Spanner

Designed to fit the octagonal nut on 'B' type rear axles. The opposite end may be used on the smaller 'A' type axles.

18G207. Bevel Pinion Bearing Preload Gauge

The movable arms of the tool are located in opposite holes of the bevel pinion flange and the weight moved along the rod to the poundage required.

18G207

18G285. Bevel Pinion Inner Race Remover and Replacer

This tool is necessary for withdrawing the inner bearing race from the pinion shaft. It can also be used for replacing the race on the shaft without damage.

This is a universal tool.

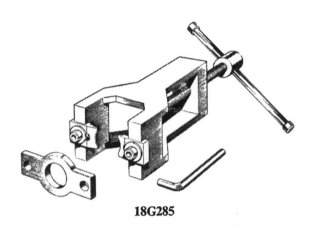

18G285

18G304, 18G304B, and 18G304J. Front and Rear Hub Plate Remover (basic tool)

The plate in combination with two bolts and plug will remove the front and rear hubs of the M.G. 'MGA' with pressed-steel wheels.

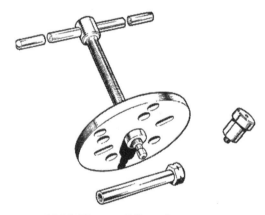

18G304 (Plate and Screw)
18G304B ($\frac{7}{16}$ in. UNF Bolt)
18G304J (Rear Axle Plug)

18G372. Torque Wrench

A universal torque spanner for use with standard sockets. This tool is essential if the recommended maximum torque for various studs is not to be exceeded.

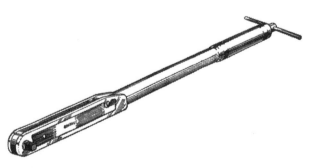

18G372

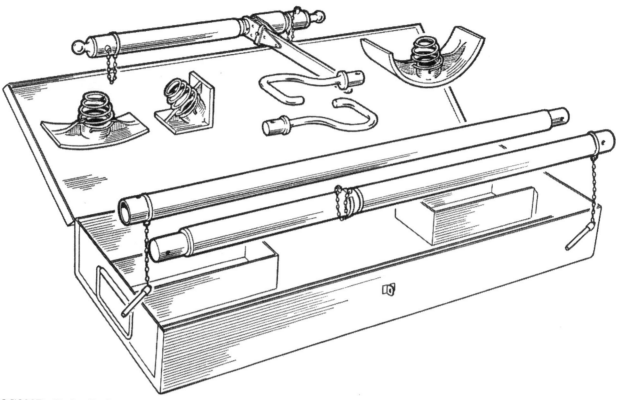

18G308B. Body Jack

The jack is a tool which has been designed to deal with repairs to bodies of all-steel construction. It is supplied in a metal case complete with the various attachments and will be found capable of dealing with all normal requirements.

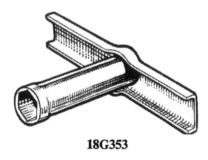

18G353

18G353. Lockheed Bleeder Screw Wrench

This specially designed tube spanner and integral tommy-bar greatly assists the brake bleeding operation. The spanner remains square on the bleeder screw without disturbing the bleed tube.

18G598. Gearbox Front Cover Centralizing Tool

In the case of front covers fitted with oil seals, the use of this tool is essential to ensure concentricity of the front cover with the mainshaft, thereby obviating any possibility of oil leaks.

18G598

18G134Q. Rear Hub Bearing Remover, Differential Bearing Replacer, and Rear Hub Assembly Replacer Adaptor

Use in conjunction with detachable handle 18G134 to replace the gearbox front cover oil seal. Its other uses are applicable to other B.M.C. models only.

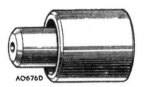

18G134Q

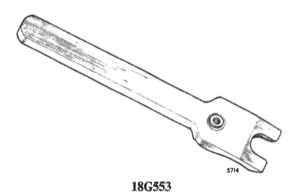

18G553. Disc Brake Piston Resetting Tool (Dunlop)

This tool is used to press the operating piston down the cylinder bore before refitting a friction pad.

18G553

18G590. Disc Brake Piston Resetting Tool (Lockheed)

This tool is used for refitting the operating piston and seals. It is also used for resetting the piston to the zero position before fitting a friction pad.

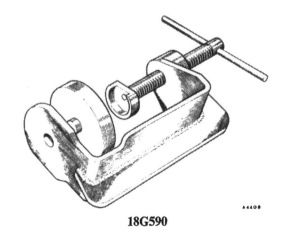

18G590

Notes

SECTION R

CHASSIS

General.

THE CHASSIS, RADIATOR, EXHAUST AND FUEL SUPPLY COMPONENTS

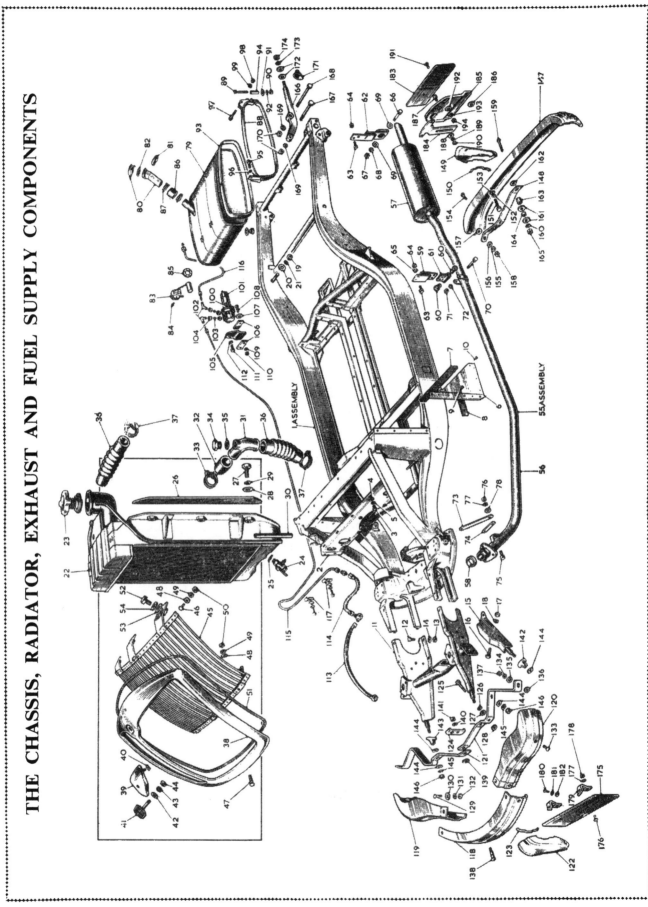

R

KEY TO THE CHASSIS, RADIATOR, EXHAUST AND FUEL SUPPLY COMPONENTS

No.	Description
1.	Chassis frame complete.
2.	Toe-board support plate assembly.
3.	Screw—toe-board plate to goalpost.
4.	Washer—plain—plate to post screw.
5.	Washer—spring—plate to post screw.
6.	Cover plate L/H—goalpost.
7.	Felt seal—cover plate—lower.
8.	Felt seal—cover plate—upper.
9.	Screw—cover plate to post (vertical).
10.	Screw—cover to post stay.
11.	Front extension assembly.
12.	Screw—extension to frame bracket.
13.	Nut—extension to frame bracket screw.
14.	Washer—spring.
15.	Bracket assembly L/H—bumper support.
16.	Screw—bracket assembly to front extension.
17.	Nut—bracket to front extension screw.
18.	Washer—spring.
19.	Nut—rebound strap spindle.
20.	Washer—plain—rebound strap spindle.
21.	Washer—spring—rebound strap spindle.
22.	Radiator block assembly.
23.	Cap—filler.
24.	Tap—drain.
25.	Washer—tap.
26.	Packing piece—block to body.
27.	Screw—block to body.
28.	Washer—plain—block to body.
29.	Washer—spring—block to body.
30.	Drain tube.
31.	Pipe—water pump connector.
32.	Hose—connector pipe.
33.	Clip—hose.
34.	Plug—connector pipe.
35.	Washer—plug.
36.	Hose—top and bottom.
37.	Clip—1⅛″ dia. hose (top and bottom).
38.	Case sub-assembly.
39.	False nose assembly.
40.	Speed fix (¼″)—nose to case.
41.	Badge.
42.	Washer—plain—badge.
43.	Washer—double coil—badge.
44.	Nut—badge.
45.	Grille assembly.
46.	Stud—grille retaining.
47.	Stud—grille lower fixing.
48.	Washer—grille lower fixing stud.
49.	Washer—spring—grille studs.
50.	Nut—grille studs.
51.	Piping—grille.
52.	Screw—case grille to body.
53.	Washer—plain—case/grille to body.
54.	Washer—spring—case/grille to body.
55.	Exhaust pipe assembly.
56.	Pipe—front.
57.	Silencer and tail pipe.
58.	Packing—exhaust pipe.
59.	Mounting—exhaust—intermediate.
60.	Bush—intermediate exhaust mounting.
61.	Housing—intermediate exhaust housing.
62.	Mounting—exhaust—rear.
63.	Screw—mountings to frame.
64.	Nut—mountings screw.
65.	Washer—spring.
66.	Bolt—rear mounting and tail pipe.
67.	Nut—rear mounting bolt.
68.	Washer—plain.
69.	Washer—plain.
70.	Bolt—bush and housing to pipe.
71.	Nut—bush and housing bolt.
72.	Washer—spring.
73.	Strap—upper front mounting.
74.	Strap—lower front mounting.
75.	Screw—strap to front pipe.
76.	Nut—strap screw.
77.	Washer—spring.
78.	Washer—plain.
79.	Petrol tank.
80.	Filler extension.
81.	Sealing ring.
82.	Ferrule—filler extension.
83.	Tank unit attachment.
84.	Screw—unit to tank.
85.	Gasket—cork—tank attachment.
86.	Hose—filler connection.
87.	Clip—hose.
88.	Strap and mounting assembly.
89.	Bolt.
90.	Nut.
91.	Washer—plain.
92.	Washer—spring.
93.	Packing strap—tank mounting.
94.	Distance-piece—rear mounting.
95.	Screw—front mounting.
96.	Washer—spring—front mounting.
97.	Bolt—rear mounting.
98.	Nut—rear mounting.
99.	Washer—spring—rear mounting.
100.	Pump.
101.	Cover—pump.
102.	Elbow—petrol pump.
103.	Olive—petrol pump.
104.	Nut—petrol pump olive.
105.	Bracket—pump to frame bracket.
106.	Rubber—bracket.
107.	Washer—special large—pump mounting.
108.	Bolt—pump mounting.
109.	Nut—pump mounting.
110.	Locknut—pump mounting.
111.	Screw—pump to bracket.
112.	Washer—spring—pump to bracket.
113.	Pipe—flexible—between carburetters.
114.	Pipe—flexible—main pipe to carburetter.
115.	Pipe assembly—pump to carburetters.
116.	Pipe assembly—tank to pump.
117.	Clip—petrol pipe to bulkhead.
118.	Centre bar.
119.	Corner bar—R/H.
120.	Corner bar—L/H.
121.	Main spring.
122.	Over-rider assembly.
123.	P.V.C. moulding.
124.	Bracket—starting handle.
125.	Bolt—over-rider to bumper bar.
126.	Washer—spring—over-rider to bumper.
127.	Washer—plain over-rider to bumper.
128.	Washer—special plain—over-rider to bumper.
129.	Bolt—corner bar to centre bar.
130.	Washer—special plain—corner to centre bar.
131.	Washer—spring—corner to centre bar.
132.	Nut—centre bar to corner bars.
133.	Bolt—mainspring to corner bar.
134.	Washer—spring—mainspring to corner bar.
135.	Washer—plain—mainspring to corner bar.
136.	Washer—special plain.
137.	Nut—mainspring to corner bar.
138.	Bolt—mainspring to centre.
139.	Washer—spring—mainspring to centre.
140.	Washer—special plain—mainspring to centre.
141.	Nut—mainspring to centre.
142.	Grommet—outer stud.
143.	Grommet—inner stud.
144.	Washer—plain.
145.	Washer—spring.
146.	Nut—bumper support stud.
147.	Bar.
148.	Mainspring.
149.	Over-rider assembly.
150.	P.V.C. moulding.
151.	Bolt—over-rider to bumper bar.
152.	Washer—spring—over-rider to bumper.
153.	Washer—special—over-rider to bumper.
154.	Bolt—outer—mainspring to bar.
155.	Washer—spring—mainspring to bar.
156.	Washer—plain—mainspring to bar.
157.	Washer—special plain—mainspring to bar.
158.	Nut—mainspring to bar.
159.	Bolt—inner—mainspring to bar.
160.	Washer—spring—mainspring to bar.
161.	Washer—plain—mainspring to bar.
162.	Washer—special plain—mainspring to bar.
163.	Distance-piece—mainspring to bar.
164.	Distance tube—mainspring to bar.
165.	Nut—mainspring to bar.
166.	Bracket L/H—rear bumper mounting.
167.	Bolt—bracket to frame.
168.	Bolt—bracket to frame.
169.	Washer—spring.
170.	Nut—bracket to frame bolt.
171.	Rubber grommet—bracket.
172.	Washer—plain—bracket and bumper.
173.	Washer—spring—bracket and bumper.
174.	Nut—bracket and bumper.
175.	Front number-plate.
176.	Screw—plate to bracket.
177.	Washer—spring—plate to bracket.
178.	Nut—plate to bracket screw.
179.	Bracket L/H—number-plate.
180.	Screw—brackets to bumper.
181.	Washer—spring—brackets to bumper.
182.	Nut—brackets to bumper.
183.	Rear number-plate.
184.	Lamp bracket.
185.	Bracket—number-plate mounting.
186.	Rubber washer—number-plate.
187.	Screw—lamp bracket to mounting bracket.
188.	Washer—plain—bracket to bracket.
189.	Washer—spring—bracket to bracket.
190.	Locknut—bracket to bracket.
191.	Screw—number-plate to bracket.
192.	Washer—plain—plate to bracket.
193.	Washer—spring—plate to bracket.
194.	Nut—plate to bracket.

GENERAL

This section deals with the repair of the M.G. type of box-sectioned chassis frame damaged in accident, where the facilities as used by frame manufacturers are not available. The manufacturers, naturally, have the benefit of their production equipment, but the methods adopted by them, particularly in regard to the use of assembly jigs and welding equipment, are out of reach of the average repair organization.

These instructions will therefore deal mainly with methods of repairing damage to chassis frames without dismantling the component parts, i.e. breaking down welds, any more than is absolutely necessary to eliminate torn or badly buckled metal or deformed cross-members which are beyond economic repair.

Repairs carried out in this manner fall into two categories:—

(a) Repair of the frame in position in the vehicle, which may be regarded as an emergency repair, and

(b) Repair of the frame out of the vehicle, in which complete rectification of the chassis frame is attempted.

In general terms it may be stated that chassis frames with considerable damage may be recovered sufficiently to be serviceable units, but, naturally, the skill and experience of the repairer and the extent of the equipment available will determine whether any particular frame is repairable, bearing in mind that there are certain fundamental accuracies to be restored, also that the cost of labour and material involved in effecting a complete repair will not always be economically justified if it exceeds the cost to the user of replacing the entire frame assembly.

Damage to a frame is usually a combination of torn and buckled metal, either in side-members or cross-members, and lateral or vertical displacement of side-members, causing misalignment. The resultant repair is concerned with elimination of the local damage by smoothing or renewal of parts—generally both—and the recovery of alignment of the frame as a whole.

In practice, this result is achieved by carrying out the local repairs to the metal and applying the necessary corrective loads to the side-members, coupled with the judicious use of heat to the strains to permit the members to recover their natural positions.

Section R.1

FRAME REPAIRS

As the heating of the frame plays a vital part in its repair, it is essential that a good gas-welding equipment is available. At certain points spot and arc welding are

preferable, but a skilled gas welder will be able to make effective welds with the use of a gas-welding torch only.

A screw or hydraulic jack is then required, with a few adaptors to make it of universal application. A selection of bending irons, some metal-straightening equipment, such as dolly blocks, spoons, levers and hammers, will complete the equipment necessary. If a suitable hydraulic press is available, this in conjunction with hardwood reaction blocks would be more preferable than some of the methods detailed later, using jacks.

When the frame is heated for straightening, the area affected should be maintained at a cherry-red throughout the entire straightening operation.

When an acetylene torch is used for heating, a " neutral " flame should be employed and played over the entire area to be heated until the metal has reached a uniform cherry-red. Never heat the metal beyond a cherry-red as it will seriously weaken the steel. It is good practice to frequently check the temperature of the heated metal with a dry pine stick, while it is being worked, to maintain it at the proper state of ductility and avoid burning. Touching a dry pine stick to metal that has been heated to a cherry-red will cause the stick to glow and char, but not to ignite. The heated area of the frame should be protected from draughts to prevent sudden cooling of the metal.

An important point to observe here, prior to commencing repairs, is in regard to the front suspension cross-member. Correct alignment of the front suspension is of such vital importance that if there is appreciable distortion of this member it should be renewed, due to the fact that it is a very difficult operation to re-form it to its exact shape.

Note.—We do not recommend that this operation be carried out by the Distributor or Dealer unless adequate assembly fixtures are available, but that the frame be returned to the Service Department, The M.G. Car Co. Ltd., Abingdon, Berks.

Section R.2

INVESTIGATION OF DAMAGED CHASSIS

Although in most cases of accident the resultant primary damage to the frame is readily apparent, there are cases where the damage may only be slight and is masked by the wings and body structure, and in such cases it may be necessary to carry out a complete check of chassis alignment, including front suspension and rear axle, to determine the full extent of the damage.

When checking cars damaged in accident, it is most essential to do the checking on a flat surface large enough to receive the complete car. It is preferable to use a large iron slab, but a concrete slab carefully

*R.*4

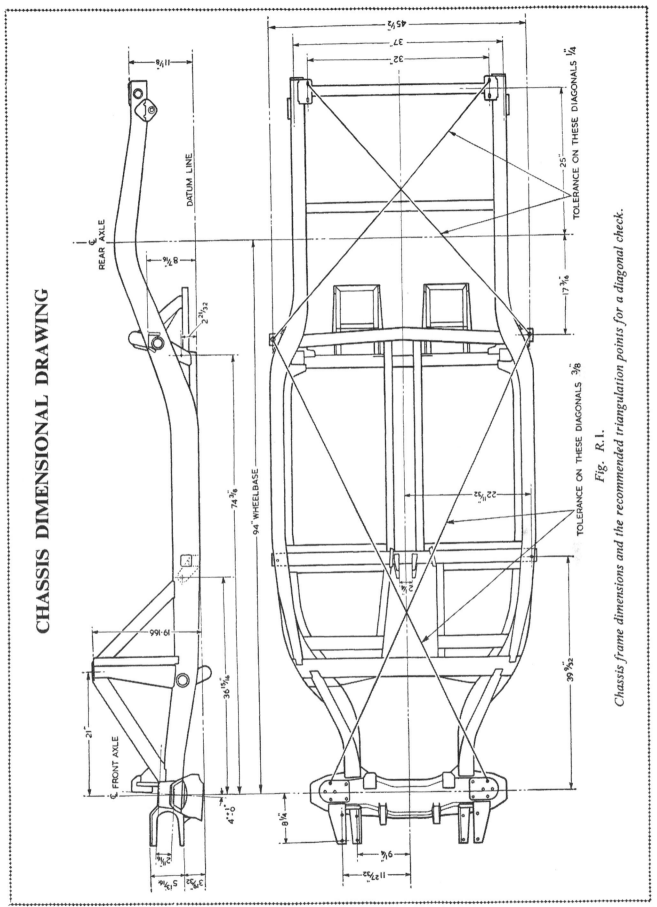

CHASSIS DIMENSIONAL DRAWING

Fig. R.1.

Chassis frame dimensions and the recommended triangulation points for a diagonal check.

prepared and hand-surfaced will be suitable. The car may then be directly checked by comparative measurements or a centre line dropped down from the front and rear centre of the frame and parallel track lines laid out. From these lines the squareness of the car may easily be checked.

Section R.3

A TWISTED FRAME

Checking the alignment of the frame bare is relatively a simple matter, especially if the frame can be set up on a large flat surface or face plate. It involves establishing a datum or centre line, from which all measurements can be taken. Diagonals are checked from suitable fixed locating points, which can be cross-checked at the centre line on which the diagonals should cross, as detailed in chassis dimensional drawing Fig. R.1. The angle of the front cross-member should be 4° but may be given an allowance of $+1°$ or $-0°$. Diagonal measurements quickly determine which section of the frame is bent.

Accuracies of side-members are usually checked with suitable straight-edges, and squareness of side rails is checked with straight-edge and square. Twist is checked visually against straight-edges laid transversely across the frame at suitable points.

In the event of the frame being twisted, this condition can be corrected by anchoring the frame to a fixed trestle and by using a suitable lever or a stout beam of timber. The frame can then be sprung back with effort applied at the end of the lever.

If necessary apply a well-spread heat at the twisted section.

The frame being completely dismantled, it is possible to remove all signs of damage by cutting out holes for access in the inner liner plates with the welding torch, hammering out bulges, dents or buckled areas, and closing the holes by welding in the piece previously removed. The welds can be cleaned up and the repair is then invisible.

When any adjustment to the frame is carried out do not forget the light gauge of the material.

Final checking of the frame should be carried out in accordance with the dimensions and diagonals indicated.

Comparative vertical measurements should reveal any frame twist.

Check the wheel camber, castor angle, king-pin angle and front wheel alignment.

Section R.4

DIAGONAL FRAME CHECK (COMPLETE CAR)

In checking the frame for distortion, diagonal measurements may be taken without removing the body from the chassis by using a plumb-bob as follows:—

Place the car on a level surface and block up the car equally at each wheel approx. 12 in. (30 cm.) high with all tyres properly inflated.

Perform the measuring with accuracy and care.

Suspend the plumb-bob from various corresponding points on the frame, such as indicated by the diagonal lines in Fig. R.1. The plumb-bob should be suspended slightly above the floor. When the plumb-bob comes to rest, mark the floor directly underneath it. The marks made on the floor will represent various points of the frame to be checked diagonally.

Measure the diagonal distance between the points; this distance should agree within $\frac{1}{4}$ to $\frac{3}{8}$ in. (6·5 to 10 mm.).

Care must be taken to see that any two diagonals compared represent exactly corresponding points on each side of the frame.

Upon the result of this preliminary investigation a decision can be taken whether the frame can be repaired in position or whether the frame must be stripped out completely.

SECTION S

BODY

Fig. S.1

Withdrawing a door lock

Section S.1

REMOVING A DOOR LOCK

Take out the three securing screws and remove the sidescreen clamp plate.

Unscrew the 10 recessed-headed screws securing the trim panel to the door. Lower the trim panel downwards to free it from the lipped rail along the bottom edge of the pocket.

Remove the nut and bolt securing the door lock cable to the bracket in the top forward corner of the pocket. Note which one of the four holes is used in the bracket to secure the cable.

Take out the four screws securing the lock to the door panel and withdraw the lock through the opening at the top of the door pocket, at the same time feeding the cable through the grommet.

Reassembly is a reversal of the dismantling procedure.

Section S.2

REMOVING THE WINDSHIELD

Unscrew the six recessed-headed screws securing the interior trim panel on either side of the car, forward of the doors.

Remove the two pieces of sealing material which cover the windshield securing bolt holes and unscrew the bolts, taking care not to drop the plain and spring washers between the two body panels.

Take out the three screws securing the windshield frame to the hand-grip on each side and lift away the windshield.

Reassembly is a reversal of the dismantling procedure. Replace the pieces of sealing material with Bostik.

Section S.3

REMOVING THE FRONT BUMPER

The front bumper is secured to four mounting brackets attached to the front frame extension assembly. Remove the four nuts and spring and plain washers.

If necessary, release one of the outer bumper mounting brackets from the frame extension by unscrewing the three nuts and bolts.

The bumper may now be withdrawn forward from the brackets.

One bolt secures each over-rider to the bumper. Unscrew the bolt and the over-rider will become detached.

Section S.4

REMOVING THE REAR BUMPER

The rear bumper is secured to two mounting brackets attached to the rear of the frame.

Remove the two bumper securing nuts and spring and plain washers.

Disconnect the wiring to the rear number-plate light. Withdraw the bumper from the mounting brackets.

Each over-rider is secured to the bumper by one bolt.

Section S.5

REMOVING THE HOOD

Place the hood in the folded position.

Remove from one side the three recessed-headed screws securing the hood frame to the body. With an assistant holding the free end, remove the three screws securing the hood on the opposite side.

The hood may now be lifted away.

Section S.6

REMOVING A REAR WING

Disconnect the wiring to the rear lamps (see Section N).

Remove from inside of the rear wing the five bolts and spring and plain washers securing the wing to the body. Five more bolts are situated behind the baffle panel inside the rear of the wing.

Remove the bolt securing the baffle panel to the wing flange.

Gently ease away the rear portion of the trim panel situated behind the seats and remove the bolt securing the upper corner of the wing.

Remove the two nuts and bolts from the forward end of the wing on the underside.

Remove the three recessed-headed screws securing the wing to the door striker panel.

Remove the wing rearwards to disengage it from the flange of the door striker panel.

Reassembly is a reversal of the dismantling procedure.

When refitting the wing ensure that the piping is correctly and neatly positioned before finally tightening the wing bolts.

Section S.7

REMOVING A FRONT WING

Disconnect the wiring to the lamps (see Section N).

Remove the four nuts and bolts from the rear underside of the front wing, also the six bolts, spring and plain washers securing the baffle panel to the body. Remove the panel.

Take out the three bolts situated just above and to the outside rear of the front bumper.

Remove from inside of the wing the nine bolts, spring and plain washers securing the wing to the body and also two more situated under the bonnet in the rear corner.

Remove the trim panel from inside the car forward of the doors (see Section S.2) and take out the two bolts situated at the top. Remove the four remaining bolts running down the side of the body panel and the wing may be lifted away.

Have an assistant to bear the weight of the wing while the securing bolts are being removed.

Reassembly is a reversal of the dismantling procedure.

When refitting the wing ensure that the piping is correctly and neatly positioned before finally tightening the wing bolts.

Fig. S.2.

Showing the rear wing attachment points.

Fig. S.3.

The front wing attachment points.

Section S.8

REMOVING THE BODY

The following items must be disconnected or dismantled when removing the body.

Wiring

Disconnect the positive lead from the battery.

Disconnect the horn wires and the wires from the dynamo and " SW " connection on the coil. Disconnect the low-tension lead from the starter solenoid and finally the snap-connectors situated at the rear of the front wheel arch.

Pipes, Controls, etc.

Unscrew the oil-gauge pipe union nut from the flexible hose adaptor.

Disconnect the hydraulic clutch pipe from the flexible hose union, and detach the brake pipe from the three-way connector on the frame.

Disconnect the speedometer cable from the gearbox, the cable clip on the engine bulkhead, and the tachometer drive cable from the engine.

Unscrew the gland-nut and remove the thermal transmitter from the engine. Release the clip securing the thermo-coupling to the right-hand side of the cylinder head.

Release the mixture control cable from the carburetters.

Unscrew the bolts and remove both the air-cleaners.

Drain and remove the radiator (see Section C.4).

Remove the front and rear bumpers (see Sections S.3 and S.4). The rear bumper support brackets must be removed by undoing the two nuts and bolts securing them to the chassis at either side. Three nuts and bolts secure each outer front bumper support bracket to the front extension, and these bolts and brackets must be removed.

Remove the fuel tank (see Section D.1).

Remove the nut and bolt securing the top steering column clamp to the body bracket and release the clamp.

Unscrew the nine bolts situated along the top forward edge of the engine bulkhead, and also the four bolts securing the brake and clutch pedal assembly bracket to the bulkhead cross-member. These four bolts are situated inside the car, two on either side of the pedals.

Body Mounting Points

Remove the front small nut and bolt securing the baffle plate to the bottom flange of the wing and the seven bolts securing each baffle plate inside the front wings.

Take out the four bolts each side securing the body valances to the frame goalpost on the chassis frame.

Undo the two nuts and bolts on each side at the front, gaining access between the radiator and the grille.

Remove the trim panels from inside of the car forward of each door, and lift off the loose trimming covers which envelop the body mounting bracket at this point. Take out the bolt each side, which secures the body bracket to the chassis frame.

Gaining access from underneath the car, remove the bolt from each side just forward of the rear wheel arch.

Working from inside the boot, remove from each rear corner the two bolts which secure the rear of the body to the chassis.

The body may now be lifted from the chassis. As the body is lifted it must be manœuvred slightly forward to disengage it from the two remaining front bumper mounting brackets which protrude through the body.

Before replacing the body by reversing the dismantling procedure, ensure that the laminated cork on each body mounting point is in good condition and squarely mounted, also check the rubber strips along the chassis longitudinal members, the engine bulkhead cross-member and the cross-member forward of the battery boxes.

Section S.9

FITTING THE OPTIONAL HEATING AND DEMISTING EQUIPMENT

Drain the water from the radiator and engine (see Section C.2).

Disconnect the battery lead from the positive terminal.

Remove the brass blanking plug from the bottom

radiator hose elbow and fit the water union and washer. Fit the hose and secure with a clip to the water union.

Slide the two cleats onto the water return pipe. Remove the two nuts and spring washers securing the accelerator cable abutment bracket to the inlet manifold and fit the water return pipe to the two studs. Slide a second hose clip over the hose on the water union, insert the water return pipe in the hose and tighten the clip. Refit the nuts and spring washers to the studs on the manifold and tighten up to secure the water return pipe.

Remove the blanking plate from the right-hand side of the cylinder block. Fit the water control valve assembly, with the hose connection pointing to the rear of the car, using the new gasket. Refit the thermocouple clip to the lower bolt.

Remove the blanking plate from the air hose connector on the right-hand side of the radiator.

Insert the air hose between the radiator and the grille and fit the chamfered end onto the connector.

Secure the front end of the air hose to the valance tie-plate behind the grille with the cleat, screw, spring washer and nut, through the hole provided.

Fit one end of the air hose to the connector and secure it in position with the clip.

Secure the hose to the valance, using the second cleat and bolt through the hole in the valance which is blanked off with a nut and bolt.

Remove the blanking plate, secured by eight screws, on the engine bulkhead shelf, and fit the heater assembly, using the rubber seal fitted under the blanking plate. Secure with the screws removed from the blanking plate.

Fit the air intake to the connector on the heater assembly.

Fit one end of the water inlet hose to the water control valve connector and secure with one of the clips. Fit the other end of the water hose to the connector on the heater assembly which is adjacent to the air intake hose and secure with a second clip.

Fig. S.4.

The position of the water valve lever with the control pulled out to the " Min." position. The air flap lever may also be seen with the flap fully open.

Fig. S.5.
A general view of the under-bonnet installation showing the arrangement of the components.

Fit one end of the water outlet hose to the return pipe on the engine and the other end to the second hose connection on the heater assembly. Secure the hose with the clips.

Fit onto the rear of the heater assembly, from inside the car, the two demister flanged tubes, securing with the six screws provided. Slide the rubber adaptors onto each of the ducts below the facia panel.

On the left-hand duct and over the adaptor fit the 14 in. (355·6 mm.) air hose. Connect the other end to the left-hand demister tube on the heater.

On the right-hand duct and over the adaptor fit the 25 in. (635·4 mm.) air hose. Connect the other end to the right-hand demister tube on the heater.

Fit one end of the temperature control cable to the quadrant on the lever control assembly, securing the outer cable under the clamp plate and using the trunnion to secure the inner cable.

Remove from the underside of the facia the two screws which secure the radio mounting rails. Having removed the fibre bolt retainers from the lever control assembly fit the assembly underneath the facia, screwing the bolts into the radio mounting rails.

Special sealing grommets are fitted in the scuttle which may easily be pierced to allow the entry of the control cables. Pass the temperature control cable through the grommet on the right-hand side of the scuttle (Fig. S.6).

With the control knob in the " Min." position, connect the inner cable to the water valve lever with the trunnion supplied and with the water valve lever right back (close to outer clamp). Secure the outer cable under the clamp.

Pass the air control cable through the special grommet on the left-hand side of the scuttle and, with the air knob pulled fully out, connect the inner cable to the air control flap lever with the trunnion. When connecting the cable, the control flap must be held fully closed with the flap lever close to the outer cable clamp. Secure the outer cable under the clamp. Strap the air control cable to the water inlet pipe, using the rubber clip.

Connect the demister control cable to the demister flap on the heater box, passing the inner cable through the hole in the flap box (close to outer cable clamp) and also through the slot in the flap. Tighten the trunnion screw with the control knob pulled fully out and with the flap tightly shut.

Under the facia panel and behind the speaker fret, two wires (green and green with red) will be found to be connected together by a snap connector. Disconnect these two wires and reconnect them to the black wires from the switch on the lever control assembly, using the existing snap-connector and the one fitted to one of the switch wires.

Connect the black wires from the blower motor to the two wires (black and green with red) situated in the centre of the scuttle under the bonnet. Snap connectors are fitted to these two wires.

Reconnect the battery lead and switch on the blower motor (with the ignition switched on) and check that it is working.

See that both drain taps are closed (handles in line with the taps) and fill the cooling system with water.

Switch on the ignition and start the engine, letting it run at a fast idling speed.

Move the temperature lever to the " Max." position.

After the engine has been run for a few minutes, both

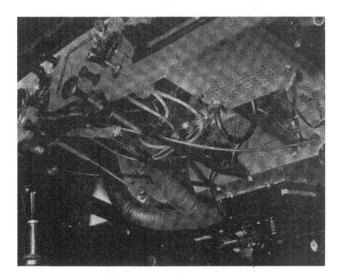

Fig. S.6.
Showing the general layout of the heater controls and demister tubes inside the car.

THE HEATER AND DEMISTER COMPONENTS

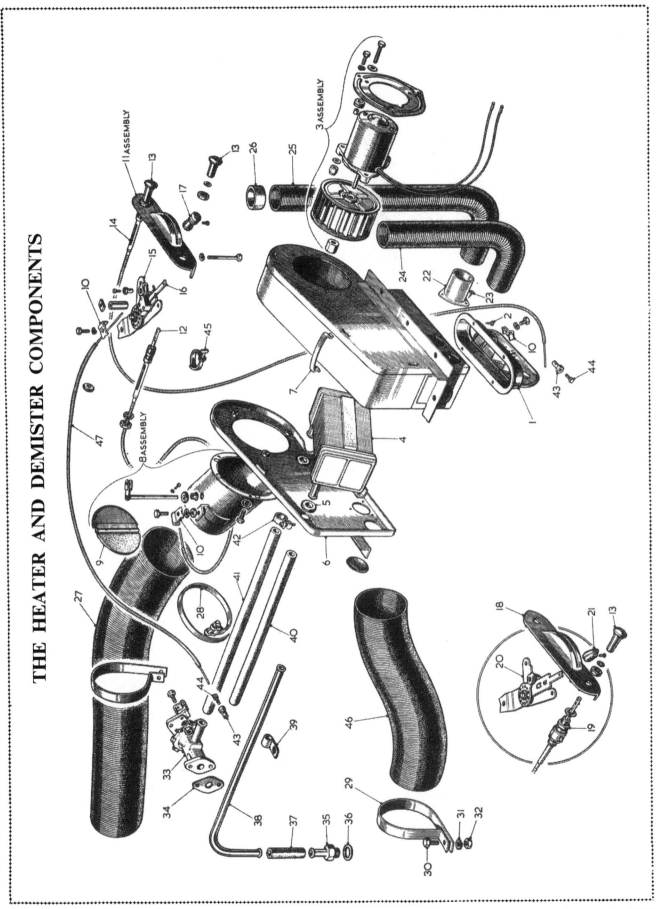

S

KEY TO THE HEATER AND DEMISTER COMPONENTS

No.	Description	No.	Description	No.	Description
1.	Outlet door assembly.	17.	Knob.	33.	Water valve assembly.
2.	Screw—outlet door.	18.	Lever control assembly.	34.	Gasket.
3.	Motor and runner.	19.	Air push cable.	35.	Water union.
4.	Radiator.	20.	Lever control.	36.	Washer.
5.	Grommet.	21.	Knob.	37.	Water pipe—2½ in. (63·5 mm.).
6.	Side cover—heater.	22.	Demister tubes.	38.	Water return pipe.
7.	Clip—side cover.	23.	Screw—cadmium-plated.	39.	Clip.
8.	Flap valve assembly.	24.	Air hose—14 in. (355·6 mm.).	40.	Water hose—12½ in. (317·5 mm.).
9.	Flap.	25.	Air hose—25 in. (635 mm.).	41.	Water hose—14 in. (355·6 mm.).
10.	Clamp—cable.	26.	Adaptors.	42.	Hose clip.
11.	Lever control assembly.	27.	Air hose—31 in. (787·4 mm.).	43.	Trunnion.
12.	Air push-pull cable.	28.	Clip.	44.	Screw—cadmium-plated.
13.	Knob.	29.	Clip—large.	45.	Cable clip.
14.	Demister cable.	30.	Screw—hexagon head.	46.	Air hose.
15.	Lever control sub-assembly.	31.	Washer—spring.	47.	Wire cable assembly.
16.	Lever and switch assembly.	32.	Nut.		

M.G. 'MGA'. Issue 5. 34279

S.7

275

the rubber pipes leading to the heater should become warm, indicating that the system is working satisfactorily.

If one or both of the pipes do not warm up this indicates an air lock in the heater circuit which can be cleared by disconnecting the flexible rubber hose connected to the rear of the copper return pipe on the engine, while the engine is ticking over.

After a few moments water should flow from the hose, which should be reconnected quickly and the hose clip tightened. If necessary, replenish the radiator with water.

The heating and demisting equipment is available as a complete kit, Part No. AHH5422.

Section S.10

COLD AIR VENTILATION EQUIPMENT

A cold air ventilation kit, Part No. AHH5532, is available which merely provides fresh cold air to the car interior.

The fittings for this installation are similar to those of the heater kit, bearing in mind the fact that there are no water or demister fitments or connections. The control switch, for the blower motor, is mounted on the fascia panel by the air control (pull/push) switch.

Section S.11

MAINTENANCE OF BODYWORK AND UPHOLSTERY

It is advisable to wash the coachwork of the car with an abundant quantity of water to remove all traces of dust, mud and traffic film. Polish the paintwork frequently with a good-quality car polish which is free from abrasive.

Metal polish must not be used to clean chromium, plastic, stainless steel, or anodized aluminium bright parts. Wash them frequently with soap and water, and when the dirt has been removed polish the surface with a clean, dry cloth or chamois-leather until bright. Never use an abrasive.

A slight tarnish may be found on stainless steel that has not been washed regularly, and this can be removed with impregnated wadding such as is used on silverware.

Surface deposits on chromium parts may be removed with a chromium cleaner.

An occasional application of wax polish or light oil to metal trim will help to preserve the finish, particularly during winter, when salt has been applied to the roads, but these protectives should not be applied to plastic finishers.

When cleaning windshields it is advisable to use methylated spirits (de-natured alcohol) to remove tar

spots and other stains. It has been found that the use of some silicon- and wax-based polishes for this purpose can be detrimental to the windshield wiper blades.

The upholstery of the car should be cleaned periodically by wiping over with a damp cloth. Accumulations of dirt, if left too long, eventually work into the pores of the leather, giving a soiled appearance. A little neutral soap may be used, but detergents, caustic soaps or spirits of any kind must **not** be used.

When necessary, the hood cloth may be cleaned with water applied with a brush. Soaps and detergents must not be used.

Section S.12

REGLAZING THE WINDSHIELD

To detach the frame from the body

Remove the three cap nuts and screws attaching each grab handle to the windshield frame flange. Remove the four screws and lift the frame forward and upward away from the body. Note that of these four screws on each side the three top screws are $\frac{5}{8}$ in. (15·9 mm.) long and the bottom screw $\frac{3}{4}$ in. (19·1 mm.) long.

To remove the glass from the frame

Remove the two screws at the top and bottom corners of the frame side rails which screw into the angle bracket in the frame channel at the mitred corners. The frame may now be pulled apart and the glazing rubber removed.

To reglaze

Check the frame top and bottom rails with the curvature of the glass and set where necessary to suit any variation. A tolerance of $\frac{1}{8}$ in. (3·2 mm.) is permitted.

The rails can be set by placing the rear face over a soft wood block. Grip the rails with the hands on each side of the block, about 12 in. (305 mm.) apart, and gently slide from side to side over the block, exerting sufficient graduated pressure to shape the rail to the desired curvature.

When the curvature of the rails is correct reassemble the windshield by reversing the order for extracting the glass, taking care that the mitred corners are correctly aligned and all the screws are tight. Replace the frame on the body, using the correct screws as detailed above.

Section S.13

REMOVING THE WINDSHIELD (COUPÉ)

Extract the two screws securing the driving mirror and remove the mirror.

Remove the windshield wiper arms.

Unscrew the seven screws at each side securing the R.H. and L.H. side fillets. Remove the fillets.

Withdraw the five screws retaining the front windshield fillet and remove the fillet.

Press the glass from the inside of the car, commencing at one corner, and carefully ease the sealing rubber from the metal edge of the windshield housing.

Before attempting to refit the windshield glass to the body it should be assembled into the rubber channel. Make sure that the glass is right home in the channel.

To facilitate the assembly of the windshield to the car body and the outside finisher to the glass, lengths of cord each about 15 feet (4·6 m.) long should be threaded around the rubber channel. Insert one length of cord into the channel to be fitted over the metal edge of the windshield housing and the other into the finisher channel on the outer side. It is convenient to have the ends of the inner cord at the bottom of the windshield and the ends of the outer cord at the top.

Threading the cords is easily carried out if one end of a cord is threaded through approximately 6 in. (15 cm.) of small-diameter tubing—brake pipe is ideal (see Fig. S.7). Radius one end of the tube inside and out and bell out the opposite end. Allow 6 in. (15 cm.) of the cord to protrude from the plain end of the tube and then press that end of the tube into the channel to which the cord is to be fitted. Run the tube around the channel, allowing the cord to flow freely through it until it surrounds the windshield and the free ends overlap and

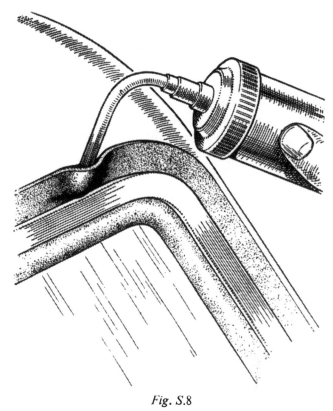

Fig. S.8

Showing the use of a pressure gun to apply Seelastik

hang from the channel. The ends should be long enough to allow a good pull when the windshield is fitted.

Insert one edge of an external finisher into the channel in the rubber, press in position, and finally position by withdrawing the string. Insert the second finisher in the same manner and fit the upper and lower cappings.

To fit the windshield to the body it must be offered to the windshield aperture from outside the car. With the assembly pressed into position from the outside the inner cord must be pulled away progressively round the aperture to draw the inside lip of the rubber channel over the flange.

Use a rubber mallet round the outside edge of the windshield to ensure complete seating of the assembly.

Seelastik sealing compound should be injected between the outer lip of the rubber seal and the body and between the seal and the glass. The application must be evenly distributed round the windshield. To ensure this the outside lip should be firmly pressed down, with the fingers or a wooden roller, to spread the sealing compound under the rubber seal.

Fig. S.8 shows the method of applying Seelastik sealing compound between the channel lip and body flange, using an Expandite pressure applicator gun, if possible, fitted with a special $\frac{3}{16}$ in. (4·5 mm.) bore brass tube nozzle.

Refit the driving mirror.

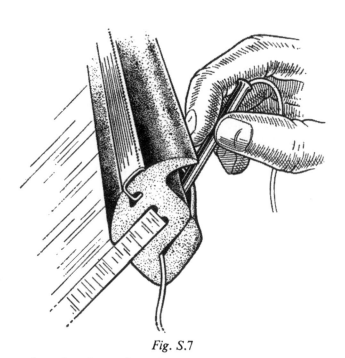

Fig. S.7

Inserting the cord in the outside lip of the rubber channel after fitting the plated finisher

Section S.14

REMOVING AND REPLACING THE REAR LIGHT (COUPÉ)

Push the glass, and rubber seal, towards the outside of the car until it is free.

To replace, fit the rubber seal round the glass. Thread a length of cord along the inner flange of the rubber seal with the ends protruding.

With the aid of a second operator to apply hand pressure to the outside of the glass draw the string from the rubber seal so that the flange is lifted over the metal edge of the window opening.

Seelastik sealing compound should be injected between the outer lip of the rubber channel and the body flange. The application must be evenly distributed right round the glass. To ensure this the outside lip should be firmly pressed down, with the fingers or a wooden roller, to spread the sealing compound under the rubber seal.

Section S.15

REMOVING A DOOR LOCK OR WINDOW REGULATOR (COUPÉ)

Extract the three securing screws and remove the top finisher panel. Remove the three retaining screws from the door-pull and plate.

Push the inner escutcheons clear of the shanks of the door lock handle and the window regulator handle and push out the exposed retaining pins to release the handles.

Remove the eight recessed-headed screws securing the trim panel to the door. Remove the door aperture sealing material.

Extract the two recessed-headed screws securing the window channel top brackets, one either side, to the top of the door. Lift the felt from the bottom of the channel and remove the screws from the bottom brackets.

With the glass fully raised, remove the self-locking nut and tension spring from the ventilator window through the aperture in the door. Lift out the ventilator window.

Unscrew the two barrel nuts securing the window frame to the door.

Withdraw the three screws securing the waist rail finisher to the outside of the door and prise the finisher up and out away from the door.

Extract the screw securing the wooden glass stop to the top rear inner face of the door. Raise the glass and withdraw the quadrant arm from the window lift channel and remove the glass.

Lift out the window frame assembly.

To remove the door lock

Take out the three screws securing the remote control to the door and the four screws in the shut face of the door securing the lock. The lock, remote control, and remote control link can then be withdrawn from the door.

To remove the window regulator

Extract the six screws securing the regulator to the inner door panel and lift out the regulator.

Reassembly is a reversal of the dismantling procedure.

Replace the piece of sealing material with rubber cement.

Section S.16

REMOVING THE HEAD LINER (COUPÉ)

Unscrew the seven screws at each side securing the R.H. and L.H. side fillets and the five screws securing the front fillet and remove the fillets.

Remove the rear light as detailed in Section S.14.

Carefully ease the head liner away from the roof above the windshield, rear light, and above the doors; the head liner is attached to the roof with rubber cement.

Extract the two screws from each side of the three head liner lists and remove the head liner complete with lists.

Reassembly is a reversal of the dismantling procedure.

When resticking the head liner to the roof with rubber cement work from the rear to the front and from the centre to the sides.

Section S.17

SEALING HOOD SEAMS

Commencing at Car No. 16101, a new hood is introduced with the seams stitched, lapped, and welded. On cars prior to this commencing number trouble may be experienced with water seeping through the stitch holes. These hood seams can be sealed with stitch sealing solution (Part No. 17H9663), available in one-pint tins.

The solution must be applied in a well-ventilated place and a dry atmosphere is essential. Normal fire precautions must be taken as the solution is inflammable.

The hood seams should be painted both on the inside and outside with two thin coatings of solution, with 10 minutes' drying time allowed between each coat. The hood should be left for 24 hours after the application of the solution before it is used.

Section S.18

SLIDING SIDESCREENS

From Car No. 68851 aluminium de-luxe sliding sidescreens are available as an optional extra on the 'MGA 1600'.

Section S.19

USE OF ADHESIVES

The following adhesives are recommended for resticking leathercloth, rubber, carpet, etc., to various surfaces. Metal must be clean and free from grease before being treated.

Particular attention must be paid to these recommendations in tropical climates, where difficulty may be experienced if alternatives are used.

Materials	Adhesive	Method of application
Rubber to bare, primed, or painted metal	Dunlop S.79	Opposing surfaces to be treated with a uniform coating of adhesive and allowed to become tacky (almost dry). Press firmly together
Carpet to primed metal		
Carpet to hardboard or millboard		
Rubber-backed needleloom felt to metal (rubber side to metal)		
Polyurethane foam to primed metal		Apply an even coating to metal. Press the foam firmly on to the metal immediately
Needleloom felt to primed metal	Goodliff G2	Apply to metal surface only and stick the felt on immediately
Leathercloth to primed or painted metal	Dunlop S758	Opposing surfaces to be treated with a uniform coating of adhesive and allowed to become tacky (almost dry). Press firmly together
Leathercloth to hardboard		
Leathercloth to leathercloth		
Leathercloth to phosphated metal		
P.V.C.-coated felt to bitumastic compound coated surfaces	Flintkote 833 N	Apply to both surfaces and press firmly together after five minutes

Section S.20

PAINT REFINISHING INSTRUCTIONS

Operation	Material	Thinning	Drying times	Application	Instructions
Stripping original paint	Water-soluble paint remover, e.g. Sunbeam Anti-corrosives 'Stripolene 799'	—	—	Brush	Remove the original finish with a scraper after allowing paint-strip 10 minutes to react (repeat if necessary). Wash off thoroughly with cold water, rubbing with wire wool. Dry. Blow out crevices with compressed air. Strip a small area at a time to enable correct neutralizing of the stripper
Metal abrazing	Emery-cloth, e.g. Howarth Blue Twill, grade 1½ M	—	—	Hand or disc	Paper thoroughly to ensure satisfactory key. Wipe with cleaner solvent or white spirits
Acid etching	Apply Deoxidine 125 (I.C.I.)	1 part Deoxidine, 1 part water	—	Brush	Apply solution generously and rub in with wire wool. Do not allow Deoxidine solution to dry off before the wash-off operation. Allow approximately five minutes to complete reaction. Wash thoroughly with cold water to remove all traces of Deoxidine solution, followed by a hot rinse. Thoroughly dry surfaces with a clean cloth and blow out crevices with compressed air
Priming	Synthetic primer G.I.P. No. S3178 or	6 to 1 with Z1048	½-hour to 4 hours	Spray	Apply one thin coat of synthetic primer (recommended for superior adhesion) or one thin coat of cellulose primer (recommended for good adhesion). The use of a primer coat enhances adhesion and gives the system a much greater safety factor
	Grey cellulose primer G.I.P. C3971 MOD	50/50 with 2045M	½-hour	Spray	
Applying stopper	Stopper Grey G.I.P. 824D or Stopper Brown G.I.P. 1543	—	6–8 hours, or overnight if possible	Glazing knife	Apply stopper in thin layers, allowing 15–20 minutes' drying between applications. Heavy layers result in insufficient drying, with subsequent risk of cracking
Filling	Primer Filler Grey G.I.P. C3663M	50/50 with 2045M	3–4 hours	Spray	Apply two or three full coats, allowing 15–25 minutes' drying time between coats

Operation	Material	Thinners	Drying time	Application	Remarks
Wet-sanding	Abrasive paper 280 grade	—	—	—	Rub down wet until smooth; a guide coat (a weak contrasting colour) may be used to ensure that the whole surface is rubbed level. Wash off thoroughly with water, sponge all sludge, wash off, dry with clean sponge. Dry off. Minimum of paint should be removed consistent with a satisfactory surface. Film thickness after rubbing should be ·0025 in. (·06 mm.) min.
Applying sealer or undercoat	Sealer Grey or Sealer White or Red undercoat (see B.M.C. Paint Scheme schedule)	50/50 with 2045M	15–20 minutes	Spray	Apply one coat, flash off
Dry-sanding or de-nibbing as required	320 grade paper	—	—	—	De-nib or dry-sand with 320 paper. Clean with white spirit. The grade of paper quoted is from the 3M Company (Minnesota Mining and Mfg. Co. Ltd.); the grade of paper may vary according to manufacture
Applying colour coats	B.M.C. body finishes (see B.M.C. Paint Scheme schedule)	50/50 with 2045M	5–10 minutes' flash between coats. Overnight dry	Spray	Apply two double coats with a 5–10-minute flash between coats. Overnight dry
Flatting	320 or 400 paper (dependent on conditions)	—	—	Hand	Flat with 320 or 400 paper, dependent on conditions
Applying final colour coat	B.M.C. body finishes (see B.M.C. Paint Scheme schedule)	50/50 with 2045M	Overnight dry	Spray	Spray final double colour coat
Polishing	Cut and polish (see B.M.C. Paint Scheme schedule)	—	—	Hand or machine	The colour coat must be thoroughly dry before polishing. After cutting, burnish to a high gloss with a clean mop, and finally clean with a liquid polish, e.g. Apollo liquid polish

NOTE.—(1) For faster drying of undercoats or local repairs G.I.P. thinners 1523 may be used.
(2) Under extreme circumstances of heat and/or humidity retarder G.I.P. Z1694 can be used added to the 2045M thinners.

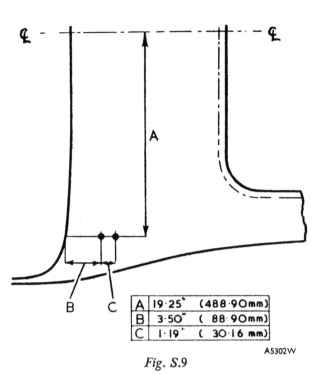

A	19·25"	(488·90 mm)
B	3·50"	(88·90 mm)
C	1·19"	(30·16 mm)

A5302W

Fig. S.9

The tonneau panel anchorage point

Section S.21

SEAT BELT ANCHORAGE FITTING INSTRUCTIONS

Seat belt kits complete with anchorage fittings are available under Part Nos. AHH6141 (R.H.) and AHH6193 (L.H.) for fitting to the driver's or passenger's seats on both Tourer and Coupé cars.

The harness comprises a long and a short belt, both of which are adjustable. When in use a tongue on the long belt engages a quick-release buckle on the short belt; the positions of the tongue and buckle may be adjusted to suit individual users.

The upper end of the long belt is fitted to the tonneau panel and the lower end of the belt to the chassis side-member. The lower end of the short belt is fitted to the side of the drive shaft tunnel adjacent to the seat being equipped.

The fitting instructions are as follows.

Tonneau panel
(1) Mark off and drill two $\frac{11}{32}$ in. (7·73 mm.) holes in the tonneau panel in the position shown in Fig. S.9.
(2) Place the two spring washers on the two pan-head screws, pass the screws through the top end bracket of the long belt, and position the bracket on top of the tonneau panel.

(3) Place the tapped reinforcement plate under the tonneau panel and secure with the two pan-head screws.

Side-member
(1) Remove the carpet from the side-member and sill. Place the anchor bracket on the side-member with the centre of the rear hole $4\frac{3}{4}$ in. (12·0 cm.) forward of the rear floor panel.
(2) Ensure that the anchor plate weld nuts face outwards, and then securely arc-weld the plate to the chassis side-member.
(3) Place the spring washers on the hexagon-headed screws and secure the quick-release bracket to the anchor plate.
(4) Secure the lower end of the long belt to the quick-release bracket with the pin provided. Ensure that the head of the pin faces the centre of the car and that it is correctly locked.

Drive shaft tunnel
(1) Remove the tunnel trim. Mark out and cut a $1\frac{1}{2}$ in. (38·1 mm.) diameter hole 2 in. (50·8 mm.) from the

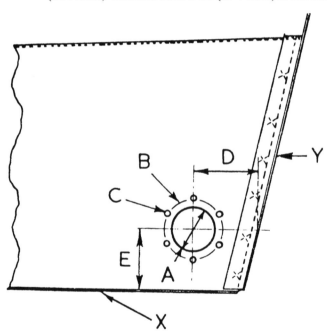

A	1·500" (38·10 mm)
B	2·00 " (50·80 mm)
C	·203 " (5·16 mm)
D	2·125 " (53·97 mm)
E	2·00 " (50·80 mm)

A5304W

Fig. S.10

The drive shaft tunnel anchorage point

floor and $2\frac{1}{8}$ in. (53·97 mm.) from the heelboard on the side of the tunnel adjacent to the seat being equipped.

(2) Using the centre bracket as a template, mark out and drill six equally spaced holes $\frac{13}{64}$ in. (5·16 mm.) diameter on a pitch circle diameter of 2 in. (50·8 mm.).

(3) Position the centre bracket inside the tunnel and secure it with the six No. 10 UNF. pan-head screws. The heads of the screws must be inside the tunnel and the spring washers and nuts inside the car.

(4) Cut a 1 in. (25·4 mm.) hole in the carpet to coincide with the middle of the centre bracket.

(5) To the $\frac{7}{16}$ in. (11·11 mm.) hexagon-headed screw fit the plain washer, anti-rattle washer (concave face to belt bracket), belt bracket, and distance piece (large diameter to tunnel). Pass the bolt through the centre-piece and secure it with the spring washer and nut.

Notes

SECTION SS

BODY

(MGA 1600 and MGA 1600 [Mk. II])

Section SS.1

FITTING A FIBREGLASS HARD-TOP

To fit the Fibreglass hard-top, fold and stow the hood or remove it completely from the car. If the hood is removed fit attachment brackets to the body sides in place of the hood frame pivot plates to form anchorage points for the hard-top retaining clips.

Lower the front of the hard-top onto the top of the windshield so that the two pins on the windshield surround locate in their respective housings in the hard-top. Allow the rear end of the hard-top to seat on the tonneau panel and ensure that the rubber moulding around the base of the hard-top is a good fit all round. Secure the front of the hard-top with the two thumbscrews provided.

Remove the cap nuts and knurled thumbscrews from the retaining clips, pass the clips through the brackets on the sides of the hard-top, and engage the hooked ends of the clips in the hood frame pivot plates or the attachment plates; secure the clips with the knurled thumbscrews and refit the cap nuts.

The aluminium de-luxe sliding sidescreens for use with the Fibreglass hard-top are then fitted in the normal way.

Section SS.2

SEAT BELTS

From Car No. 100352 the body of the car incorporates anchorage points to facilitate the fitting of B.M.C. seat belts (Part No. AHH6122) to the driver's seat and to the passenger's seat. The fitting of the belts to the car should be carried out by an Authorized M.G. Distributor or Dealer.

The anchorage points are located one on each frame side-member, one on each side of the tunnel, and one

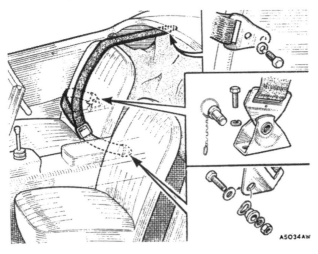

Fig. SS.1

The seat belt attachment points and fittings (Tourer)

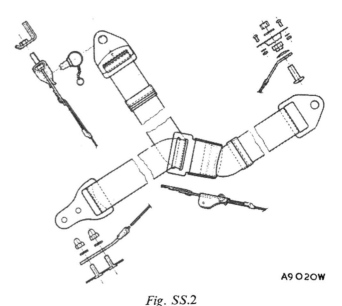

A9020W

Fig. SS.2

The seat belt and anchorage fittings (Coupe)

each under the left-hand and right-hand sides respectively of the tonneau panel (Tourer) and rear wheel arch (Coupé). The seat belt is made up of a long and a short belt, each of which is adjustable. When in use the two belts are connected by a quick-release buckle.

To fit the belts to the car, bolt the end bracket of the long belt (having two holes) to the under side of the tonneau panel, using the spring washers and set screws provided. Cut the trim covering the frame member locating points to obtain metal-to-metal contact and use the spring washers and set screws provided to secure the anchor plate to the side-member. Secure the remaining end bracket of the long belt to the anchor plate with the quick-release pin and circlip; insert the pin from the inside of the bracket. Cut the trim covering the tunnel locating point and remove the grommet from the tunnel. Place the anti-rattle washer and the under side of the short belt end bracket over the distance piece and pass the set screw through the plain washer, distance piece, and the tunnel so that the large diameter of the distance piece is adjacent to the tunnel. Secure the set screw with the spring washer and nut.

NOTE.—It is important that the short belt is secured to the side of the tunnel next to the wearer.

Section SS.3

SEAT BELT ANCHORAGE FITTING INSTRUCTIONS

Seat belt kits complete with anchorage fittings are available under Part No. AHH6141 (R.H.) and AHH6193 (L.H.) for Tourer cars up to Car No. 100351 and Part No. AHH6175 (R.H.) and AHH6194 (L.H.) for Coupé cars up to Car No. 100351.

The description and fitting instructions are as detailed in Section S.21, with the exception of the upper end of the long belt, which is fitted to the rear wheel arch instead of the tonneau panel on Coupé cars only.

The fitting instructions are as follows.

Rear wheel arch

(1) Remove the rear road wheel.

(2) Using the vertical spot-welds securing the bulkhead to the wheel arch as a datum line, mark out and drill an $\frac{11}{32}$ in. (8·73 mm.) hole $\frac{5}{8}$ in. (15·87 mm.) forward of the datum mark and $2\frac{1}{2}$ in. (6·35 cm.) up from the spot-welds securing the floor to the wheel arch.

(3) Mark out a second hole $1\frac{1}{4}$ in. (31·75 mm.) forward from the datum line and $3\frac{1}{2}$ in. (8·89 cm.) up from the spot-welds securing the floor to the wheel arch.

(4) Position the reinforcement plate over the first hole and drill a second hole $\frac{11}{32}$ in. (8·73 mm.) diameter.

(5) Pass the bolts attached to the reinforcement plate through the holes in the wheel arch and secure the

upper end of the long belt to the bolts with the spring washers and cap nuts provided.

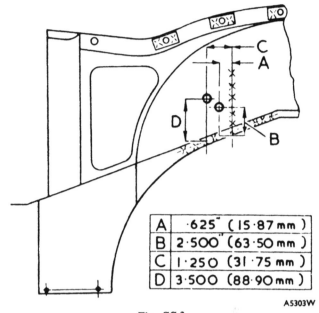

A	·625" (15·87 mm)
B	2·500" (63·50 mm)
C	1·250 (31·75 mm)
D	3·500 (88·90 mm)

A5303W

Fig. SS.3

The rear wheel arch anchorage point (Coupe only)

THE BODY SHELL (COUPÉ)

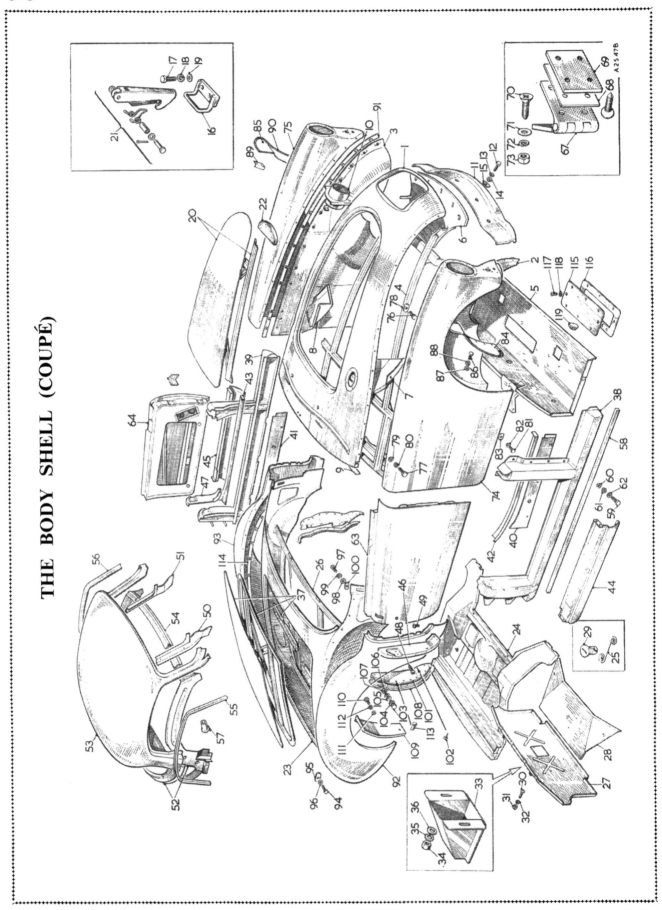

KEY TO THE BODY SHELL (COUPE)

No.	Description	No.	Description	No.	Description
1.	Panel assembly—bonnet surround.	40.	Panel—sill inner—R.H.	81.	Screw—front wing to sill.
2.	Front side assembly—R.H.	41.	Panel—sill inner—L.H.	82.	Washer—spring—for screw.
3.	Front side assembly—L.H.	42.	Plate—sill sealing—R.H.	83.	Washer—plate—for screw.
4.	Bonnet surround reinforcement.	43.	Plate—sill sealing—L.H.	84.	Splash plate—R.H. front wing.
5.	Panel assembly—front bulkhead.	44.	Panel—sill outer—R.H.	85.	Splash plate—L.H. front wing.
6.	Panel—radiator duct.	45.	Panel—sill outer—L.H.	86.	Screw fixing plate.
7.	Box assembly—R.H. air duct.	46.	Panel—R.H. shut pillar facing.	87.	Washer—plain—for screw.
8.	Box assembly—L.H. air duct.	47.	Panel—L.H. shut pillar facing.	88.	Washer—spring—for screw.
9.	Strip—front bulkhead packing.	48.	Screw—facing panel—side.	89.	Bracket assembly—splash plate.
10.	Tube—air duct.	49.	Screw—facing panel—bottom.	90.	Seal—L.H. front splash plate (rubber).
11.	Panel—front valance.	50.	Pillar reinforcement—windscreen—R.H.	91.	Piping—front wing.
12.	Screw fixing panel.	51.	Pillar reinforcement—windscreen—L.H.	92.	Wing assembly—R.H. rear.
13.	Washer—spring—for screw.	52.	Roof assembly.	93.	Wing assembly—L.H. rear.
14.	Washer—plain—for screw.	53.	Panel assembly—roof.	94.	Screw—rear wing to body.
15.	Washer—plate—for screw.	54.	Reinforcement assembly—roof panel.	95.	Washer—plate—for screw.
16.	Bracket—bonnet lock safety catch.	55.	Dripway—door opening—R.H.	96.	Washer—plain—for screw.
17.	Screw fixing bracket.	56.	Dripway—door opening—L.H.	97.	Screw—rear wing to sill.
18.	Washer—spring—for screw.	57.	Rivet—dripway to roof panel.	98.	Washer—plain—for screw.
19.	Washer—plain—for screw.	58.	Finisher—sill lower.	99.	Washer—spring—for screw.
20.	Panel assembly—bonnet lid.	59.	Bolt for sill finisher (special).	100.	Washer—plate—for screw.
21.	Catch assembly—safety.	60.	Nut for bolt.	101.	Splash plate—rear wing—front—L.H.
22.	Batten—bonnet lid stiffening.	61.	Lock washer for nut.	102.	Screw fixing plate.
23.	Tonneau panel.	62.	Washer—plain—for nut.	103.	Screw fixing plate.
24.	Floor assembly—boot.	63.	Door assembly—R.H.	104.	Washer—plain—for screw.
25.	Dzus spring.	64.	Door assembly—L.H.	105.	Washer—plain—for screw.
26.	Reinforcement assembly—tonneau.	67.	Hinge—door.	106.	Washer—spring—for screw.
27.	Panel assembly—rear bulkhead.	68.	Packing—door hinge.	107.	Nut for screw.
28.	Panel assembly—battery cover.	69.	Plate—tapping.	108.	Seal—R.H. rear splash plate (rubber).
29.	Dzus fastener.	70.	Screw fixing hinge.	109.	Splash plate—rear wing—rear—R.H.
30.	Screw—rear bulkhead panel.	71.	Washer—plain—for screw.	110.	Screw fixing plate.
31.	Nut for screw.	72.	Washer—spring—for screw.	111.	Washer—plain—for screw.
32.	Washer—spring—for screw.	73.	Nut for screw.	112.	Washer—spring—for screw.
33.	Bracket assembly—spare wheel.	74.	Wing assembly—R.H. front.	113.	Rivet—rubber seal to plate.
34.	Nut for bracket.	75.	Wing assembly—L.H. front.	114.	Piping—rear wing.
35.	Washer—spring—for nut.	76.	Screw—front wing to body.	115.	Cover-plate—heater aperture.
36.	Washer—plain—for nut.	77.	Screw—front wing to body.	116.	Gasket—heater sealing.
37.	Lid assembly—boot.	78.	Washer—plate—for screw.	117.	Screw—cover-plate.
38.	Sill reinforcement and pillar assembly—R.H.	79.	Washer—plain—for screw.	118.	Washer—spring—for screw.
39.	Sill reinforcement and pillar assembly—L.H.	80.	Washer—spring—for screw.	119.	Grommet—blanking heater valve control hole.

THE DOOR FITTINGS

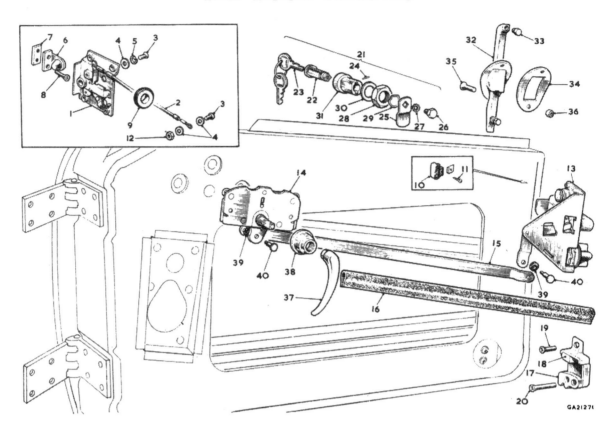

GA21271

No.	Description	No.	Description
Tourer		20.	Screw fixing striker.
1.	Lock assembly—door—R.H.	21.	Lock assembly—private.
2.	Cable—operating.	22.	Cylinder—lock—with keys.
3.	Screw fixing lock.	23.	Key.
4.	Washer—plain—for screw.	24.	Peg—retaining.
5.	Washer—spring—for screw.	25.	Latch.
6.	Plate—striking.	26.	Bolt—latch.
7.	Plate—striker tapping.	27.	Washer for bolt.
8.	Screw fixing striker.	28.	Locknut.
9.	Grommet—door release (rubber).	29.	Washer—waved.
10.	Buffer assembly—door.	30.	Washer—seating.
11.	Screw fixing buffer.	31.	Housing—lock.
12.	Nut—door lock cable.	32.	Handle—door.
		33.	Insert—door handle (rubber).
Coupé		34.	Washer—seating.
13.	Lock—R.H.	35.	Screw fixing handle.
14.	Remote-control mechanism—R.H.	36.	Nut for screw.
15.	Link—connecting.	37.	Handle—remote-control.
16.	Sleeve—anti-rattle.	38.	Escutcheon assembly—handle.
17.	Plate—striker—R.H.	39.	Washer (rubber).
18.	Plate—striker tapping.	40.	Rivet.
19.	Screw fixing striker.		

THE HEATER COMPONENTS

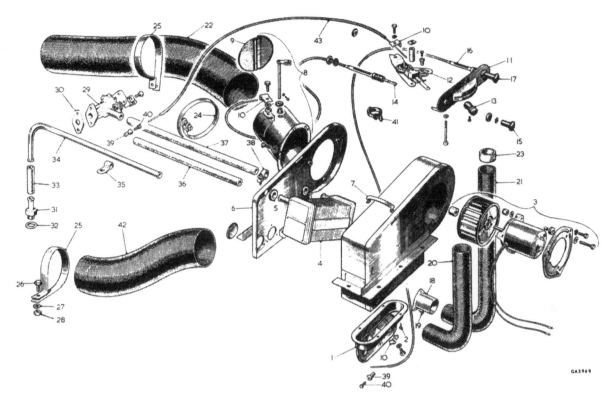

No.	Description	No.	Description
1.	Door assembly—outlet.	23.	Adaptor.
2.	Screw fixing outlet door.	24.	Clip.
3.	Motor and runner.	25.	Clip (large).
4.	Radiator.	26.	Screw for clip.
5.	Grommet.	27.	Washer—spring—for screw.
6.	Cover (side).	28.	Nut for screw.
7.	Clip—side cover.	29.	Valve assembly—water.
8.	Valve assembly—flap.	30.	Gasket—water valve.
9.	Flap.	31.	Union—water.
10.	Clamp for cable.	32.	Washer.
11.	Escutcheon—heater control unit.	33.	Pipe—water ($2\frac{1}{2}$ in.).
12.	Lever control sub-assembly.	34.	Pipe return—water.
13.	Knob.	35.	Clip.
14.	Control cable and knob—air.	36.	Hose—water ($12\frac{1}{2}$ in.).
15.	Knob.	37.	Hose—water (14 in.).
16.	Control cable and knob—demist.	38.	Clip—hose.
17.	Knob.	39.	Trunnion.
18.	Tube—demister.	40.	Screw for trunnion.
19.	Screw—fixing demister tube.	41.	Clip—cable.
20.	Hose—air (14 in.).	42.	Hose—air.
21.	Hose—air (25 in.).	43.	Cable assembly.
22.	Hose—air (31 in.).		

THE FASCIA PANELS

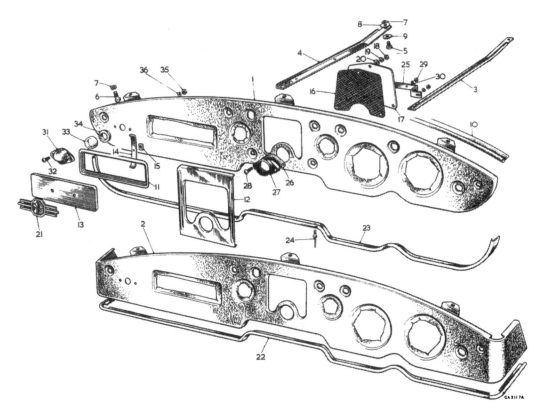

No.	Description	No.	Description
1.	Panel assembly—fascia.	19.	Washer—spring—for nut.
2.	Panel assembly—fascia.	20.	Washer—plain—for nut.
3.	Support angle assembly—R.H.	21.	Motif—'M.G.'.
4.	Support angle assembly—L.H.	22.	Finisher—fascia panel lower.
5.	Screw—support angle.	23.	Finisher—fascia panel lower.
6.	Screw—support angle.	24.	Rivet—finisher and radio bezel.
7.	Washer—spring—for screw.	25.	Bracket—horn-push.
8.	Washer—plain—for screw.	26.	Insulating piece—horn-push.
9.	Washer—plain—for screw.	27.	Horn-push.
10.	Piping—fascia panel.	28.	Screw—push to bracket.
11.	Bezel—radio aperture.	29.	Nut for screw.
12.	Bezel—radio speaker.	30.	Washer—spring—for screw.
13.	Plate—blanking—radio aperture.	31.	Cover assembly—map light.
14.	Clip—blanking plate to fascia.	32.	Screw—map light cover to fascia.
15.	Nut—clip (Spire).	33.	Glass—map light.
16.	Grille—radio speaker.	34.	Rubber—map light glass.
17.	Plate—blanking—radio speaker grille.	35.	Nut for screw.
18.	Nut—blanking plate to panel.	36.	Washer—spring—for screw.

LUBRICATION   CHART

KEY TO LUBRICATION CHART

DAILY

(1) ENGINE. Check the oil level with the dipstick, and replenish if necessary with oil to Ref. A.

EVERY 3,000 MILES (5000 Km.)

(2) GEARBOX. Check the oil level with the dipstick, and replenish if necessary with oil to Ref. A.

(3) REAR AXLE. Remove the filler plug and replenish to the filler plug level with oil to Ref. B.

(4) STEERING. Give three or four strokes of the gun filled to Ref. C to the nipples on the steering joints.

(5) PROPELLER SHAFT. Give the shaft coupling and sliding joint nipples three or four strokes with the gun filled to Ref. C. Earlier cars have two propeller shaft nipples only.

(6) HAND BRAKE. Give the cable nipples three or four strokes of the gun filled with grease to Ref. C.

(7) CARBURETTERS. Remove the cap from the top of each suction chamber and add a few drops of oil to Ref. D.

(8) AIR CLEANERS. Wash the filter elements and allow to dry. Re-oil the elements with oil to Ref. A and allow to drain before reassembly.

(9) ENGINE. Drain off the old oil from the sump and refill with fresh oil to Ref. A.

EVERY 6,000 MILES (10000 Km.)

(10) DISTRIBUTOR. Withdraw the rotor arm and add a few drops of oil to Ref. D to the cam bearing and to the advance mechanism through the gap round the cam spindle. Smear the cam lightly with grease to Ref. C and add a spot of oil to the contact breaker pivot.

(11) OIL FILTER. Release the filter bowl, wash in fuel, and fit a new element.

(12) GEARBOX. Drain off the old oil and refill with fresh oil to Ref. A.

(13) REAR AXLE. Drain off the old oil and refill with fresh oil to Ref. B.

(14) DYNAMO. Add a few drops of oil to Ref. D through the oil hole in the commutator end bearing.

EVERY 12,000 MILES (20000 Km.)

(15) STEERING. Give up to 10 strokes to the nipple on the steering gearbox and two strokes only to the pinion shaft nipple. Use oil to Ref. B.

(16) WATER PUMP. Remove the plug from the water pump and lubricate sparingly with grease to Ref. C.

MULTIGRADE MOTOR OILS

In addition to the recommended lubricants listed in the Manual we approve the use of these motor oils, as produced by the oil companies shown in our publications, for all climatic conditions unless the engine is in poor mechanical condition.

NOTE.—Oil and grease references are to be found on page P.2.

THE 'MGA' AND 'MGA 1600' LUBRICATION CHART

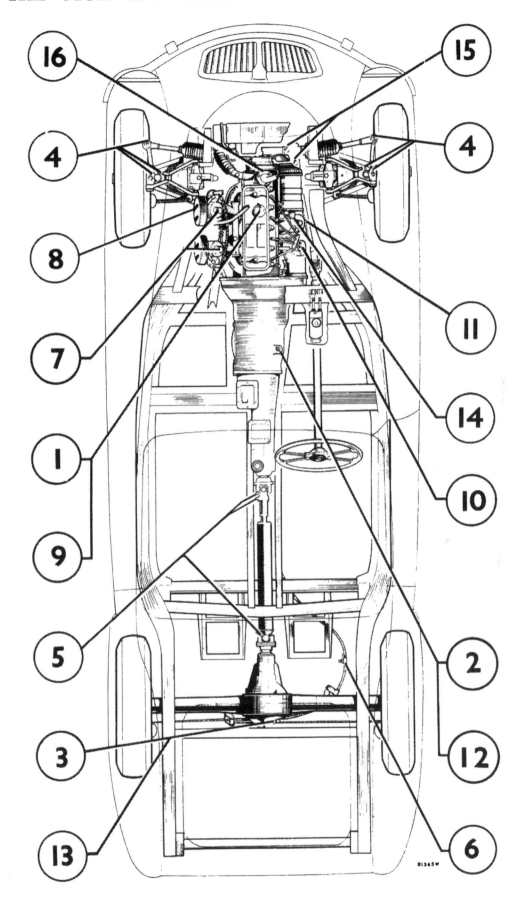

M.G. 'MGA'. Issue 5. 48539

OFFICIAL TECHNICAL BOOKS

Brooklands Technical Books has been formed to supply owners, restorers and professional repairers with official factory literature.

Workshop Manuals

Midget Instruction Manual		9781855200739
Midget TD & TF	AKD580A	9781870642552
MGA 1500 1600 & 1600 Mk. 2	AKD600D	9781869826307
MGA Twin Cam	AKD926B	9781855208179
Austin-Healey Sprite Mk. 2, Mk. 3 & Mk. 4 and		
MG Midget Mk. 1, Mk. 2 & Mk. 3		
	AKD4021	9781855202818
Midget 1500	AKM4071B	9781855201699
MGB & MGB GT	AKD3259 & AKD4957	9781855201743
MGB GT V8 Supplement		9781855201859
MGB, MGB GT and MGB GT V8		9781783180578
MGC	AKD 7133	9781855201828
Rover 25 & MG ZR 1999-2005		
	RCL0534ENGBB	9781855208834
Rover 75 & MG ZT 1999-2005		
	RCL0536ENGBB	9781855208841
MGF - 1.6 MPi, 1.8 MPi, 1.8VVC		
	RCL 0051ENG, RCL0057ENG	
	& RCL0124	9781855207165
MGF Electrical Manual 1996-2000 MY		
	RCL0341	9781855209077
MG TF	RCL0493	9781855207493

Parts Catalogues

MGA 1500	AKD1055	9781870642569
MGA 1600 Mk. 1 & Mk. 2	AKD1215	9781870642613
Austin-Healey Sprite Mk. 1 & Mk. 2 and		
MG Midget Mk. 1 (Mechanical & Body Edition)		
	AKD3566 & AKD3567	9781783180509
Austin-Healey Sprite Mk. 3 & Mk. 4 and		
MG Midget Mk. 2 & Mk. 3 (Mechanical & Body		
Edition 1969)	AKD3513 & AKD3514	9781783180554
Austin-Healey Sprite Mk. 3 & Mk. 4 and		
MG Midget Mk. 2 & Mk. 3 (Feb 1977 Edition)		
	AKM0036	9780948207419
MGB up to Sept 1976	AKM0039	9780948207068
MGB Sept 1976 on	AKM0037	9780948207440

Owners Handbooks

Midget Series TD		9781870642910
Midget TF and TF 1500		
Operation Manual	AKD658A	9781870642934
MGA 1500	AKD598G	9781855202924
MGA 1600	AKD1172C	9781855201668
MGA 1600 Mk. 2	AKD1958A	9781855201675
MGA Twin Cam (Operation)	AKD879	9781855207929
MGA Twin Cam (Operation)	AKD879B	9781855207936
MGA 1500 Special Tuning	AKD819A	9781783181728
MGA 1500 and 1600 Mk. 1 Special Tuning		
	AKD819B	9781783181735
Midget TF and TF 1500	AKD210A	9781855202979
Midget Mk. 3 (GB 1967-74)	AKD7596	9781855201477
Midget (Pub 1978)	AKM3229	9781855200906
Midget Mk. 3 (US 1967-74)	AKD7883	9781855206311
Midget Mk. 3 (US 1976)	AKM3436	9781855201767
Midget Mk. 3 (US 1979)	AKM4386	9781855201774
MGB Tourer (Pub 1965)	AKD3900C	9781869826741

MGB Tourer & GT (Pub 1969)	AKD3900J	9781855200609
MGB Tourer & GT (Pub 1974)	AKD7598	9781869826727
MGB Tourer & GT (Pub 1976)	AKM3661	9781869826703
MGB GT V8	AKD8423	9781869826710
MGB Tourer & GT (US 1968)	AKD7059B	9781870642514
MGB Tourer & GT (US 1971)	AKD7881	9781870642521
MGB Tourer & GT (US 1973)	AKD8155	9781870642538
MGB Tourer (US 1975)	AKD3286	9781870642545
MGB (US 1979)	AKM8098	9781855200722
MGB Tourer & GT Tuning	CAKD4034L	9780948207051
MGB Special Tuning 1800cc	AKD4034	9780948207006
MGC	AKD4887B	9781869826734
MGF (Modern shape)	RCL0332ENG	9781855208339

Owners Workshop Manuals - Autobooks

MGA & MGB & GT 1955-1968	
(Glove Box Autobooks Manual)	9781855200937
MGA & MGB & GT 1955-1968	
(Autobooks Manual)	9781783180356
Austin-Healey Sprite Mk. 1, 2, 3 & 4 and	
MG Midget Mk. 1, 2, 3 & 1500 1958-1980	
(Glove Box Autobooks Manual)	9781855201255
Austin-Healey Sprite Mk. 1, 2, 3 & 4 and	
MG Midget Mk. 1, 2, 3 & 1500 1958-1980	
(Autobooks Manual)	9781783180332
MGB & MGB GT 1968-1981	
(Glove Box Autobooks Manual)	9781855200944
MGB & MGB GT 1968-1981	
(Autobooks Manual)	9781783180325

Carburetters

SU Carburetters Tuning Tips & Techniques	
	9781855202559
Solex Carburetters Tuning Tips & Techniques	
	9781855209770
Weber Carburettors Tuning Tips and Techniques	
	9781855207592

Restoration Guide

MG T Series Restoration Guide	9781855202115
MGA Restoration Guide	9781855203020
Restoring Sprites & Midgets	9781855205987
Practical Classics On MGB Restoration	9780946489428

MG - Road Test Books

MG Gold Portfolio 1929-1939	9781855201941
MG TA & TC GOLD PORT 1936-1949	9781855203150
MG TD & TF Gold Portfolio 1949-1955	9781855203167
MG Y-Type & Magnette Road Test Portfolio	9781855208629
MGB & MGC GT V8 GP 1962-1980	9781855200715
MGA & Twin Cam Gold Portfolio 1955-1962	9781855200784
MGB Roadsters 1962-1980	9781869826109
MGC & MGB GT V8 LEX	9781855203631
MG Midget Road Test Portfolio 1961-1979	9781855208957
MGF & TF Performance Portfolio 1995-2005	9781855207073
Road & Track On MG Cars 1949-1961	9780946489398
Road & Track On MG Cars 1962-1980	9780946489817

From MG specialists, Amazon and all good motoring bookshops.

Brooklands Books Ltd., P.O. Box 904, Amersham, Bucks, HP6 9JA, England, UK

www.brooklandsbooks.com

Brooklands MG 'Road Test' & Restoration Titles

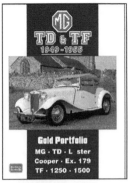

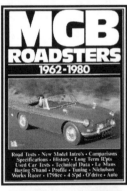

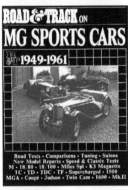

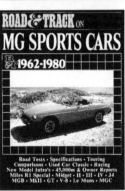

www.brooklandsbooks.com

Brooklands Books Ltd., PO Box 904,
Amersham, Bucks, HP6 9JA, UK
www.brooklandsbooks.com

Part Number: AKD4021 (15 Edition)

ISBN 9781869826307   Ref: MG31WH   4W4/2986

Printed in Great Britain
by Amazon

46568334R00165